An Introduction to
Categorical Data Analysis

An Introduction to
Categorical Data Analysis

ALAN AGRESTI

University of Florida

A Wiley-Interscience Publication

JOHN WILEY & SONS, INC.

New York • Chichester • Brisbane • Toronto • Singapore

Copyright © 1996 by John Wiley & Sons, Inc.

Library of Congress Cataloging-in-Publication Data:

Agresti, Alan.
     An introduction to categorical data analysis / Alan Agresti.
       p.   cm. — (Wiley series in probability and
  statistics. Applied probability and statistics)
     "A Wiley-Interscience publication."
     Includes bibliographical references (p.   –   ) and index.
     ISBN 0-471-11338-7
     1. Multivariate analysis.   I. Title.   II. Series.
  QA278.A355   1996
  519.5'35—dc20   95-21928

Printed in the United States of America

10 9 8 7 6 5 4 3 2

Contents

Preface

In recent years, the use of specialized statistical methods for categorical data has increased dramatically, particularly for applications in the biomedical and social sciences. Partly this reflects the development during the past few decades of sophisticated methods for analyzing categorical data. It also reflects the increasing methodological sophistication of scientists and applied statisticians, most of whom now realize that it is unnecessary and often inappropriate to use methods for continuous data with categorical responses.

This book presents the most important methods for analyzing categorical data. It summarizes methods that have long played a prominent role, such as chi-squared tests and measures of association. It gives special emphasis, however, to logistic regression and loglinear modeling techniques. These methods share many features with linear model methods for continuous variables.

The presentation in this book has a low technical level and does not require familiarity with advanced mathematics such as calculus or matrix algebra. Readers should possess a background that includes material from a two-semester statistical methods sequence for undergraduate or graduate nonstatistics majors. This background should include estimation and hypothesis testing and exposure to regression modeling and the analysis of variance.

This book is designed for students taking an introductory course in categorical data analysis. But I also have written it for applied statisticians and practicing scientists involved in data analyses. I hope that the book will be helpful to analysts dealing with categorical response data in the social, behavioral, and biomedical sciences, as well as in public health, marketing, education, biological and agricultural sciences, and industrial quality control.

The material in Chapters 1–6 forms the heart of an introductory course in categorical data analysis. Chapters 2 and 3 survey standard descriptive and inferential methods for contingency tables, such as odds ratios, tests of independence, and Cochran–Mantel–Haenszel procedures. I feel that an understanding of methods is enhanced, however, by viewing them in the context of statistical models. Thus, the major focus of the text is the modeling of categorical responses. Chapter 4 introduces generalized linear models for binary data and count data. Chapter 5 discusses logistic regression for binomial (binary) data, and Chapter 6 discusses loglinear models for

Poisson (count) data. Later chapters connect these model types and show extensions for a variety of problems. I believe that logistic regression is more important than loglinear models, since most applications with categorical responses have a single binomial or multinomial response variable. Thus, I have given more attention to this topic than most introductory books do.

I prefer to teach categorical data methods by unifying their logistic regression and loglinear models with ordinary regression and ANOVA models. Chapter 4 does this under the umbrella of generalized linear models. Some instructors might prefer to cover this chapter rather lightly, using it primarily to introduce logistic regression models for binomial data and loglinear models for Poisson data. Instructors can choose sections from Chapters 7–9 to supplement the basic topics in Chapters 1–6. Chapter 7 covers topics bearing on the construction of both logit and loglinear models, such as graphical representations, the use of ordinal information, and effects of sparse data. Chapter 8 introduces logit models for multinomial responses, both nominal and ordinal. Chapter 9 presents methods for matched-pairs data. Courses having a biostatistical orientation might put more emphasis on Chapters 8 and 9 and relatively less on Chapters 6 and 7. Courses having a social science orientation might prefer the reverse. Throughout the text, asterisks appended to titles indicate sections and subsections that are less important at first introduction.

This book is of a lower technical level than my book *Categorical Data Analysis* (Wiley, 1990). I hope that it will appeal to readers who prefer a more applied introduction than the earlier book provides. For instance, this book does not attempt to derive likelihood equations, prove asymptotic distributions, discuss current research work, or present a complete bibliography. Some features of this new book not contained in the 1990 text include an appendix showing the use of a new SAS procedure (GENMOD) for generalized linear modeling that can conduct nearly all methods presented in this book, a section on association graphs (Section 7.1), and a chapter providing a historical perspective of the development of the methods (Chapter 10). In addition, this text maintains many features of the 1990 text, including

1. a unified presentation based on generalized linear models,
2. methods for ordinal data,
3. exact, small-sample methods,
4. models for multicategory responses,
5. methods for matched pairs,
6. more than 200 exercises, and
7. about 100 examples of real data sets.

Most methods presented in this text require extensive computations. For the most part, I have avoided details about complex calculations, feeling that computing software should relieve this drudgery. Software for categorical data analyses is widely available in most large commercial packages. I recommend that readers of this text use software wherever possible in answering homework problems and checking text examples. The Appendix discusses SPSS for categorical data analyses and contains examples of the use of SAS (particularly PROC GENMOD) for nearly all methods discussed in the text. For ease of reference, the methods are organized there according to the text chapters in which they occur. The tables in the Appendix and many of the data sets analyzed in the book are available from StatLib. This file is available on the World Wide Web at http://lib.stat.cmu.edu/datasets/agresti.

I thank those individuals who commented on parts of the manuscript or who made suggestions about examples or material to cover. These include Patricia Altham, Scott Beck, James Booth, Jane Brockmann, Brent Coull, Al DeMaris, Harry Khamis, Svend Kreiner, Stephen Stigler, Joffre Swait, Larry Winner, and an anonymous reviewer. I owe special thanks to Joan Hilton and Peter Imrey, who spent substantial time reviewing the first draft and who provided numerous helpful comments. Atalanta Ghosh helped to obtain some data sets and Brent Coull helped with graphical software. Many thanks to Kate Roach at Wiley for her continuing interest and to the Wiley staff for their usual high-quality support. As always, special thanks to my wife, Jacki Levine, for her advice and constant encouragement.

Finally, a truly nice by-product of this profession is the opportunity to meet people around the world. Many of these professional colleagues have become valued friends. It is to them that I dedicate this book.

ALAN AGRESTI

London, England
October 1995

CHAPTER 1

Introduction

From helping to assess the value of new medical treatments to evaluating the factors that affect our opinions on various controversial issues, scientists today are finding myriad uses for methods of analyzing categorical data. It's primarily for these scientists and their collaborating statisticians—as well as those training to perform these roles—that this book was written. The book provides an applied introduction to methods for analyzing categorical data, emphasizing uses and interpretations of the methods rather than the theory behind them.

This first chapter introduces fundamental statistical prerequisites. It reviews two key probability distributions for categorical data, the *binomial* and the *Poisson*. It also introduces *maximum likelihood*, the most popular method for estimating parameters for categorical data, and illustrates its use for analyzing proportion data. We begin by discussing the major types of categorical data and describing the book's outline.

1.1 CATEGORICAL RESPONSE DATA

Let's first define categorical data. A *categorical* variable is one for which the measurement scale consists of a set of categories. For instance, political philosophy may be measured as "liberal," "moderate," or "conservative"; choice of breakfast cereal might use categories "hot," "cold," "none"; a diagnostic test for Alzheimer's disease might use categories "symptoms present," "symptoms absent." One and only one category should apply to each subject, unlike the set of categories (liberal, Christian, Republican).

Categorical scales are pervasive in the social sciences for measuring attitudes and opinions on various issues. Categorical scales also occur frequently in the health sciences, for measuring such responses as whether a patient survives an operation (yes, no), severity of an injury (none, mild, moderate, severe), and stage of a disease (initial, advanced).

Though categorical variables are common in the social and health sciences, they are by no means restricted to those areas. They frequently occur in the behavioral sciences (e.g., categories "schizophrenia," "depression," "neurosis" for diagnosis of type of mental illness), public health (e.g., categories "yes" and "no" for whether awareness of AIDS has led to increased use of condoms), zoology (e.g., categories "fish," "invertebrate," "reptile" for alligators' primary food preference), education (e.g., categories "correct" and "incorrect" for students' responses to an exam question), and marketing (for example, categories "Brand A," "Brand B," and "Brand C" for consumers' preference among three leading brands of a product). They even occur in highly quantitative fields such as engineering sciences and industrial quality control, when items are classified according to whether or not they conform to certain standards.

1.1.1   Response/Explanatory Variable Distinction

Many statistical analyses distinguish between *response* variables and *explanatory* variables. For instance, regression models describe how the distribution of a continuous response variable, such as annual income, changes according to levels of explanatory variables, such as number of years of education and number of years of job experience. The response variable is sometimes called the *dependent variable* or *Y variable*, and the explanatory variable is sometimes called the *independent variable* or *X variable*.

The subject of this text is the analysis of categorical response variables. The categorical variables listed in the previous subsection, such as political philosophy, are response variables; in some studies, they might also serve as explanatory variables. Statistical models for categorical response variables analyze how such responses are influenced by explanatory variables. For example, one might study how political philosophy depends on factors such as annual income, attained education, religious affiliation, age, gender, and race. The explanatory variables can be categorical or continuous.

1.1.2   Nominal/Ordinal Scale Distinction

There are two primary types of measurement scales for categorical variables. Many categorical scales have a natural ordering. Examples are attitude toward legalization of abortion (disapprove in all cases, approve only in certain cases, approve in all cases), appraisal of a company's inventory level (too low, about right, too high), response to a medical treatment (excellent, good, fair, poor), and diagnosis of whether a patient is mentally ill (certain, probable, unlikely, definitely not). Categorical variables having ordered scales are called *ordinal* variables.

Categorical variables having unordered scales are called *nominal* variables. Examples are religious affiliation (categories Catholic, Jewish, Protestant, other), mode of transportation to work (automobile, bicycle, bus, subway, walk), favorite type of music (classical, country, folk, jazz, rock), and choice of residence (apartment, condominium, house, other).

For nominal variables, the order of listing the categories is irrelevant, and the statistical analysis should not depend on that ordering. Methods designed for nominal variables give the same results no matter in what order the categories are listed.

Methods designed for ordinal variables utilize the category ordering. Whether we list the categories from low to high or from high to low is usually irrelevant, but results of ordinal analyses would change if the categories were reordered in any other way.

Methods designed for ordinal variables cannot be used with nominal variables, since nominal variables do not have ordered categories. Methods designed for nominal variables can be used with nominal or ordinal variables, since they only require a categorical scale. When used with ordinal variables, however, they do not use the information about that ordering. This can result in serious loss of power. It is usually best to apply methods appropriate for the actual scale.

Categorical variables are often referred to as *qualitative*, to distinguish them from numerical-valued or *quantitative* variables such as weight, age, income, and number of times arrested. However, it is often advantageous to treat ordinal data in a quantitative manner, for instance, by assigning ordered scores to the categories.

1.1.3   Organization of This Book

The chapters of this book divide into three parts. The first part, consisting of Chapters 1–3, describes some standard methods of categorical data analysis developed prior to about 1960. These include basic ways of assessing association between two categorical variables and analyzing the effect on that association of controlling for another variable.

The second part of the book, Chapters 4–7, introduces models for categorical responses. These models resemble regression models for continuous response variables, but they assume binomial or Poisson, rather than normal, distributions. We present two types of models in detail, *logistic regression* (also called *logit*) *models* and *loglinear models*. Logit models apply to binomial responses, whereas loglinear models apply to Poisson responses. Connections exist between them, and Chapter 4 shows they are special cases of a generalized class of models that also contains the usual normal-distribution-based regression models for continuous responses.

The third part of the book, Chapters 8 and 9, discusses logit and loglinear models for two specialized situations. Chapter 8 presents extensions of binomial-response models to *multinomial* responses, which have several outcome categories, rather than two; included there are special models for ordinal responses. Chapter 9 presents models for matched pairs; these apply, for instance, when we observe a categorical response for the same subjects at two separate times. The book concludes (Chapter 10) with a historical overview of the development of categorical data methods.

Many methods for analyzing categorical data require extensive computations. The Appendix discusses the use of SAS and SPSS statistical software and provides SAS code for performing nearly all the analyses presented in this book.

1.2   SAMPLING MODELS

The analysis of categorical data, or of any type of data for that matter, requires assumptions about the random mechanism that generated the data. For regression and analysis of variance (ANOVA) models for continuous data, the normal distribution plays a central role. This section discusses two key distributions for categorical data: the *Poisson* and *binomial* distributions.

1.2.1 Poisson Sampling

In England, the M1 is a primary north–south motorway, heavily used by commercial as well as private vehicles. A group of British transportation researchers plan to study the rate of fatal automobile accidents on a rural stretch of that motorway. Their study will catalog for the next year all accidents resulting in a fatality on this part of the M1. The data will consist of weekly counts of the number of fatal accidents.

The *Poisson* distribution is a potential probability model for the number of fatal accidents in any given week. It is indexed by a parameter μ, its mean. Let y denote a possible outcome for a Poisson variate. The formula for Poisson probabilities is

$$P(y) = \frac{e^{-\mu}\mu^{y}}{y!} \qquad y = 0, 1, 2, \ldots. \tag{1.2.1}$$

The possible outcomes for y are the nonnegative integers. The term $y!$, called y *factorial*, denotes the product $1 \times 2 \times \cdots \times y$, with $0! = 1$. The term $e^{-\mu}$ denotes the *exponential function*, evaluated at $-\mu$. This is sometimes expressed as $\exp(-\mu)$.

The exponential function is available on many pocket calculators; one enters the value for $-\mu$ (call it x) and presses the e^x key. The exponential function is the antilog for the logarithm using the *natural log* scale. This means that $e^a = b$ is equivalent to $\log(b) = a$, where log denotes the natural log. For instance, $e^0 = \exp(0) = 1$ corresponds to $\log(1) = 0$; similarly, $e^{.7} = \exp(.7) = 2.0$ corresponds to $\log(2) = .7$. All logarithms in this text use this natural log scale, which has $e = e^1 = 2.718$ as the base.

Suppose that fatal accidents on the M1 occur at the average rate of 2 per week. For the Poisson model with $\mu = 2$, the probability of 0 fatal accidents ($y = 0$) in a given week equals

$$P(0) = \frac{(e^{-2})(2^0)}{0!} = e^{-2} = .135.$$

(Recall that a number raised to the zero power equals 1, and $0! = 1$.) Similarly, $P(1) = (e^{-2})(2^1)/1! = 2e^{-2} = .271$. Table 1.1 lists the probabilities for all y, and Figure 1.1 plots them. Though the distribution applies to all nonnegative integer y values, when $\mu = 2$ practically all the probability falls between y values of 0 and 8.

Over a period of t weeks having constant average rate of two fatal accidents per week, the Poisson model for the total number of fatal accidents has a mean of $2t$. For instance, for a three-week period, the Poisson distribution has $\mu = 6$. The probability of no fatalities in a given three-week period is $(e^{-6})(6^0)/0! = e^{-6} = .002$. This equals $(.135)^3$, the probability of no fatalities in every one of the three one-week periods. Figure 1.1 also plots the Poisson distribution having this mean; it is slightly more spread out than the one having a mean of 2.

Table 1.1 **Poisson and Binomial Distributions with
Means of 2.0**

	P(y)	
y	Poisson	Binomial
0	.135	.107
1	.271	.268
2	.271	.302
3	.180	.201
4	.090	.088
5	.036	.027
6	.012	.005
7	.004	.001
8	.001	.000
9	.000	.000
10	.000	.000
≥ 11	.000	—

Note: Binomial has $N = 10$ and $\pi = .2$.

Let $E(Y)$ denote the expected value of a variable Y, which is the mean of its probability distribution. Let $\text{Var}(Y) = \sigma^2(Y)$ denote its variance, and $\sigma(Y)$ its standard deviation. Unlike the normal distribution, the Poisson distribution (1.2.1) does not have a separate parameter describing variation. In fact, the Poisson mean parameter μ is also the variance of the distribution. That is,

$$E(Y) = \text{Var}(Y) = \mu, \qquad \sigma(Y) = \sqrt{\mu}.$$

The distribution in Figure 1.1 having mean equal to 2 has standard deviation equal to $\sqrt{2} = 1.4$. If the fatal accident rate truly stays constant over time with weekly expectation of 2, then over a long period the weekly fatal-accident counts should have a sample mean of about 2 and a sample standard deviation of about 1.4. The distribution in Figure 1.1 having mean equal to 6 has standard deviation equal to $\sqrt{6} = 2.4$.

An important sampling model for categorical data treats each category count as an independent Poisson observation. This sampling scheme is called *Poisson sampling*. A key feature of the Poisson distribution is that its variance increases as the mean does. Sample counts tend to vary more when their average level is higher. When the mean number of fatal accidents equals 20 per week, we observe greater variability in the weekly counts than when the mean equals 2 per week.

In practice, count observations often have variance exceeding the mean, rather than equaling the mean as the Poisson requires. This phenomenon is called *overdispersion*. Our example assumed that the expected number of fatal accidents was the same (namely, 2) in each week of the year. Traffic has seasonal fluctuations in intensity, however, probably producing weekly variation in this expectation. Such variation

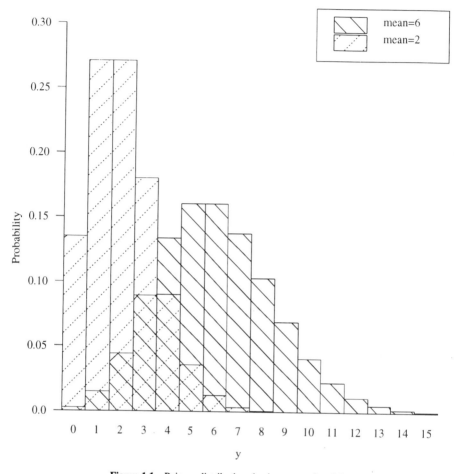

Figure 1.1 Poisson distributions having means 2 and 6.

would cause fatality counts to display somewhat more variation than predicted by the Poisson model with a constant expectation.

The assumption of Poisson sampling is often too simplistic, because of factors such as overdispersion. Nevertheless, Poisson sampling assumptions produce useful results, albeit in an approximate manner, in a wide variety of categorical data analyses.

1.2.2 Binomial Sampling

In the auto-fatality example, the weekly number of fatal accidents is random, rather than fixed. The weekly number of accidents itself, whether fatal or not, is also random. Before any particular week, we do not know the number of accidents. Many applications, however, have a sample size index such as this that is fixed, rather than random. For instance, suppose the researchers plan to classify outcomes of all accidents until N occur, for the purpose of estimating the *proportion* π of accidents that result in a fatality. The total sample size is then fixed at N. For this design the number of fatal accidents cannot have a Poisson distribution. Poisson probabilities are positive for all nonnegative integers, yet the count of fatal accidents cannot exceed N in this design.

To illustrate, suppose the number of fatal accidents over t weeks has a Poisson distribution with mean $2t$, and the number of nonfatal accidents has a Poisson distribution with mean $8t$. The researchers observe a total number N of accidents and then stop the observation process. The counts of fatal and nonfatal accidents sum to N and then have a *binomial distribution*. The number of fatal accidents has a binomial distribution with N "trials" and parameter $\pi = 2t/(2t + 8t) = .2$, which is the probability that any given accident is fatal. This probability equals the Poisson mean for fatal accidents divided by the sum of the Poisson means for fatal and nonfatal accidents. The number of nonfatal accidents equals (N − number fatal accidents) and has a binomial distribution with N trials and parameter .8.

We now more formally define the binomial distribution. Consider N independent and identical trials with two possible outcomes for each, referred to as "success" and "failure." These are simply generic labels, and the success outcome need not refer to a preferred result. In this example, the two outcomes are "fatal accident" and "nonfatal accident." *Identical trials* means that the probability of success is the same for each trial. *Independent trials* means that one can represent them by independent random variables; in particular, the outcome of one trial does not affect the outcome of another. These are often called *Bernoulli trials*. Let π denote the probability of success for a given trial. Let Y denote the number of successes out of the N trials. The variate Y has the binomial distribution with N trials and parameter π.

You are probably already familiar with the binomial distribution, but we review it briefly here. The probability of outcome y for Y equals

$$P(y) = \frac{N!}{y!\,(N-y)!}\pi^y(1-\pi)^{N-y}, \qquad y = 0, 1, 2, \ldots, N. \qquad (1.2.2)$$

To illustrate, let Y denote the number of fatal accidents out of 10 accidents, when the chance of fatality is .2 for each accident; then $N = 10$ and $\pi = .2$. The probability of $y = 0$ fatal accidents, and hence $N - y = 10$ nonfatal accidents, equals

$$P(0) = [10!/(0!\,10!)](.2)^0(.8)^{10} = (.8)^{10} = .107;$$

the probability of 1 fatal accident equals

$$P(1) = [10!/(1!\,9!)](.2)^1(.8)^9 = 10(.2)(.8)^9 = .268.$$

Table 1.1 tabulates the entire distribution.

The binomial distribution for N trials with parameter π has mean and variance

$$E(Y) = N\pi, \qquad \mathrm{Var}(Y) = N\pi(1 - \pi).$$

Unlike the Poisson variance, the binomial variance is smaller than the mean. The binomial distribution in Table 1.1 has a mean of $10(.2) = 2.0$ and a variance of $10(.2)(.8) = 1.6$. The binomial mean is the same as the Poisson mean in Table 1.1, but the binomial variability is slightly less. Both the binomial and Poisson distributions become more bell-shaped as their mean increases, and when the mean is large they are often approximated by normal distributions.

Some trials have more than two possible outcomes. For instance, one might summarize the outcome for the driver in each accident using the categories "uninjured," "injury not requiring hospitalization," "injury requiring hospitalization," "fatality." The distribution of counts in the various categories is then the *multinomial*, and we say there is *multinomial sampling*. The binomial distribution is the special case of the multinomial with only two possible outcomes for each trial. We study the multinomial distribution in Chapter 8, which deals with modeling multicategory response variables.

Standard procedures for categorical data assume a Poisson, binomial, or multinomial sampling model. Most inferential analyses have the pleasing result that parameter estimates and standard errors are identical for Poisson and (binomial/multinomial) sampling schemes. In particular, this happens for parameters in the logistic regression and loglinear models that are the primary focus of this text. For completeness, we have presented the formulas for Poisson and binomial probabilities, but there is little need for them in the remainder of the text. The calculations for most parameter estimates and standard errors based on these sampling models are complex, and widely available statistical software computes them for us.

1.3 INFERENCE FOR A PROPORTION

In practice, sampling models such as the Poisson and binomial have unknown parameter values. Using sample data, we estimate the parameters. This section introduces the estimation method used in this text, called *maximum likelihood*. We illustrate this method by applying it to inference for the binomial parameter π.

1.3.1 Maximum Likelihood Estimation

For a particular sampling model, we can substitute the sample data into the probability function and then view that probability as a function of the unknown parameter value. For instance, in $N = 10$ trials, suppose a binomial count equals $y = 0$. From the binomial formula (1.2.2) with parameter π, the probability of this outcome equals

$$P(0) = [10!/(0!)(10!)]\pi^0(1 - \pi)^{10} = (1 - \pi)^{10}.$$

This probability is defined for all the potential values of π, namely the values between 0 and 1.

The probability of the observed data, expressed as a function of the parameter, is called a *likelihood function*. The binomial likelihood function for $y = 0$ successes

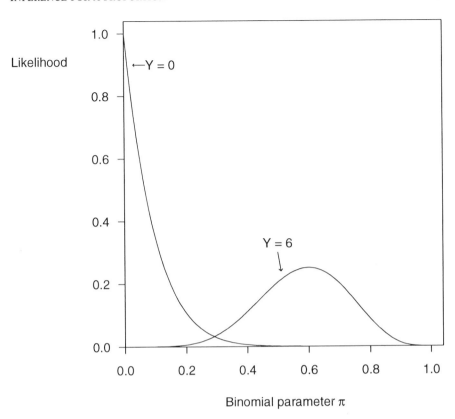

Likelihood

Figure 1.2 Binomial likelihood functions for $Y = 0$ successes and for $Y = 6$ successes in 10 trials.

in 10 trials is $l(\pi) = (1 - \pi)^{10}$, defined for π between 0 and 1. Figure 1.2 shows a plot of this likelihood function. From this function, for instance, the probability that $Y = 0$ is $l(.4) = (1 - .4)^{10} = .006$ if $\pi = .4$, $l(.2) = (1 - .2)^{10} = .107$ if $\pi = .2$, and $l(.0) = (1 - .0)^{10} = 1.0$ if $\pi = .0$.

The *maximum likelihood estimate* of the parameter is defined to be the parameter value for which the probability of the observed data takes its greatest value. This is the parameter value at which the likelihood function takes its maximum. Figure 1.2 shows that the likelihood function $l(\pi) = (1 - \pi)^{10}$ has its maximum at $\pi = 0.0$. Thus, when 10 trials have 0 successes, the maximum likelihood estimate of π equals 0.0.

Generally, the binomial outcome of y successes in N trials has maximum likelihood estimate of π equal to $p = y/N$. This is the sample proportion of successes for the N trials. If the binomial outcome for $N = 10$ trials equals $y = 6$, then the maximum likelihood estimate of π equals $p = 6/10 = .6$. Figure 1.2 also plots the likelihood function when $N = 10$ with $y = 6$, which from formula (1.2.2) equals $l(\pi) = [10!/(6!)(4!)]\pi^6(1 - \pi)^4$. The maximum value occurs when $\pi = .6$. That is, the result $y = 6$ in $N = 10$ trials is more likely to occur when $\pi = .6$ that when π equals any other value.

Denote each success by a 1 and each failure by a 0. Then the sample proportion equals the sample mean of the results of the individual trials. For instance, for 4 failures and 6 successes in 10 trials, the data are $0, 0, 0, 0, 1, 1, 1, 1, 1, 1$, and the sample mean of $p = (0 + 0 + 0 + 0 + 1 + 1 + 1 + 1 + 1 + 1)/10 = .6$ estimates π.

The Poisson distribution also has maximum likelihood estimate of the mean parameter μ equal to the sample mean of the data. If in 10 successive weeks, the numbers of M1 motorway traffic accidents resulting in fatalities are 3, 1, 2, 0, 4, 2, 2, 5, 0, 1, then the maximum likelihood estimate of μ equals $(3 + 1 + 2 + 0 + 4 + 2 + 2 + 5 + 0 + 1)/10 = 2.0$. The Poisson probability of observing a sample like this is higher when $\mu = 2.0$ than when μ equals any other value.

The abbreviation ML symbolizes the term *maximum likelihood*. The ML estimate is often denoted by the parameter symbol with a ˆ (a "hat") over it. The ML estimate of the Poisson parameter μ, for instance, is denoted by $\hat{\mu}$, called *mu-hat*.

Before we observe the data, the value of an estimate is unknown, and it is a variate having some sampling distribution. We refer to this variate as an *estimator*, and its value for observed data as an *estimate*. Estimators based on the method of maximum likelihood are popular because they have good large-sample behavior. Most importantly, it is not possible to find estimators that are more precise, in terms of having smaller large-sample standard errors. Also, large-sample distributions of ML estimators are approximately normal. All estimators reported in this text use this method.

1.3.2 Test about a Proportion

We now use the ML estimator p in statistical inference for the binomial parameter π. The sampling distribution of the sample proportion p has mean and standard error

$$E(p) = \pi, \qquad \sigma(p) = \sqrt{\frac{\pi(1 - \pi)}{N}}.$$

As the number of trials N increases, the standard error of p decreases toward zero; that is, the sample proportion tends to be closer to the parameter value π.

The approximate normality of the distribution of p, for large N, suggests simple large-sample inferential methods for π. First, we test the null hypothesis $H_0 : \pi = \pi_0$ that the parameter equals some fixed value, π_0. The test statistic

$$z = \frac{p - \pi_0}{\sqrt{\frac{\pi_0(1 - \pi_0)}{N}}}$$

divides the difference between the sample proportion p and the null hypothesis value π_0 by the *null* standard error of p. The null standard error is the one computed under the assumption that the null hypothesis is true. For large samples, the null sampling distribution of the z test statistic is the standard normal, having a mean of 0 and standard deviation of 1. The z statistic represents the number of standard errors that the sample proportion falls from the null hypothesized proportion.

To illustrate, we analyze whether a majority or minority of adults in the United States believes that a pregnant woman should be able to obtain an abortion. Let π denote the proportion of the American adult population that responds "yes" to the question, "Please tell me whether or not you think it should be possible for a pregnant woman to obtain a legal abortion if she is married and does not want any more children." We test $H_0 : \pi = .5$ against the two-sided alternative hypothesis, $H_a : \pi \neq .5$. This item was one of many included in the 1991 General Social Survey, conducted by the National Opinion Research Center (NORC) at the University of Chicago. This survey, conducted annually, asks about 1500 adult American subjects their opinions about a wide variety of issues. The NORC makes available computer tapes and disks and a cumulative codebook summarizing responses to surveys since 1972. Of 950 subjects who responded to this abortion item in 1991, 424 replied "yes" and 526 replied "no."

The sample proportion of "yes" responses was $424/950 = .446$. For a sample of size $N = 950$, the null standard error of p equals $\sqrt{(.5)(.5)/950} = .016$. The test statistic equals $z = (.446 - .5)/.016 = -3.31$. The two-sided P-value is the probability that the absolute value of a standard normal variate exceeds 3.31, which equals $P = .001$. There is strong evidence that, in 1991, $\pi < .5$; that is, that fewer than half of Americans favored legal abortion in this situation. In some other situations, such as when the mother's health was endangered, an overwhelming majority favored legal abortion. Responses also depended strongly on the question wording.

1.3.3 Confidence Interval for a Proportion

A significance test simply indicates whether a particular value for a parameter (such as .5) is plausible. In nearly all applications we learn much more by estimating the parameter, using a confidence interval to determine the range of plausible values. A large-sample 95% confidence interval for π has the formula

$$p \pm 1.96\hat{\sigma}(p), \quad \text{where } \hat{\sigma}(p) = \sqrt{p(1 - p)/N}. \tag{1.3.1}$$

This formula substitutes the sample proportion p for the unknown parameter π in the standard error formula for $\sigma(p)$.

For the example just discussed, this interval equals $.446 \pm 1.96\sqrt{(.446)(.554)/950}$, or $.446 \pm .032$, or $(.414, .478)$. We can be 95% confident that the population proportion of Americans favoring legalized abortion for married pregnant women who do not want more children is between .414 and .478.

Though formula (1.3.1) is simple, when $\pi < .20$ or $\pi > .80$, it does not work well. The actual coverage probability may not be near the nominal level, even for moderately large sample sizes. A better way to construct confidence intervals uses a duality with significance tests. This confidence interval consists of all values π_0 for the null hypothesis parameter that are judged plausible in the test. A 95% confidence interval contains all values π_0 for which the two-sided P-value exceeds .05; that is, it contains all values that are "not rejected" at the .05 significance level. These are the null values that have test statistic z less than 1.96 in absolute value. This alternative method does not require estimating π in the standard error, since the test statistic's standard error uses the null value π_0.

For a sample proportion of $p = .446$ based on $N = 950$, the value $\pi_0 = .478$ for the null hypothesis parameter leads to the test statistic value $z = (.446 - .478)/\sqrt{(.478)(.522)/950} = -1.96$ and a two-sided P-value of $P = .05$; the value $\pi_0 = .415$ leads to $z = (.446 - .415)/\sqrt{(.415)(.585)/950} = 1.96$ and $P = .05$. (We explain in the following paragraph how we got .478 and .415.) All π_0 values between .415 and .478 give $|z| < 1.96$ and $P > .05$, so the 95% confidence interval equals $(.415, .478)$. Here, the sample size is large, and this method gives essentially the same result as the simpler method (1.3.1).

For given p and N, one can determine the π_0 values that yield the test statistic value $z = \pm 1.96$ by solving the equation $|p - \pi_0|/\sqrt{\pi_0(1 - \pi_0)/N} = 1.96$ for π_0. This equation is quadratic in π_0 (see Problem 1.15). Alternatively, one can determine the limits by trial and error, using the endpoints of interval (1.3.1) as initial guesses.

As a rough guideline, the large-sample test and the confidence interval based on that test perform well when $N\pi > 5$ and $N(1 - \pi) > 5$; that is, when the expected number of outcomes of the two types both exceed 5. Testing $H_0 : \pi = .5$ requires only $N > 10$, whereas testing $H_0 : \pi = .1$ (or $H_0 : \pi = .9$) requires $N > 50$. The sample size requirement reflects the increasingly skewed behavior of the sampling distribution of p as π approaches 0 or 1, with larger samples needed before a symmetric normal shape occurs.

For samples that are too small for the normal approximation to be adequate, one can use the binomial distribution directly in calculating P-values. For instance, for testing $H_0 : \pi = .5$ against $H_a : \pi > .5$, suppose there are $Y = 4$ successes in $N = 5$ trials. The P-value for this one-sided alternative is the right-tail probability, $P(4) + P(5)$, for a binomial distribution with $N = 5$ and $\pi = .5$. Using formula (1.2.2), this equals .188.

In summary, this chapter has introduced two key distributions for categorical data analysis: the binomial and the Poisson. It has also introduced maximum likelihood estimation and illustrated its use for proportion data. The rest of the text uses ML inference for binomial and Poisson parameters in a wide variety of contexts.

PROBLEMS

1.1. In the following examples, identify the response variable and the explanatory variables.

    a. Attitude toward gun control (favor, oppose), Gender (female, male), Mother's education (high school, college).

    b. Heart disease (yes, no), Blood pressure, Cholesterol level.

    c. Race (white, nonwhite), Religion (Catholic, Jewish, Protestant), Vote for President (Democrat, Republican, Other), Annual income.

1.2. Which scale of measurement is most appropriate for the following variables— nominal, or ordinal?

    a. Political party affiliation (Democrat, Republican, unaffiliated).

    b. Highest degree obtained (none, high school, bachelor's, master's, doctorate).

    c. Patient condition (good, fair, serious, critical).

    d. Hospital location (London, Boston, Madison, Rochester, Toronto).

    e. Favorite beverage (beer, juice, milk, soft drink, wine, other).

1.3. Silicon wafers for computer chips manufactured by a high-tech company have an average number of defects of 1.0 per wafer. If the number of defects has the Poisson distribution, find the probability that a wafer has (a) 0 defects, (b) 1 defect, (c) at least 2 defects.

1.4. Each of 100 multiple-choice questions on an exam has four possible answers but one correct response. For each question, a student randomly selects one response as the answer. Specify the distribution of the student's number of correct answers on the exam. Based on the mean and standard deviation of that distribution, would it be surprising if the student made at least 50 correct responses?

1.5. A balanced coin is flipped twice. Let Y = number of heads obtained.

    a. Specify the probabilities for the possible values for Y, and give the distribution's mean and variance.

    b. Calculate probabilities for the Poisson distribution having the same mean as the distribution in **(a)**. How does its variance compare to that in **(a)**?

    c. For each flip of a possibly unbalanced coin, let π denote the probability of a head. Suppose there are 0 heads in 2 flips. Find the ML estimate of π. Does this estimate seem "reasonable"? (The Bayesian estimator, an alternative one that combines one's prior beliefs about parameter values with the sample data, provides a nonzero estimate of π; it equals $(Y + 1)/(N + 2) = .25$ when the prior belief is that π is equally likely to be anywhere between 0 and 1.)

1.6. Refer to the previous problem.

    a. Calculate the binomial probabilities for $N = 2$ when the probability of a head for each flip equals (i) $\pi = .6$, (ii) $\pi = .4$.

    b. Suppose we observe $Y = 1$. Calculate and sketch the likelihood function.

    c. Using the plotted likelihood function from **(b)**, show that the ML estimate of π equals .5.

1.7. In his autobiography *A Sort of Life*, British author Graham Greene described a period of severe mental depression during which he played Russian Roulette. This "game" consists of putting a bullet in one of the six chambers of a pistol, spinning the chambers to select one at random, and then firing the pistol once at one's head.

    a. Greene played this game six times, and was lucky that none of them resulted in a bullet firing. Find the probability of this outcome.

    b. Suppose one kept playing this game until the bullet fires. Let Y denote the number of the game on which the bullet fires. Argue that the probability of the outcome y equals $(5/6)^{y-1}(1/6)$, for $y = 1, 2, 3, \ldots$. (This is called the *geometric distribution.*)

1.8. A sample of women suffering from dysmenorrhea have been taking an analgesic designed to diminish the effects. A new analgesic is claimed to provide greater relief. After trying the new analgesic, 40 women reported greater relief with the standard analgesic, and 60 reported greater relief with the new one.

a. Test the hypothesis that the probability of greater relief with the standard analgesic is the same as the probability of greater relief with the new analgesic. Report and interpret the P-value for the two-sided alternative.

b. Construct and interpret a 95% confidence interval for the probability of greater relief with the new analgesic.

1.9. Refer to the previous problem. The researchers wanted a sufficiently large sample to be able to estimate the probability of preferring the new analgesic to within .08, with confidence .95. If the true probability is .75, how large a sample is needed to achieve this accuracy? (*Hint:* How large should N be so that a 95% confidence interval has plus and minus term equal to .08?)

1.10. *Newsweek* magazine (March 27, 1989) reported results of a poll about religious beliefs, conducted by the Gallup Organization. Of 750 American adults, 24% believed in reincarnation. Treating this as a random sample, construct and interpret a 95% confidence interval for the true proportion of American adults believing in reincarnation.

1.11. A criminologist wants to estimate the proportion of U.S. citizens who live in a home in which firearms are available. The 1991 General Social Survey asked respondents, "Do you have in your home any guns or revolvers?" Of the respondents, 393 answered "yes" and 583 answered "no." Construct a 90% confidence interval for the true proportion of "yes." Interpret.

1.12. If Y is a variate and c is a positive constant, then the standard deviation of the distribution of cY equals $c\sigma(Y)$. Suppose Y is a binomial variate, and let $p = Y/N$. Show that $\sigma(p) = \sqrt{\pi(1-\pi)/N}$. Explain why it is easier to get a close estimate of π when it is near 0 or 1 than when it is near $\frac{1}{2}$.

1.13. A variate has a Poisson distribution, with unknown parameter μ. The sole observation equals 0.

a. Find and plot the likelihood function over the space of potential values for μ.

b. What is the ML estimate of μ? (Recall: The ML estimate of μ equals the sample mean.)

1.14. Using calculus, it is easier to derive the maximum of the log of the likelihood function, $L = \log l$, than the likelihood function l itself. Both functions have maximum at the same value, so it is sufficient to do either.

a. Calculate the log likelihood $L(\pi)$ for the binomial distribution (1.2.2).

b. One can usually determine the point at which the maximum of a log likelihood L occurs by solving the *likelihood equation*. This is the equation resulting from differentiating L with respect to the parameter, and setting the derivative equal to zero. Find the likelihood equation for the binomial distribution, and solve it to show that the ML estimate equals $p = y/N$.

1.15. Show that a value π_0 for which the statistic $z = (p - \pi_0)/\sqrt{\pi_0(1 - \pi_0)/N}$ takes some fixed value z_0 is a solution to the equation $(1 + z_0^2/N)\pi_0^2 + (-2p - z_0^2/N)\pi_0 + p^2 = 0$. Hence, using the formula $x = (-b \pm \sqrt{b^2 - 4ac})/2a$ for solving the quadratic equation $ax^2 + bx + c = 0$, obtain the limits for the 95% confidence interval in Section 1.3.3 for the proportion of Americans favoring legalized abortion.

CHAPTER 2

Two-Way Contingency Tables

Table 2.1 cross classifies a sample of Americans according to their gender and their opinion about an afterlife. For the females in the sample, for instance, 435 said they believed in an afterlife and 147 said they did not or were undecided. For such data, we might study whether an association exists between gender and belief in an afterlife. Is one sex more likely than the other to believe in an afterlife, or is belief in an afterlife independent of gender?

Analyzing associations is at the heart of most multivariate statistical analyses. This chapter deals with associations between two categorical variables. We introduce parameters that describe the association and present inferential methods for those parameters.

Many applications involve comparing two groups with respect to the relative numbers of observations in two categories. For Table 2.1, one might compare the proportions of males and females who believe in an afterlife. For such data, Section 2.2 presents methods for analyzing differences and ratios of proportions. Section 2.3 presents another measure, the *odds ratio*, that plays a key role for several methods discussed in this text. Sections 2.4 and 2.5 describe large-sample significance tests about whether an association exists between two categorical variables; Section 2.4 presents tests for nominal variables, and Section 2.5 presents an alternative test for ordinal variables. Section 2.6 discusses small-sample analyses. First, Section 2.1 introduces terminology and notation.

2.1 PROBABILITY STRUCTURE FOR CONTINGENCY TABLES

Categorical data consist of frequency counts of observations occurring in the response categories. Let X and Y denote two categorical variables, X having I levels and Y having J levels. We display the IJ possible combinations of outcomes in a rectangular table having I rows for the categories of X and J columns for the categories of Y. The *cells* of the table represent the IJ possible outcomes. A table of this form in

16

Table 2.1 Cross Classification of Belief in Afterlife by Gender

	Belief in Afterlife	
Gender	Yes	No or Undecided
Females	435	147
Males	375	134

Source: Data from 1991 General Social Survey.

which the cells contain frequency counts of outcomes is called a *contingency table*. A contingency table that cross classifies two variables is called a *two-way table*; one that cross classifies three variables is called a *three-way table*, and so forth. A two-way table having I rows and J columns is called an $I \times J$ (read I-by-J) table. Table 2.1, for instance, is a 2×2 table.

2.1.1 Joint, Marginal, and Conditional Probabilities

Probability distributions for contingency tables relate to the sampling scheme, as we shall discuss in Section 2.1.4. We first present the fundamental types of probabilities for two-way contingency tables. Suppose first that each subject in a sample is randomly chosen from some population of interest, and then classified on two categorical responses, X and Y. Let $\pi_{ij} = P(X = i, Y = j)$ denote the probability that (X, Y) falls in the cell in row i and column j. The probabilities $\{\pi_{ij}\}$ form the *joint distribution* of X and Y. They satisfy $\sum_{i,j} \pi_{ij} = 1$.

The *marginal distributions* are the row and column totals of the joint probabilities. These are denoted by $\{\pi_{i+}\}$ for the row variable and $\{\pi_{+j}\}$ for the column variable, where the subscript "+" denotes the sum over the index it replaces. For instance, for 2×2 tables,

$$\pi_{1+} = \pi_{11} + \pi_{12} \quad \text{and} \quad \pi_{+1} = \pi_{11} + \pi_{21}.$$

We use similar notation for samples, with Roman p in place of Greek π. For instance, $\{p_{ij}\}$ denotes the sample joint distribution. These are the sample cell proportions. The cell counts are denoted by $\{n_{ij}\}$, with $n = \sum_{i,j} n_{ij}$ denoting the total sample size. The cell proportions and cell counts are related by

$$p_{ij} = \frac{n_{ij}}{n}.$$

The marginal frequencies are the row totals $\{n_{i+}\}$ and the column totals $\{n_{+j}\}$.

In many contingency tables, one variable (say, the column variable, Y) is a response variable and the other (the row variable, X) is an explanatory variable. Then it is informative to construct a separate probability distribution for Y at each level of X. Such a distribution consists of *conditional probabilities* for Y, given the level of X, and is called a *conditional distribution*.

Table 2.2   Notation for Table 2.1

| Gender | Belief in Afterlife | | Total |
	Yes	No or Undecided	
Females	$n_{11} = 435$	$n_{12} = 147$	$n_{1+} = 582$
Males	$n_{21} = 375$	$n_{22} = 134$	$n_{2+} = 509$
Total	$n_{+1} = 810$	$n_{+2} = 281$	$n = 1091$

2.1.2   Belief in Afterlife Example

Table 2.1, a 2×2 contingency table, cross classifies $n = 1091$ respondents to the 1991 General Social Survey by their gender and their belief in an afterlife. Table 2.2 illustrates the cell count notation for these data. For instance, $n_{11} = 435$, and the corresponding sample joint proportion is $p_{11} = 435/1091 = .399$.

In Table 2.1, belief in the afterlife is a response variable and gender is an explanatory variable. We therefore study the conditional distributions of belief in the afterlife, given gender. For females, the proportion of "yes" responses was $435/582 = .747$, and the proportion of "no" responses was $147/582 = .253$. The proportions $(.747, .253)$ are females' sample conditional distribution of belief in the afterlife. For males, the sample conditional distribution is $(.737, .263)$. (Problem 2.11 requests further analyses of these data using methods of this chapter.)

2.1.3   Independence

Two variables are said to be *statistically independent* if the conditional distributions of Y are identical at each level of X. When two variables are independent, the probability of any particular column response j is the same in each row. Belief in an afterlife is independent of gender, for instancé, if the actual probability of believing in an afterlife equals .740 both for females and for males.

When both variables are response variables, one can describe their relationship using their joint distribution, or the conditional distribution of Y given X, or the conditional distribution of X given Y. Statistical independence is, equivalently, the property that all joint probabilities equal the product of their marginal probabilities,

$$\pi_{ij} = \pi_{i+}\pi_{+j} \quad \text{for } i = 1, \ldots, I \quad \text{and} \quad j = 1, \ldots, J.$$

That is, the probability that X falls in row i and Y falls in column j is the product of the probability that X falls in row i with the probability that Y falls in column j.

2.1.4   Poisson, Binomial, and Multinomial Sampling

The sampling models introduced in Section 1.2 extend to cell counts in contingency tables. For instance, the Poisson sampling model for a 2×2 table treats each of the four cell counts in the table as an independent Poisson variate.

When the rows of a contingency table refer to different groups, the sample sizes for those groups are often fixed by the sampling design. The next section contains a 2×2 table of this type. The first row refers to 11,000 subjects receiving one treatment, the second row refers to 11,000 separate subjects receiving a different treatment, and each subject is measured on a categorical response variable. When the marginal totals for the levels of X are fixed rather than random, a joint distribution for X and Y is no longer meaningful, but conditional distributions for Y at each level of X are. When there are two response categories, we assume a binomial distribution for the sample in each row, with number of trials equal to the fixed row total. When the samples in the rows are independent, such as separate random samples, this sampling scheme is called *independent binomial sampling*.

When the total sample size in the table is fixed but not the row or column totals, a *multinomial* sampling model applies, in which the cells are the possible outcomes. For instance, Table 2.1 cross classifies a random sample of 1091 subjects according to gender and belief in afterlife. The four cell counts are sample values from a multinomial distribution having four categories.

For many multinomial samples over the cells of a contingency table, the columns are a response variable and the rows are an explanatory variable. Then, to describe the data, it is sensible to divide the cell counts by the row totals to form conditional distributions on the response. In doing so, we inherently treat the row totals as fixed and analyze the data the same way as if they formed separate independent samples. In Table 2.1, for instance, we might treat the results for females as a binomial sample with outcome categories "yes" and "no or undecided" for belief in an afterlife, and the results for males as an independent binomial sample on the same response variable. If there were more than two response categories, such as ("yes," "no," "undecided"), we would treat the samples as independent multinomial samples.

For most analyses, one need not worry about which sampling model makes the most sense. For the primary inferential methods in this text, the same results occur for Poisson, multinomial, and independent binomial/multinomial sampling models.

2.2   COMPARING PROPORTIONS IN TWO-BY-TWO TABLES

Response variables having two categories are called *binary variables*. For instance, "belief in afterlife" is binary when measured using the categories (yes, no). Many studies compare two groups on a binary response, Y. The data can be displayed in a 2×2 contingency table, in which the rows are the two groups and the columns are the response levels of Y. This section presents measures for comparing two groups on binary responses.

2.2.1   Difference of Proportions

We use the generic terms *success* and *failure* for the response categories of a binary variable. For subjects in row 1, let π_1 denote the probability of a success, with $1 - \pi_1$ the probability of a failure. The probabilities $(\pi_1, 1 - \pi_1)$ form the conditional distribution of Y in row 1. For subjects in row 2, let π_2 denote the probability of success.

The *difference of proportions* $\pi_1 - \pi_2$ compares the success probabilities in the two rows. This difference falls between -1 and $+1$. It equals zero when $\pi_1 = \pi_2$; that is, when the response is independent of the group classification.

Let p_1 and p_2 denote *sample* proportions of successes for the two rows. The sample difference $p_1 - p_2$ estimates $\pi_1 - \pi_2$. For instance, in row 1 of Table 2.2, $p_1 = n_{11}/n_{1+} = 435/582 = .747$ is the number of "yes" responses divided by the sample size in that row; $p_2 = 375/509 = .737$ is the corresponding sample proportion in row 2. The sample difference of proportions is $.747 - .737 = .010$.

For simplicity, we denote the sample sizes for the two groups (i.e., the row totals n_{1+} and n_{2+}) by N_1 and N_2. When the counts in the two rows are independent binomial samples, the estimated standard error of $p_1 - p_2$ is

$$\hat{\sigma}(p_1 - p_2) = \sqrt{\frac{p_1(1 - p_1)}{N_1} + \frac{p_2(1 - p_2)}{N_2}}. \tag{2.2.1}$$

The standard error decreases, and hence the estimate of $\pi_1 - \pi_2$ improves, as the sample sizes increase. A large-sample $100(1 - \alpha)\%$ confidence interval for $\pi_1 - \pi_2$ is

$$(p_1 - p_2) \pm z_{\alpha/2}\,\hat{\sigma}(p_1 - p_2), \tag{2.2.2}$$

where $z_{\alpha/2}$ denotes the standard normal percentile having right-tail probability equal to $\alpha/2$ (e.g., for a 95% interval, $\alpha = .05$, $z_{\alpha/2} = z_{.025} = 1.96$).

2.2.2 Aspirin and Heart Attacks Example

Table 2.3 is taken from a report on the relationship between aspirin use and myocardial infarction (heart attacks) by the Physicians' Health Study Research Group at Harvard Medical School. The Physicians' Health Study was a five-year randomized study testing whether regular intake of aspirin reduces mortality from cardiovascular disease. Every other day, physicians participating in the study took either one aspirin tablet or a placebo. The study was blind—the physicians in the study did not know which type of pill they were taking.

Table 2.3   Cross Classification of Aspirin Use and Myocardial Infarction (MI)

Group	Myocardial Infarction		Total
	Yes	No	
Placebo	189	10,845	11,034
Aspirin	104	10,933	11,037

Source: Preliminary Report: Findings from the Aspirin Component of the Ongoing Physicians' Health Study. *N. Engl. J. Med., 318*: 262–264 (1988).

We treat the two rows in Table 2.3 as independent binomial samples. Of the $N_1 = 11,034$ physicians taking placebo, 189 suffered myocardial infarction (MI) over the course of the study, a proportion of $p_1 = 189/11,034 = .0171$. Of the $N_2 = 11,037$ physicians taking aspirin, 104 suffered MI, a proportion of $p_2 = .0094$. The sample difference of proportions is $.0171 - .0094 = .0077$. From (2.2.1), this difference has an estimated standard error of

$$\sqrt{\frac{(.0171)(.9829)}{11,034} + \frac{(.0094)(.9906)}{11,037}} = 0.0015.$$

A 95% confidence interval for the true difference $\pi_1 - \pi_2$ is $.0077 \pm 1.96(0.0015)$, or $.008 \pm 0.003$, or $(.005, .011)$. Since this interval contains only positive values, we conclude that $\pi_1 - \pi_2 > 0$; that is, $\pi_1 > \pi_2$, so taking aspirin appears to diminish the risk of MI.

2.2.3 Relative Risk

A difference between two proportions of a certain fixed size may have greater importance when both proportions are near 0 or 1 than when they are near the middle of the range. Consider a comparison of two drugs on the proportion of subjects who have adverse reactions when using the drug. The difference between .010 and .001 is the same as the difference between .410 and .401, namely .009. The first difference seems more noteworthy, since ten times as many subjects have adverse reactions with one drug as the other. In such cases, the ratio of proportions is also a useful descriptive measure.

In 2×2 tables, the *relative risk* is the ratio of the "success" probabilities for the two groups,

$$\frac{\pi_1}{\pi_2}. \tag{2.2.3}$$

It can be any nonnegative real number. The proportions .010 and .001 have a relative risk of $.010/.001 = 10.0$, whereas the proportions .410 and .401 have a relative risk of $.410/.401 = 1.02$. A relative risk of 1.00 occurs when $\pi_1 = \pi_2$; that is, when response is independent of group.

Two groups with *sample* proportions of p_1 and p_2 have a sample relative risk of p_1/p_2. Its sampling distribution can be highly skewed unless the sample sizes are quite large, so its confidence interval formula is rather complex (Problem 2.12).

For Table 2.3, the sample relative risk is $p_1/p_2 = .0171/.0094 = 1.82$. The sample proportion of MI cases was 82% higher for the group taking placebo. Using computer software (SAS-PROC FREQ), we find that a 95% confidence interval for the true relative risk is $(1.43, 2.30)$. We can be 95% confident that, after five years, the proportion of MI cases for physicians taking placebo is between 1.43 and 2.30 times the proportion of MI cases for physicians taking aspirin.

The confidence interval for the relative risk indicates that the risk of MI is at least 43% higher for the placebo group. The confidence interval $(.005, .011)$ for the difference of proportions makes it seem as if the two groups differ by a trivial amount, but the relative risk shows that the difference may have important public health implications. Using the difference of proportions alone to compare two groups can be somewhat misleading when the proportions are both close to zero.

It is sometimes informative to compute also the ratio of "failure" probabilities, $(1 - \pi_1)/(1 - \pi_2)$. This takes a different value than the ratio of the success probabilities. When one of the two outcomes has small probability, normally one computes the ratio of the probabilities for that outcome.

2.3 THE ODDS RATIO

We next present another measure of association for 2×2 contingency tables, called the *odds ratio*. This is a fundamental parameter for models presented in later chapters.

In 2×2 tables, the probability of "success" is π_1 in row 1 and π_2 in row 2. Within row 1, the *odds* of success are defined to be

$$\text{odds}_1 = \frac{\pi_1}{(1 - \pi_1)}.$$

For instance, if $\pi_1 = .75$, then the odds of success equal $.75/.25 = 3$.

The odds are nonnegative, with value greater than 1.0 when a success is more likely than a failure. When odds $= 4.0$, a success is four times as likely as a failure. The probability of success is .8, the probability of failure is .2, and the odds equal $.8/.2 = 4$. We then expect to observe four successes for every one failure. When odds $= \frac{1}{4}$, a failure is four times as likely as a success; we expect to observe one success for every four failures.

Within row 2, the odds of success equal

$$\text{odds}_2 = \frac{\pi_2}{1 - \pi_2}.$$

In either row, the success probability is the function of the odds,

$$\pi = \frac{\text{odds}}{\text{odds} + 1}.$$

For instance, when odds $= 4$, then $\pi = 4/(4 + 1) = .8$. When the conditional distributions are identical in the two rows (i.e., $\pi_1 = \pi_2$), the odds satisfy $\text{odds}_1 = \text{odds}_2$. The variables are then independent.

The ratio of odds from the two rows,

$$\theta = \frac{\text{odds}_1}{\text{odds}_2} = \frac{\pi_1/(1 - \pi_1)}{\pi_2/(1 - \pi_2)}, \tag{2.3.1}$$

is called the *odds ratio*. Whereas the relative risk is a ratio of two probabilities, the odds ratio θ is a ratio of two odds.

2.3.1 Properties of the Odds Ratio

The odds ratio can equal any nonnegative number. When X and Y are independent, $\pi_1 = \pi_2$, so that $odds_1 = odds_2$ and $\theta = odds_1/odds_2 = 1$. The value $\theta = 1$ corresponding to independence serves as a baseline for comparison. Odds ratios on each side of 1 reflect certain types of associations. When $1 < \theta < \infty$, the odds of success are higher in row 1 than in row 2. For instance, when $\theta = 4$, the odds of success in row 1 are four times the odds of success in row 2. Thus, subjects in row 1 are more likely to have successes than are subjects in row 2; that is, $\pi_1 > \pi_2$. When $0 < \theta < 1$, a success is less likely in row 1 than in row 2; that is, $\pi_1 < \pi_2$.

Values of θ farther from 1.0 in a given direction represent stronger levels of association. An odds ratio of 4 is farther from independence than an odds ratio of 2, and an odds ratio of 0.25 is farther from independence than an odds ratio of 0.50. Two values for θ represent the same level of association, but in opposite directions, when one value is the inverse of the other. When $\theta = 0.25$, for instance, the odds of success in row 1 are 0.25 times the odds of success in row 2, or equivalently $1/0.25 = 4.0$ times as high in row 2 as in row 1. When the order of the rows is reversed or the order of the columns is reversed, the new value of θ is the inverse of the original value. This ordering is usually arbitrary, so whether we get 4.0 or 0.25 for the odds ratio is simply a matter of how we label the rows and columns.

The odds ratio does not change value when the orientation of the table reverses so that the rows become the columns and the columns become the rows. The same value occurs when we treat the columns as the response variable and the rows as the explanatory variable, or the rows as the response variable and the columns as the explanatory variable. Since the odds ratio treats the variables symmetrically, it is unnecessary to identify one classification as a response variable in order to calculate it. By contrast, the relative risk requires this, and its value also depends on whether we apply it to the first or second response category.

When both variables are responses, the odds ratio can be defined using joint probabilities as

$$\theta = \frac{\pi_{11}/\pi_{12}}{\pi_{21}/\pi_{22}} = \frac{\pi_{11}\pi_{22}}{\pi_{12}\pi_{21}}. \tag{2.3.2}$$

The odds ratio is also called the *cross-product ratio*, since it equals the ratio of the products $\pi_{11}\pi_{22}$ and $\pi_{12}\pi_{21}$ of cell probabilities from diagonally opposite cells.

The sample odds ratio equals the ratio of the sample odds in the two rows,

$$\hat{\theta} = \frac{p_1/(1 - p_1)}{p_2/(1 - p_2)} = \frac{n_{11}/n_{12}}{n_{21}/n_{22}} = \frac{n_{11}n_{22}}{n_{12}n_{21}}. \tag{2.3.3}$$

For the standard sampling schemes, this is the ML estimator of the true odds ratio.

2.3.2 Odds Ratio for Aspirin Study

To illustrate the odds ratio, we revisit Table 2.3 from Section 2.2.2 on aspirin use and myocardial infarction (MI). For the physicians taking placebo, the estimated odds of MI equal $n_{11}/n_{12} = 189/10{,}845 = 0.0174$. The value 0.0174 means there were 1.74 "yes" responses for every 100 "no" responses. The estimated odds equal $104/10{,}933 = 0.0095$ for those taking aspirin, or 0.95 "yes" responses per every 100 "no" responses.

The sample odds ratio equals $\hat{\theta} = 0.0174/0.0095 = 1.832$. This also equals the cross-product ratio $(189)(10{,}933)/(10{,}845)(104)$. The estimated odds of MI for physicians taking placebo equal 1.832 times the estimated odds for physicians taking aspirin. The estimated odds were 83% higher for the placebo group.

2.3.3   Inference for Odds Ratios and Log Odds Ratios

For small to moderate sample sizes, the sampling distribution of the odds ratio is highly skewed. When $\theta = 1$, for instance, $\hat{\theta}$ cannot be much smaller than θ (since $\hat{\theta} \geq 0$), but it could be much larger with nonnegligible probability.

Because of this skewness, statistical inference for the odds ratio uses an alternative but equivalent measure: its natural logarithm, $\log(\theta)$. Independence corresponds to $\log(\theta) = 0$. That is, an odds ratio of 1.0 is equivalent to a log odds ratio of 0.0. An odds ratio of 2.0 has a log odds ratio of 0.7. The log odds ratio is symmetric about zero, in the sense that reversal of rows or reversal of columns changes its sign. Two values for $\log(\theta)$ that are the same except for sign, such as $\log(2.0) = 0.7$ and $\log(0.5) = -0.7$, represent the same level of association. Doubling a log odds ratio corresponds to squaring an odds ratio. For instance, log odds ratios of $2(0.7) = 1.4$ and $2(-0.7) = -1.4$ correspond to odds ratios of $2^2 = 4$ and $0.5^2 = 0.25$.

The log transform of the sample odds ratio, $\log \hat{\theta}$, has a less skewed sampling distribution that is closer to normality. Its large-sample approximating normal distribution has a mean of $\log \theta$ and a standard deviation, referred to as an *asymptotic standard error* and denoted by *ASE*, of

$$ASE(\log \hat{\theta}) = \sqrt{\frac{1}{n_{11}} + \frac{1}{n_{12}} + \frac{1}{n_{21}} + \frac{1}{n_{22}}}. \tag{2.3.4}$$

The *ASE* value decreases as the cell counts increase. Because this sampling distribution is closer to normality, it is best to construct confidence intervals for $\log \theta$ and then transform back (i.e., take antilogs, using the exponential function) to form a confidence interval for θ. A large-sample confidence interval for $\log \theta$ is

$$\log \hat{\theta} \pm z_{\alpha/2} ASE(\log \hat{\theta}).$$

Exponentiating endpoints of this confidence interval yields one for θ.

For Table 2.3, the natural log of $\hat{\theta}$ equals $\log(1.832) = 0.605$. The *ASE* (2.3.4) of $\log \hat{\theta}$ equals $(1/189 + 1/10{,}933 + 1/10{,}845 + 1/104)^{1/2} = 0.123$. For the population this sample represents, a 95% confidence interval for $\log \theta$ equals $0.605 \pm 1.96(0.123)$, or $(0.365, 0.846)$. The corresponding confidence interval for θ is $[\exp(0.365), \exp(0.846)] = (e^{0.365}, e^{0.846}) = (1.44, 2.33)$. Since the confidence interval for θ does not contain 1.0, the true odds of MI seem different for the two groups. The interval predicts that the odds of MI are at least 44% higher for subjects taking placebo than for subjects taking aspirin. The endpoints of the interval are not equally distant from $\hat{\theta} = 1.83$, because the sampling distribution of $\hat{\theta}$ is skewed to the right.

The sample odds ratio $\hat{\theta}$ equals 0 or ∞ if any $n_{ij} = 0$, and it is undefined if both entries in a row or column are zero. The slightly amended estimator

$$\tilde{\theta} = \frac{(n_{11} + 0.5)(n_{22} + 0.5)}{(n_{12} + 0.5)(n_{21} + 0.5)},$$

corresponding to adding $\frac{1}{2}$ to each cell count, does not have this problem. It is preferred when the cell counts are very small or any zero cell counts occur. In that case, the *ASE* formula (2.3.4) replaces $\{n_{ij}\}$ by $\{n_{ij} + 0.5\}$. For Table 2.3, $\tilde{\theta} = (189.5)(10,933.5)/(10,845.5)(104.5) = 1.828$ is close to $\hat{\theta} = 1.832$, since no cell count is especially small.

2.3.4   Relationship Between Odds Ratio and Relative Risk

A sample odds ratio of 1.83 does *not* mean that p_1 is 1.83 times p_2; that would be the interpretation of a *relative risk* of 1.83, since that measure deals with proportions rather than odds. Instead, $\hat{\theta} = 1.83$ means that the *odds* value $p_1/(1-p_1)$ is 1.83 times the odds value $p_2/(1 - p_2)$. From (2.3.3) and from the sample analog of definition (2.2.3),

$$\text{Odds ratio} = \frac{p_1/(1 - p_1)}{p_2/(1 - p_2)} = \text{Relative risk} \times \left(\frac{1 - p_2}{1 - p_1}\right).$$

When the proportion of successes is close to zero for both groups, the fraction in the last term of this expression equals approximately 1.0. The odds ratio and relative risk then take similar values. Table 2.3 illustrates this similarity. For each group, the sample proportion of MI cases is close to zero. Thus, the sample odds ratio of 1.83 is similar to the sample relative risk of 1.82 obtained in Section 2.2.3. In such a case, an odds ratio of 1.83 *does* mean that p_1 is about 1.83 times p_2.

This relationship between the odds ratio and the relative risk is useful. For some data sets calculation of the relative risk is not possible, yet one can calculate the odds ratio and use it to approximate the relative risk. Table 2.4 is an example of this type. These data refer to a study that investigated the relationship between myocardial infarction and smoking. The first column refers to 262 young and middle-aged women (age <69) admitted to 30 coronary care units in northern Italy with acute MI during the period 1983–1988. Each case was matched with two control patients admitted to the same hospitals with other acute disorders. The controls fall in the second column of the table. All subjects were classified according to whether they had ever been smokers. The "yes" group consists of women who were current smokers or ex-smokers, whereas the "no" group consists of women who never were smokers. We refer to this variable as *smoking status*.

Table 2.4   Cross Classification of Smoking Status and Myocardial Infarction (MI)

Ever Smoker	Myocardial Infarction	Controls
Yes	172	173
No	90	346

Source: A. Gramenzi et al., *J. Epidemiol. and Commun. Health, 43*: 214–217 (1989).
Reprinted with permission of BMJ Publishing Group.

We would normally regard MI as a response variable and smoking status as an explanatory variable. In this study, however, the marginal distribution of MI is fixed by the sampling design, there being two controls for each case. The outcome measured for each subject is whether she ever was a smoker. The study, which uses a *retrospective* design to "look into the past," is called a *case-control study*. Such studies are common in health-related applications, for instance, to ensure a sufficiently large sample of subjects having the disease studied.

We might wish to compare ever-smokers with nonsmokers in terms of the proportion who suffered MI. These proportions refer to the conditional distribution of MI, given smoking status. We cannot estimate such proportions for this data set. For instance, about a third of the sample suffered MI. This is because the study matched each MI case with two controls, and it does not make sense to use $\frac{1}{3}$ as an estimate of the probability of MI. We *can* compute proportions in the reverse direction, for the conditional distribution of smoking status, given myocardial infarction status. For women suffering MI, the proportion who ever were smokers was $172/262 = .656$, while it was $173/519 = .333$ for women who had not suffered MI.

When the sampling design is retrospective, one can construct conditional distributions for the explanatory variable, within levels of the fixed response. It is usually not possible to estimate the probability of the response outcome of interest, or to compute the difference of proportions or relative risk for that outcome. Using Table 2.4, for instance, we cannot estimate the difference between nonsmokers and ever smokers in the probability of suffering MI. We can compute the odds ratio, however. This is because the odds ratio takes the same value when it is defined using the conditional distribution of X given Y as it does when defined (as in (2.3.1)) using the distribution of Y given X; that is, it treats the variables symmetrically. The odds ratio is determined by the conditional distributions in *either* direction, and can be calculated even if we have a study design that measures a response on X within each level of Y. In Table 2.4, the sample odds ratio is $[.656/(1 - .656)]/[.333/(1 - .333)] = (172 \times 346)/(173 \times 90) = 3.82$. The estimated odds of ever being a smoker were about 2 for the MI cases (i.e., $.656/.344$) and about $\frac{1}{2}$ for the controls (i.e., $.333/.667$), yielding an odds ratio of about $2/(1/2) = 4$.

We noted that when the probability that $Y = 1$ is small for each value of X, the odds ratio and relative risk take similar values. Even if we can estimate only conditional probabilities of X given Y, if we expect $P(Y = 1 \mid X)$ to be small, then we can use the sample odds ratio to provide a rough indication of the relative risk. For Table 2.4, we cannot estimate the relative risk of MI or the difference of proportions suffering MI. Since the probability of young or middle-aged women suffering MI is probably small regardless of smoking status, however, the odds ratio value of 3.82 is also a rough estimate of the relative risk. We estimate that women who ever smoked were nearly four times as likely to suffer MI as women who never smoked.

In Table 2.4, it makes sense to treat each column, rather than each row, as a binomial sample. Because of the matching that occurs in case-control studies, however, the binomial samples in the two columns are *dependent* rather than independent. Each observation in column 1 is naturally paired with two of the observations in column 2. Chapter 9 presents specialized methods for analyzing dependent binomial samples.

2.3.5   Types of Observational Studies*

By contrast to the study summarized by Table 2.4, imagine a study where we follow a sample of women for the next 20 years, observing the rates of MI for smokers and nonsmokers. Such a sampling design is *prospective*. There are two types of prospective studies. In *cohort studies*, the subjects make their own choice about which group to join (e.g., whether to be a smoker), and we simply observe in future time who suffers MI. In *clinical trials*, we randomly allocate subjects to the two groups of interest, such as in the aspirin study described in Section 2.2.2, again observing in future time who suffers MI. Yet another approach, a *cross-sectional design*, samples women and classifies them simultaneously on the group classification and their current response. As in a case-control study, we can then get the data at once, rather than waiting for future events.

Case-control, cohort, and cross-sectional studies are called *observational studies*. We observe who chooses each group and who has the outcome of interest. By contrast, a clinical trial is an *experimental study*, the investigator having control over which subjects enter each group, for instance, which subjects take aspirin and which take placebo. Clinical trials have fewer potential pitfalls, because of the use of randomization, but observational studies are often more practical for biomedical and social science research.

2.4   CHI-SQUARED TESTS OF INDEPENDENCE

We next show how to test the null hypothesis (H_0) that cell probabilities equal certain fixed values $\{\pi_{ij}\}$. For a sample of size n with cell counts $\{n_{ij}\}$, the values $\{\mu_{ij} = n\pi_{ij}\}$ are called *expected frequencies*. They represent the values of the expectations $\{E(n_{ij})\}$ when H_0 is true.

This notation refers to two-way tables, but similar notions apply to multiway tables or to a set of counts for a single categorical variable. To illustrate, for n flips of a coin, let π denote the probability of a head and $1 - \pi$ the probability of a tail on each flip. The null hypothesis that the coin is balanced corresponds to $\pi = 1 - \pi = .5$. The expected frequency of heads equals $\mu = n\pi = n/2$, which also equals the expected frequency of tails. If H_0 is true, we expect to observe about half heads and half tails.

We compare sample cell counts to the expected frequencies to judge whether the data contradict H_0. If H_0 is true for a two-way table, n_{ij} should be close to μ_{ij} in each cell. The larger the differences $\{n_{ij} - \mu_{ij}\}$, the stronger the evidence against H_0. The test statistics used to make such comparisons have large-sample chi-squared distributions.

2.4.1 Pearson Statistic and the Chi-Squared Distribution

The *Pearson chi-squared statistic* for testing H_0 is

$$X^2 = \sum \frac{(n_{ij} - \mu_{ij})^2}{\mu_{ij}}. \tag{2.4.1}$$

It was proposed in 1900 by Karl Pearson, the British statistician known also for the Pearson product-moment correlation, among his many contributions. This statistic takes its minimum value of zero when all $n_{ij} = \mu_{ij}$. For a fixed sample size, greater differences between $\{n_{ij}\}$ and $\{\mu_{ij}\}$ produce larger X^2 values and stronger evidence against H_0.

Since larger X^2 values are more contradictory to H_0, the P-value of the test is the null probability that X^2 is at least as large as the observed value. The X^2 statistic has approximately a chi-squared distribution for large sample sizes. It is difficult to specify what "large" means, but $\{\mu_{ij} \geq 5\}$ is sufficient. The P-value is the chi-squared right-hand tail probability above the observed X^2 value.

The chi-squared distribution is specified by its *degrees of freedom*, denoted by df. The mean of the chi-squared distribution equals df, and its standard deviation equals $\sqrt{2\,df}$. As df increases, the distribution concentrates around larger values and is more spread out. It is defined only for nonnegative values and is skewed to the right, but becomes more bell-shaped (normal) as df increases. Figure 2.1 displays the shapes of chi-squared densities having $df = 1, 5, 10,$ and 20. The df value equals the difference between the number of parameters in the alternative and null hypotheses, as explained later in this section.

2.4.2 Likelihood-Ratio Statistic

An alternative statistic for testing H_0 results from the likelihood-ratio method for significance tests. The test determines the parameter values that maximize the likelihood function under the assumption that H_0 is true. It also determines the values that maximize it under the more general condition that H_0 may or may not be true. The test is based on the ratio of the maximized likelihoods,

$$\Lambda = \frac{\text{maximum likelihood when parameters satisfy } H_0}{\text{maximum likelihood when parameters are unrestricted}}.$$

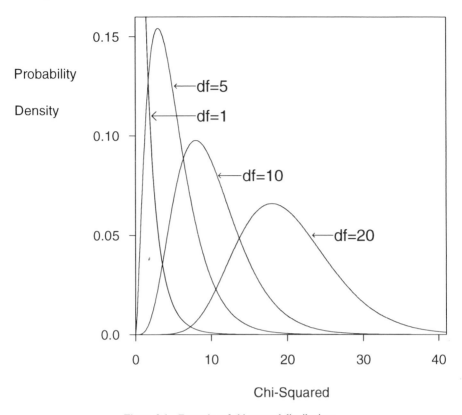

Figure 2.1 Examples of chi-squared distributions.

This ratio cannot exceed 1. If the maximized likelihood is much larger when the parameters are not forced to satisfy H_0, then the ratio Λ is far below 1 and there is strong evidence against H_0.

The test statistic for a likelihood-ratio test equals $-2\log(\Lambda)$. This value is non-negative, and "small" values of Λ yield "large" values of $-2\log(\Lambda)$. The reason for the log transform is to yield an approximate chi-squared sampling distribution. For two-way contingency tables, this statistic simplifies to the formula

$$G^2 = 2\sum n_{ij}\log\left(\frac{n_{ij}}{\mu_{ij}}\right). \qquad (2.4.2)$$

The statistic G^2 is called the *likelihood-ratio chi-squared statistic*. Like the Pearson statistic, G^2 takes its minimum value of 0 when all $n_{ij} = \mu_{ij}$, and larger values provide stronger evidence against H_0.

Though the Pearson X^2 and likelihood-ratio G^2 provide separate test statistics, they share many properties and commonly yield the same conclusions. When H_0 is true and the sample cell counts are large, the two statistics have the same chi-squared distribution, and their numerical values are similar. Each statistic has advantages and disadvantages, which we allude to later in this section and in Sections 7.3.1 and 7.4.3.

2.4.3 Tests of Independence

In two-way contingency tables, the null hypothesis of statistical independence of two responses has the form

$$H_0 : \pi_{ij} = \pi_{i+} \pi_{+j}$$

for all i and j. The marginal probabilities then specify the joint probabilities. To test H_0, we identify $\mu_{ij} = n\pi_{ij} = n\pi_{i+}\pi_{+j}$ as the expected frequency. Here, μ_{ij} is the expected value of n_{ij} assuming independence. Usually, $\{\pi_{i+}\}$ and $\{\pi_{+j}\}$ are unknown, as is this expected value.

We estimate the expected frequencies by substituting sample proportions for the unknown probabilities, giving

$$\hat{\mu}_{ij} = np_{i+}p_{+j} = n\frac{n_{i+}}{n}\frac{n_{+j}}{n} = \frac{n_{i+}n_{+j}}{n}.$$

The $\{\hat{\mu}_{ij}\}$ are called *estimated expected frequencies*. They have the same row and column totals as the observed counts, but they display the pattern of independence.

For testing independence in $I \times J$ contingency tables, the Pearson and likelihood-ratio statistics equal

$$X^2 = \sum \frac{(n_{ij} - \hat{\mu}_{ij})^2}{\hat{\mu}_{ij}}, \qquad G^2 = 2 \sum n_{ij} \log\left(\frac{n_{ij}}{\hat{\mu}_{ij}}\right). \qquad (2.4.3)$$

Their large-sample chi-squared distributions have $df = (I-1)(J-1)$. This means the following: Under H_0, $\{\pi_{i+}\}$ and $\{\pi_{+j}\}$ determine the cell probabilities. There are $I-1$ nonredundant row probabilities; since they sum to 1, the first $I-1$ determine the last one through $\pi_{I+} = 1 - (\pi_{1+} + \cdots + \pi_{I-1,+})$. Similarly, there are $J-1$ nonredundant column probabilities, for a total of $(I-1) + (J-1)$ parameters. The alternative hypothesis does not specify the IJ cell probabilities. They are then solely constrained to sum to 1, so there are $IJ - 1$ nonredundant parameters. The value for df is the difference between the number of parameters under the alternative and null hypotheses, or

$$(IJ - 1) - [(I-1) + (J-1)] = IJ - I - J + 1 = (I-1)(J-1).$$

2.4.4 Gender Gap Example

We illustrate chi-squared tests of independence using Table 2.5, from the 1991 General Social Survey. The variables are gender and party identification. Subjects indicated whether they identified more strongly with the Democratic or Republican party

Table 2.5 Cross Classification of Party Identification by Gender

	Party Identification			
Gender	Democrat	Independent	Republican	Total
Females	279	73	225	577
	(261.4)	(70.7)	(244.9)	
Males	165	47	191	403
	(182.6)	(49.3)	(171.1)	
Total	444	120	416	980

Note: Estimated expected frequencies for hypothesis of independence in parentheses.
Source: Data from 1991 General Social Survey.

or as Independents. Table 2.5 also contains estimated expected frequencies for H_0 : independence. For instance, the first cell has $\hat{\mu}_{11} = n_{1+}n_{+1}/n = (577 \times 444)/980 = 261.4$.

The chi-squared test statistics are $X^2 = 7.01$ and $G^2 = 7.00$, based on $df = (I-1)(J-1) = (2-1)(3-1) = 2$. The reference chi-squared distribution has a mean of $df = 2$ and a standard deviation of $\sqrt{2\,df} = \sqrt{4} = 2$, so a value of 7.0 is fairly far out in the right-hand tail. Each statistic has a P-value of .03. This evidence of association would be rather unusual if the variables were truly independent. Both test statistics suggest that party identification and gender are associated.

Most major statistical software packages have routines for calculating X^2, G^2, and their P-values. These P-values are approximations for true P-values, since the chi-squared distribution is an approximation for the true sampling distribution. Thus, it would be overly optimistic for us to report P-values to the 4 or 5 decimal places that software provides them. If we are lucky, the P-value approximation is good to the second decimal place, so it makes more sense to report it as .03 (or, at best, .028) rather than .02837. In any case, a P-value simply summarizes the strength of evidence against the null hypothesis, and accuracy to two or three decimal places is sufficient for this purpose.

2.4.5 Residuals

A test statistic and its P-value simply describe the evidence against the null hypothesis. A cell-by-cell comparison of observed and estimated expected frequencies helps us better understand the nature of the evidence. Larger differences between n_{ij} and $\hat{\mu}_{ij}$ tend to occur for cells that have larger expected frequencies, so the raw difference $n_{ij} - \hat{\mu}_{ij}$ is insufficient. For the test of independence, useful cell residuals have the form

$$\frac{n_{ij} - \hat{\mu}_{ij}}{\sqrt{\hat{\mu}_{ij}(1 - p_{i+})(1 - p_{+j})}} \tag{2.4.4}$$

These are called *adjusted residuals*.

When the null hypothesis is true, each adjusted residual has a large-sample standard normal distribution. An adjusted residual that exceeds about 2 or 3 in absolute value

Table 2.6 Adjusted Residuals (in Parentheses) for Testing Independence in Table 2.5

	Party Identification		
Gender	Democrat	Independent	Republican
Females	279	73	225
	(2.29)	(0.46)	(−2.62)
Males	165	47	191
	(−2.29)	(−0.46)	(2.62)

indicates lack of fit of H_0 in that cell. Table 2.6 shows the adjusted residuals for testing independence in Table 2.5. For the first cell, for instance, $n_{11} = 279$ and $\hat{\mu}_{11} = 261.4$. The first row and first column marginal proportions equal $p_{1+} = 577/980 = .589$ and $p_{+1} = 444/980 = .453$. Substituting into (2.4.4), the adjusted residual for this cell equals

$$\frac{279 - 261.4}{\sqrt{261.4(1 - .589)(1 - .453)}} = 2.29.$$

This cell shows a greater discrepancy between n_{11} and $\hat{\mu}_{11}$ than one would expect if the variables were truly independent.

Table 2.6 shows large positive residuals for female Democrats and male Republicans, and large negative residuals for female Republicans and male Democrats. Thus, there were significantly more female Democrats and male Republicans and fewer female Republicans and male Democrats than the hypothesis of independence predicts. An odds ratio describes this evidence of a gender gap. The 2×2 table of Democrat and Republican identifiers has a sample odds ratio of $(279)(191)/(225)(165) = 1.44$. Of those subjects identifying with one of the two parties, the estimated odds of identifying with the Democrats rather than the Republicans were 44% higher for females than males.

For each party, Table 2.6 shows that there is only one nonredundant residual; the one for females is the negative of the one for males. The observed counts and the estimated expected frequencies have the same row and column totals. Thus, in a given column, if $n_{ij} > \hat{\mu}_{ij}$ in one cell, the reverse must happen in the other cell. The differences $n_{1j} - \hat{\mu}_{1j}$ and $n_{2j} - \hat{\mu}_{2j}$ have the same magnitude but different signs, implying the same pattern for their adjusted residuals.

2.4.6 Partitioning Chi-Squared

Chi-squared statistics have a reproductive property. If one chi-squared statistic has $df = df_1$ and a separate, independent, chi-squared statistic has $df = df_2$, then their sum is chi-squared with $df = df_1 + df_2$. For instance, if we had a table of form Table 2.5 for college-educated subjects and a separate one for subjects not having a college education, the sum of the X^2 values or the sum of the G^2 values from the two tables would be a chi-squared statistic with $df = 2 + 2 = 4$.

Similarly, chi-squared statistics having $df > 1$ can be broken into components with fewer degrees of freedom. For instance, a statistic having $df = 2$ can be partitioned into two independent components each having $df = 1$. Another supplement to a test of independence partitions its chi-squared test statistic so that the components represent certain aspects of the association. A partitioning may show that an association primarily reflects differences between certain categories or groupings of categories.

We illustrate with a partitioning of G^2 for testing independence in $2 \times J$ tables. The test statistic then has $df = (J - 1)$, and we partition it into $J - 1$ components. The jth component is G^2 for testing independence in a 2×2 table, where the first column combines columns 1 through j of the original table, and the second column uses column $j + 1$ of the original table. That is, G^2 for testing independence in a $2 \times J$ table equals the sum of a G^2 statistic that compares the first two columns, plus a G^2 statistic for the 2×2 table that combines the first two columns and compares them to the third column, and so on, up to a G^2 statistic for the 2×2 table that combines the first $J - 1$ columns and compares them to the last column. Each component G^2 statistic has $df = 1$.

Consider again Table 2.5. The first two columns of this table form a 2×2 table with cell counts, by row, of $(279, 73/165, 47)$. For this component table, $G^2 = 0.16$, with $df = 1$. Of those subjects who identify either as Democrats or Independents, there is little evidence of a difference between females and males in the relative numbers in the two categories. We form the second 2×2 table by combining these columns and comparing them to the Republican column, giving the table with rows $(279 + 73, 225/165 + 47, 191) = (352, 225/212, 191)$. This table has $G^2 = 6.84$, based on $df = 1$. There is strong evidence of a difference between females and males in the relative numbers identifying as Republican instead of Democrat or Independent. Note that $0.16 + 6.84 = 7.00$; that is, the sum of these G^2 components equals G^2 for the test of independence for the complete 2×3 table. This overall statistic primarily reflects differences between genders in choosing between Republicans and Democrats/Independents.

It might seem more natural to compute G^2 for separate 2×2 tables that pair each column with a particular one, say the last. Though this is a reasonable way to investigate association in many data sets, these component statistics are not independent and do not sum to G^2 for the complete table. Certain rules determine ways of forming tables so that chi-squared partitions, but they are beyond the scope of this text (see, e.g., Agresti (1990), p. 53, for rules and references). A necessary condition is that the G^2 values for the component tables sum to G^2 for the original table.

The G^2 statistic has exact partitionings. The overall Pearson X^2 statistic does not equal the sum of the X^2 values for the separate tables in a partition. However, it is valid to use the X^2 statistics for the separate tables in the partition; they simply do not provide an exact algebraic partitioning of the X^2 statistic for the overall table.

2.4.7 Comments on Chi-Squared Tests

Chi-squared tests of independence, like any significance tests, have serious limitations. They simply indicate the degree of evidence for an association. They are rarely adequate for answering all questions we have about a data set. Rather than relying solely on results of these tests, one should study the nature of the association. It is sensible to decompose chi-squared into components, study residuals, and estimate parameters such as odds ratios that describe the strength of association.

The X^2 and G^2 chi-squared tests also have limitations in the types of data sets for which they are applicable. For instance, they require large samples. The sampling distributions of X^2 and G^2 get closer to chi-squared as the sample size n increases, relative to the number of cells IJ. The convergence is quicker for X^2 than G^2. The chi-squared approximation is often poor for G^2 when $n/IJ < 5$. When I or J is large, it can be decent for X^2 when some expected frequencies are as small as 1. Section 7.4.3 provides further guidelines, but these are not crucial since small-sample procedures are available whenever we question whether n is sufficiently large. Section 2.6 discusses these.

The $\{\hat{\mu}_{ij} = n_{i+}n_{+j}/n\}$ used in X^2 and G^2 depend on the row and column marginal totals, but not on the order in which the rows and columns are listed. Thus, X^2 and G^2 do not change value with arbitrary reorderings of rows or of columns. This means that these tests treat both classifications as nominal. We ignore some information when we use them to test independence between ordinal classifications. When at least one variable is ordinal, more powerful tests of independence usually exist. The next section presents such a test.

2.5   TESTING INDEPENDENCE FOR ORDINAL DATA

The chi-squared test of independence using test statistic X^2 or G^2 treats both classifications as nominal. When the rows and/or the columns are ordinal, test statistics that utilize the ordinality are usually more appropriate.

2.5.1   Linear Trend Alternative to Independence

When the row variable X and the column variable Y are ordinal, a "trend" association is quite common. As the level of X increases, responses on Y tend to increase toward higher levels, or responses on Y tend to decrease toward lower levels. One can use a single parameter to describe such an ordinal trend association. The most common analysis assigns scores to categories and measures the degree of *linear trend* or correlation.

We next present a test statistic that is sensitive to positive or negative linear trends in the relationship between X and Y. It utilizes correlation information in the data. Let $u_1 \leq u_2 \leq \cdots \leq u_I$ denote scores for the rows, and let $v_1 \leq v_2 \leq \cdots \leq v_J$ denote scores for the columns. The scores have the same ordering as the category levels and are said to be *monotone*. The scores reflect distances between categories, with greater distances between categories treated as farther apart.

The sum $\sum_{i,j} u_i v_j n_{ij}$, which weights cross-products of scores by the frequency of their occurrence, relates to the covariation of X and Y. For the chosen scores, the Pearson product-moment correlation between X and Y equals the standardization of this sum,

$$r = \frac{\sum_{i,j} u_i v_j n_{ij} - \left(\sum_i u_i n_{i+}\right)\left(\sum_j v_j n_{+j}\right)/n}{\sqrt{\left[\sum_i u_i^2 n_{i+} - \frac{\left(\sum_i u_i n_{i+}\right)^2}{n}\right]\left[\sum_j v_j^2 n_{+j} - \frac{\left(\sum_j v_j n_{+j}\right)^2}{n}\right]}}.$$

Alternative formulas exist for r, and one can compute it using standard software, entering for each subject their score on the row classification and their score on the column classification. The correlation falls between -1 and $+1$. Independence between the variables implies that its true value equals zero. The larger the correlation is in absolute value, the farther the data fall from independence in this linear dimension.

A statistic for testing the null hypothesis of independence against the two-sided alternative hypothesis of nonzero true correlation is given by

$$M^2 = (n - 1)r^2. \tag{2.5.1}$$

This statistic increases as the sample correlation r increases in magnitude and as the sample size n grows. For large samples, it has approximately a chi-squared distribution with $df = 1$. Large values contradict independence, so, as with X^2 and G^2, the P-value is the right-tail probability above the observed value. The square root, $M = \sqrt{n-1}\, r$, has approximately a standard normal null distribution. It applies to directional alternatives, such as positive correlation between the classifications.

Tests using M^2 treat the variables symmetrically. If one interchanges the rows with the columns and their scores in an $I \times J$ table, M^2 takes identical value for the corresponding $J \times I$ table.

2.5.2 Alcohol and Infant Malformation Example

Table 2.7 refers to a prospective study of maternal drinking and congenital malformations. After the first three months of pregnancy, the women in the sample completed a questionnaire about alcohol consumption. Following childbirth, observations were recorded on presence or absence of congenital sex organ malformations. Alcohol

Table 2.7 Infant Malformation and Mother's Alcohol Consumption

Alcohol Consumption	Malformation		Total	Percentage Present	Adjusted Residual
	Absent	Present			
0	17,066	48	17,114	0.28	−0.18
< 1	14,464	38	14,502	0.26	−0.71
1–2	788	5	793	0.63	1.84
3–5	126	1	127	0.79	1.06
≥ 6	37	1	38	2.63	2.71

Source: B. I. Graubard and E. L. Korn, *Biometrics 43:* 471–476 (1987).
Reprinted with permission of the Biometric Society.

consumption, measured as average number of drinks per day, is an explanatory variable with ordered categories. Malformation, the response variable, is nominal. When a variable is nominal but has only two categories, statistics (such as M^2) that treat the variable as ordinal are still valid. For instance, we could artificially regard malformation as ordinal, treating "absent" as "low" and "present" as "high." Any choice of two scores yields the same value of M^2, and we simply use 0 for "absent" and 1 for "present."

Table 2.7 has a mixture of very small, moderate, and extremely large counts. Even though the sample size is large ($n = 32,574$), in such cases the actual sampling distributions of X^2 or G^2 may not be close to chi-squared. For these data, having $df = 4$, $G^2 = 6.2$ ($P = .19$) and $X^2 = 12.1$ ($P = .02$), so they provide mixed signals. In any case, they ignore the ordinality of alcohol consumption.

Table 2.7 lists the percentage of malformation cases at each level of alcohol consumption. These percentages show roughly an increasing trend. The first two are similar and the next two are also similar, however, and any of the last three percentages changes dramatically with the addition or deletion of one malformation case. Table 2.7 also reports adjusted residuals for the "present" category in this table. They are negative at low levels of alcohol consumption and positive at high levels of consumption, though most are small, and they also change substantially with slight changes in the data. The sample percentages and the adjusted residuals both suggest a possible tendency for malformations to be more likely at higher levels of alcohol consumption.

The ordinal test statistic M^2 requires scores for levels of alcohol consumption. It seems sensible to use scores that are midpoints of the categories; that is, $v_1 = 0$, $v_2 = 0.5$, $v_3 = 1.5$, $v_4 = 4.0$, $v_5 = 7.0$, the last score being somewhat arbitrary. One can calculate r and M^2 using software (e.g., PROC FREQ in SAS; see Table A.2 in the Appendix). The sample correlation between alcohol consumption and malformation is $r = .014$, and $M^2 = (32,573)(.014)^2 = 6.6$. The P-value of .01 suggests strong evidence of a nonzero correlation. The standard normal statistic $M = 2.56$ has $P = .005$ for the one-sided alternative of a positive correlation.

For the chosen scores, the correlation value of .014 seems weak. However, r has limited use as a descriptive measure for tables, such as this one, that are highly discrete and unbalanced. Future chapters present tests such as M^2 as part of a model-based analysis. For instance, Section 4.2 presents a model in which the probability of malformation changes linearly according to alcohol consumption. Model-based approaches yield estimates of the size of the effect as well as smoothed estimates of cell probabilities. These estimates are more informative than mere significance tests.

2.5.3 Extra Power with Ordinal Test

For testing independence, X^2 and G^2 refer to the most general alternative hypothesis possible, whereby cell probabilities exhibit *any* type of statistical dependence. Their df value of $(I - 1)(J - 1)$ reflects an alternative hypothesis that has $(I - 1)(J - 1)$ more parameters than the null hypothesis. These statistics are designed to detect any type of pattern for the additional parameters. In achieving this generality, they sacrifice sensitivity for detecting particular patterns.

When the row and column variables are ordinal, one can attempt to describe the association using a single extra parameter. For instance, the test statistic M^2 is based on a correlation measure of linear trend. When a test statistic refers to a single parameter, it has $df = 1$.

When the association truly has a positive or negative trend, the ordinal test using M^2 has a power advantage over the tests based on X^2 or G^2. Since df equals the mean of the chi-squared distribution, a relatively large M^2 value based on $df = 1$ falls farther out in its right-hand tail than a comparable value of X^2 or G^2 based on $df = (I - 1)(J - 1)$; falling farther out in the tail produces a smaller P-value. When there truly is a linear trend, M^2 tends to have similar size as X^2 or G^2, so it tends to have greater power in terms of yielding smaller P-values. In attempting to detect any type of dependence, the X^2 and G^2 statistics lose power relative to statistics designed to detect a particular type of dependence if that type of dependence truly occurs.

Another advantage of chi-squared tests having small df values relates to the accuracy of chi-squared approximations. For small to moderate sample sizes, the true sampling distributions tend to be closer to chi-squared when df is smaller. When several cell counts are small, the chi-squared approximation is likely to be worse for X^2 or G^2 than it is for M^2.

2.5.4 Choice of Scores

For most data sets, the choice of scores has little effect on the results. Different choices of monotone scores usually give similar results. This may not happen, however, when the data are very unbalanced, such as when some categories have many more observations than other categories. Table 2.7 illustrates this. For the equally-spaced row scores (1, 2, 3, 4, 5), the test statistic equals $M^2 = 1.83$, giving a much weaker conclusion ($P = .18$). The magnitudes of r and M^2 do not change with transformations of the scores that maintain the same relative spacings between the categories. For instance, scores $(1, 2, 3, 4, 5)$ yield the same correlation as scores $(0, 1, 2, 3, 4)$ or $(2, 4, 6, 8, 10)$ or $(10, 20, 30, 40, 50)$.

An alternative approach avoids the responsibility of selecting scores and uses the data to form them automatically. Specifically, one assigns ranks to the subjects and uses them as the category scores. For all subjects in a category, one assigns the average of the ranks that would apply for a complete ranking of the sample from 1 to n. These are called *midranks*. We illustrate by assigning midranks to the levels of alcohol consumption in Table 2.7. The 17,114 subjects at level 0 for alcohol consumption share ranks 1 through 17,114. We assign to each of them the average of these ranks, which is the midrank $(1 + 17,114)/2 = 8557.5$. The 14,502 subjects at level < 1 for alcohol consumption share ranks 17,115 through $17,114 + 14,502 = 31,616$, for a midrank of $(17,115 + 31,616)/2 = 24,365.5$. Similarly the midranks for the last three categories are 32,013.0, 32,473.0, and 32,555.5. These scores yield $M^2 = 0.35$ and a weaker conclusion yet: ($P = .55$).

Why does this happen? Adjacent categories having relatively few observations necessarily have similar midranks. For instance, the midranks $(8557.5, 24,365.5, 32,013.0, 32,473.0, 32,555.5)$ for Table 2.7 are similar for the final three categories, since those categories have considerably fewer observations than the first two categories. A consequence is that this scoring scheme treats alcohol consumption level 1–2 (category 3) as much closer to consumption level ≥ 6 (category 5) than to consumption level 0 (category 1). This seems inappropriate. It is usually better to use one's judgment by selecting scores that reflect distances between categories. When uncertain about this choice, perform a sensitivity analysis. Select two or three "sensible" choices and check that the results are similar for each. Equally-spaced scores often provide a reasonable compromise when the category labels do not suggest any obvious choices, such as the categories (liberal, moderate, conservative) for political philosophy.

When X and Y are both ordinal, one can use midrank scores for each. The M^2 statistic is then sensitive to detecting nonzero values of a nonparametric form of correlation called *Spearman's rho*. Alternative ordinal tests for $I \times J$ tables utilize versions of other ordinal association measures. For instance, *gamma* and *Kendall's tau-b* are contingency table generalizations of the ordinal measure called *Kendall's tau*. The sample value of any such measure divided by its standard error has a large-sample standard normal distribution for testing independence, and the square of the statistic is chi-squared with $df = 1$. Like the test based on M^2, these tests share the potential power advantage that results from using a single parameter to describe the association.

2.5.5 Trend Tests for *I*-by-2 and 2-by-*J* Tables

We now study how M^2 utilizes the sample data when X or Y has only two levels. Suppose the row variable X is an explanatory variable, and the column variable Y is a response variable.

When X is binary, the table has size $2 \times J$. Tables of this size occur in comparisons of two groups, such as when the rows represent two treatments. Using scores ($u_1 = 0$, $u_2 = 1$) for levels of X in this case, we see that the covariation measure $\sum_{i,j} u_i v_j n_{ij}$ on which M^2 is based simplifies to $\sum_j v_j n_{2j}$. This term sums the scores on Y for all subjects in row 2. Divided by the number of subjects in row 2, it gives the mean score for that row. In fact, when the columns (Y) are ordinal with scores $\{v_j\}$, the M^2 statistic for $2 \times J$ tables is directed toward detecting differences between the two row means of the scores on Y. In testing independence using M^2, small P-values suggest that the true difference in row means is nonzero.

When we use midrank scores for Y, the test for $2 \times J$ tables is sensitive to differences in mean ranks for the two rows. This test is called the *Wilcoxon* or *Mann–Whitney test*. Most nonparametric statistics texts present this test for fully-ranked response data, whereas the $2 \times J$ table is an extended case in which sets of subjects at the same level of Y are tied and use midranks. The large-sample version of that nonparametric test uses a standard normal z statistic. The square of the z statistic is equivalent to M^2, using arbitrary scores (such as 0, 1) for the rows and midranks for the columns.

Tables of size $I \times 2$, such as Table 2.7, have a binary response variable rather than a binary explanatory variable. It is then natural to focus on how the proportion classified in a given response category of Y varies across the levels of X. For ordinal

X with monotone row scores and arbitrary scores for the two columns, M^2 focuses on detecting a linear trend in this proportion and relates to models presented in Section 4.2. In testing independence using M^2, small P-values suggest that the slope for this linear trend is nonzero. This $I \times 2$ version of the ordinal test is called the *Cochran-Armitage trend test*.

2.5.6 Nominal-Ordinal Tables

The test statistic (2.5.1) treats both classifications as ordinal. When one variable (say X) is nominal but has only two categories, we can still use it. When X is nominal with more than two categories, it is inappropriate, and we use a different statistic. It is based on calculating a mean response on the ordinal variable in each row and considering the variation among the row means. The statistic is rather complex computationally, and we defer discussion of it to Section 7.3.6. It has a large-sample chi-squared distribution with $df = (I - 1)$. When $I = 2$, it is identical to M^2, which then compares the two row means.

2.6 EXACT INFERENCE FOR SMALL SAMPLES

The confidence intervals and tests presented so far in this chapter are large-sample methods. As the sample size n grows, the cell counts grow, and "chi-squared" statistics such as X^2, G^2, and M^2 have distributions that are more nearly chi-squared. When the sample size is small, one can perform inference using *exact* distributions rather than large-sample approximations. This section discusses exact inference for two-way contingency tables.

2.6.1 Fisher's Exact Test

We first study the 2×2 case. The null hypothesis of independence corresponds to an odds ratio of $\theta = 1$. A small-sample probability distribution for the cell counts is defined for the set of tables having the same row and column totals as the observed data. Under Poisson, binomial, or multinomial sampling assumptions for the cell counts, the distribution that applies to this restricted set of tables fixing the row and column totals is called the *hypergeometric*.

For given row and column marginal totals, the value for n_{11} determines the other three cell counts. Thus, the hypergeometric formula expresses probabilities for the four cell counts in terms of n_{11} alone. When $\theta = 1$, the probability of a particular value n_{11} for that count equals

$$P(n_{11}) = \frac{\binom{n_{1+}}{n_{11}} \binom{n_{2+}}{n_{+1} - n_{11}}}{\binom{n}{n_{+1}}}. \tag{2.6.1}$$

The binomial coefficients equal

$$\binom{a}{b} = \frac{a!}{b!\,(a-b)!}.$$

To test independence, the P-value is the sum of hypergeometric probabilities for outcomes at least as favorable to the alternative hypothesis as the observed outcome. We illustrate for $H_a : \theta > 1$. Given the marginal totals, tables having larger n_{11} values also have larger sample odds ratios $\hat{\theta} = (n_{11}n_{22})/(n_{12}n_{21})$, and hence provide stronger evidence in favor of this alternative. The P-value equals the right-tail hypergeometric probability that n_{11} is at least as large as the observed value. This test for 2×2 tables, proposed by the eminent British statistician R. A. Fisher in 1934, is called *Fisher's exact test*.

2.6.2 Fisher's Tea Taster

To illustrate this test in his 1935 text, *The Design of Experiments*, Fisher described the following experiment: A colleague of Fisher's at Rothamsted Experiment Station near London claimed that, when drinking tea, she could distinguish whether milk or tea was added to the cup first. To test her claim, Fisher designed an experiment in which she tasted eight cups of tea. Four cups had milk added first, and the other four had tea added first. She was told there were four cups of each type, so that she should try to select the four that had milk added first. The cups were presented to her in random order.

Table 2.8 shows a potential result of the experiment. We conduct Fisher's exact test of $H_0 : \theta = 1$ against $H_a : \theta > 1$. The null hypothesis states that Fisher's colleague's guess was independent of the actual order of pouring; the alternative hypothesis reflects her claim, predicting a positive association between true order of pouring and her guess. For this experimental design, the column margins are identical to the row margins $(4, 4)$, since she knew that four cups had milk added first. Both marginal distributions are naturally fixed.

The null distribution of n_{11} is the hypergeometric distribution defined for all 2×2 tables having row and column margins $(4, 4)$. The potential values for n_{11} are $(0, 1, 2, 3, 4)$. The observed table, three correct guesses of the four cups having milk

Table 2.8 Fisher's Tea-Tasting Experiment

	Guess Poured First		
Poured First	Milk	Tea	Total
Milk	3	1	4
Tea	1	3	4
Total	4	4	

added first, has null probability

$$P(3) = \frac{\binom{4}{3}\binom{4}{1}}{\binom{8}{4}} = \frac{[4!/(3!)(1!)][4!/(1!)(3!)]}{[8!/(4!)(4!)]} = \frac{16}{70} = .229.$$

The only table that is more extreme, for the alternative $H_a : \theta > 1$, consists of four correct guesses. It has $n_{11} = n_{22} = 4$ and $n_{12} = n_{21} = 0$, and a probability of

$$P(4) = \frac{\binom{4}{4}\binom{4}{0}}{\binom{8}{4}} = \frac{1}{70} = .014.$$

Table 2.9 summarizes the possible values of n_{11} and their probabilities.

The P-value for the one-sided alternative $H_a : \theta > 1$ equals the right-tail probability that n_{11} is at least as large as observed; that is, $P = P(3) + P(4) = .243$. This is not much evidence against the null hypothesis of independence. The experiment did not establish an association between the actual order of pouring and the guess. Of course, it is difficult to show effects with such a small sample. If the tea taster had guessed all cups correctly (i.e., $n_{11} = 4$), the observed result would have been the most extreme possible in the right-hand tail of the hypergeometric distribution; then, $P = P(4) = .014$, giving some reason to believe her claim. For the potential n_{11} values, Table 2.9 shows P-values for the alternative $H_a : \theta > 1$.

2.6.3 P-values and Type I Error Probabilities

The two-sided alternative $H_a : \theta \neq 1$ refers to the general alternative of statistical dependence used in chi-squared tests. Its exact P-value is usually defined as the two-tailed sum of the probabilities of tables no more likely than the observed table. To calculate it, one adds the hypergeometric probabilities of all outcomes y for the first cell count for which $P(y) \leq P(n_{11})$, where n_{11} is the observed count. For Table 2.8, summing all probabilities that are no greater than the probability $P(3) = .229$ of the

Table 2.9 Hypergeometric Distribution for Tables with Margins of Table 2.8

n_{11}	Probability	P-value	X^2
0	.014	1.000	8.0
1	.229	.986	2.0
2	.514	.757	0.0
3	.229	.243	2.0
4	.014	.014	8.0

Note: P-value refers to right-tail probability for one-sided alternative.

observed table gives $P = P(0) + P(1) + P(3) + P(4) = .486$. When the row or column marginal totals are equal, the hypergeometric distribution is symmetric, and the two-sided P-value doubles the one-sided one.

An alternative two-sided P-value sums the probabilities of those tables for which the Pearson X^2 statistic is at least as large as the observed value. That is, it uses the exact small-sample distribution of X^2 rather than its large-sample chi-squared distribution. Table 2.9 shows the X^2 values for the five tables having the margins of Table 2.8. The statistic can assume only three distinct values, so its highly discrete distribution is far from the continuous chi-squared distribution. Figure 2.2 plots this exact small-sample distribution of X^2. It equals 0.0 with probability .514, 2.0 with probability .458, and 8.0 with probability .028. The observed table has $X^2 = 2.0$, and the P-value equals the null probability of a value this large or larger, or $.458 + .028 = .486$. For these data, this P-value based on X^2 is identical to the one based solely on probabilities.

Computations for the hypergeometric distribution are rather messy. One can sidestep this distribution and approximate the exact P-value for X^2 by obtaining

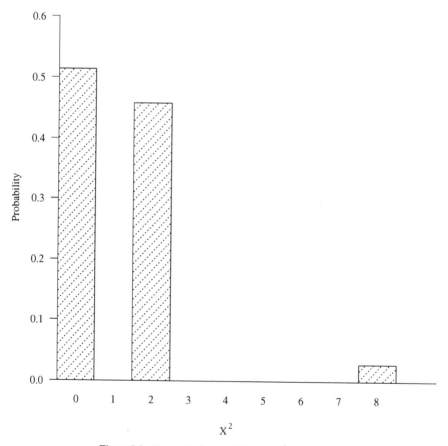

Figure 2.2 Exact distribution of Pearson X^2 for Table 2.8.

a P-value from the chi-squared distribution for an adjustment of the Pearson statistic using the *Yates continuity correction*. There is no longer any reason to use this approximation, however, since modern software makes it possible to conduct Fisher's exact test even for fairly large samples with hypergeometric P-values based on the X^2 or probability criteria.

For small samples, the exact distribution (2.6.1) is highly discrete, in the sense that n_{11} can assume relatively few values. The P-value also has a small number of possible values. For Table 2.8, it can assume five values for the one-sided test and three values for the two-sided test. This has an impact on error rates in hypothesis testing. Suppose we make a formal decision about the null hypothesis using a supposed Type I error probability such as .05. That is, we reject the null hypothesis if the P-value is less than or equal to .05. Because of the test's discreteness, it is usually not possible to achieve that level exactly. For the one-sided alternative, the tea-tasting experiment yields a P-value below .05 only when $n_{11} = 4$, in which case $P = .014$. When H_0 is true, the probability of this outcome is .014, so the actual Type I error probability would be .014, not .05. The test is said to be *conservative*, since the actual error rate is smaller than the intended one. (The approximation of exact tests using the Yates continuity correction is also conservative.)

This illustrates an awkwardness with formal decision-making at "sacred" levels such as .05 when the test statistic is discrete. For test statistics having a *continuous* distribution, the P-value has a *uniform* null distribution over the interval [0, 1]. That is, P is equally likely to fall anywhere between 0 and 1, so the probability that P falls below a fixed level α equals α, and the expected value of P is .5. For test statistics having discrete distributions, the null distribution of the P-value is discrete and has expected value greater than .5. For instance, for the one-sided test with the tea-tasting data, the P-value equals .014 with probability $P(0) = .014$, it equals .243 with probability $P(1) = .229$, and so forth; from Table 2.9, the expected value of the P-value is

$$\sum P \times \text{Prob}(P) = .014(.014) + .243(.229) + .757(.514) + .986(.229) + 1.0(.014)$$

$$= .685.$$

In this average sense, P-values for discrete distributions tend to be too large.

To diminish the conservativeness of tests for discrete data, one can use a slightly different definition of P-value. The *mid P-value* equals *half* the probability of the observed result, plus the probability of more extreme results. It has a null expected value of .5, the same as the regular P-value for continuous variates. For the tea-tasting data, with an observed value of 3 for n_{11}, the one-sided mid P-value equals $P(3)/2 + P(4) = .229/2 + .014 = .129$, compared to .243 for the ordinary P-value. The mid P-value for the two-sided test based on the X^2 statistic equals $P(X^2 = 2)/2 + P(X^2 = 8) = .257$, compared to .486 for the ordinary P-value.

Unlike an exact test with ordinary P-value, a test using the mid P-value does not guarantee that the Type I error rate falls below a fixed value (see Problem 2.27). However, it usually performs well and is less conservative than Fisher's exact test. For either P-value, rather than reducing the data to the extreme binary decision (reject

H_0, do not reject H_0), it is better simply to report the P-value, using it as a measure of the weight of evidence against the null hypothesis.

In Table 2.8, both margins are naturally fixed. When only one set is fixed, such as when rows totals are fixed with independent binomial samples, alternative exact tests exist that are less conservative than Fisher's exact test. These are beyond the scope of this text, but the reader can refer to a recent article by R. Berger and D. Boos (*J. Am. Statist. Assoc.*, 1994, p. 1012).

2.6.4 Small-Sample Confidence Interval for Odds Ratio

Exact inference is not limited to testing. One can also construct small-sample confidence intervals for the odds ratio. They correspond to a generalization of Fisher's exact test that tests an arbitrary value, $H_0 : \theta = \theta_0$. A 95% confidence interval contains all values of θ_0 for which the exact test of $H_0 : \theta = \theta_0$ yields $P > .05$; that is, for which one would not reject the null hypothesis at the .05 level.

As happens with exact tests, the discreteness makes these confidence intervals conservative. The true confidence level can be no smaller than the nominal one, but it may actually be considerably larger. For instance, a nominal 95% confidence interval may have true confidence level 98%. Moreover, the true level is unknown. The difference between the nominal and true levels can be considerable when the sample size is small. To reduce the conservativeness, one can construct the interval corresponding to the test using a mid P-value. The confidence interval consists of all θ_0 values for which the mid P-value exceeds .05. This interval is shorter. Though its actual confidence level is not guaranteed to be at least the nominal level, it tends to be close to that level. Computations for either of these types of confidence intervals are complex and require specialized software (e.g., StatXact, Cytel Software, Cambridge, MA).

For the tea-tasting data (Table 2.8), the "exact" 95% confidence interval for the true odds ratio equals $(0.21, 626.17)$. The interval based on the test using the mid P-value equals $(0.31, 308.55)$. Both intervals are very wide, because the sample size is so small.

2.6.5 Exact Tests of Independence for Larger Tables*

Exact tests of independence for tables of size larger than 2×2 use a multivariate version of the hypergeometric distribution. This distribution also applies to the set of all tables having the same row and column margins as the observed table. The exact tests are not practical to compute by hand or calculator but are feasible using computers. One selects a test statistic that describes the distance of the observed data from H_0. One then computes the probability of the set of tables for which the test statistic is at least as great as the observed one. For instance, for nominal variables, one could use X^2 as the test statistic. The P-value is then the null probability that X^2 is at least as large as the observed value, the calculation being done using the exact distribution rather than the large-sample chi-squared distribution.

Table 2.10   Example of 3 × 9 Table for Small-Sample Test

0	7	0	0	0	0	0	1	1
1	1	1	1	1	1	1	0	0
0	8	0	0	0	0	0	0	0

Recently developed software makes exact tests feasible for tables for which large-sample approximations are invalid. The software StatXact performs many exact inferences for categorical data. To illustrate, Table 2.10 is a 3 × 9 table having many zero entries and small counts. For it, $X^2 = 22.3$ with $df = 16$. The chi-squared approximation for the distribution of X^2 gives $P = .13$. Because the cell counts are so small, the validity of this approximation is suspect. Using StatXact to generate the exact sampling distribution of X^2, we obtain an exact P-value of .001, quite different from the result using the large-sample approximation.

For another example, we return to the analysis in Section 2.5 of Table 2.7, on the potential effect of maternal alcohol consumption on infant sex organ malformation. For testing independence, the values of $X^2 = 12.1$ and $G^2 = 6.2$ yield P-values from a chi-squared distribution with $df = 4$ of .02 and .19, respectively. Because of the imbalance in the table counts and the presence of some small counts, we could instead use exact tests for these statistics. The P-values using the exact distributions of X^2 and G^2 are .03 and .13, respectively. These are closer together but still give differing evidence about the association.

The columns of Table 2.7 are ordinal, and Section 2.5 presented a large-sample ordinal test based on a statistic M^2 (formula 2.5.1) that assigns scores to rows and columns. For ordinal data, exact tests exist using this statistic or using M for one-sided alternatives. For the one-sided alternative of a positive association, the exact P-value equals .02 for the midpoint scores $(0, 0.5, 1.5, 4, 7)$, .10 for the equally spaced scores $(0, 1, 2, 3, 4)$ and .29 for the midrank scores. For these data, the result depends greatly on the choice of scores.

PROBLEMS

2.1. A Swedish study considered the effect of low-dose aspirin on reducing the risk of stroke and heart attacks among people who have already suffered a stroke (*Lancet 338*: 1345–1349 (1991)). Of 1360 patients, 676 were randomly assigned to the aspirin treatment (one low-dose tablet a day) and 684 to a placebo treatment. During a follow-up period averaging about three years, the number of deaths due to myocardial infarction were 18 for the aspirin group and 28 for the placebo group.

    a. Calculate and interpret the difference of proportions, relative risk of death, and the odds ratio.

    b. Conduct an inferential analysis for these data. Interpret results.

2.2. In the United States, the estimated annual probability that a woman over the age of 35 dies of lung cancer equals .001304 for current smokers and .000121 for nonsmokers (M. Pagano and K. Gauvreau, *Principles of Biostatistics*, Belmont, CA: Duxbury Press, 1993, p. 134).

a. Calculate and interpret the difference of proportions and the relative risk. Which of these measures is more informative for these data? Why?

b. Calculate and interpret the odds ratio. Explain why the relative risk and odds ratio take similar values.

2.3. A newspaper article preceding the 1994 World Cup semifinal match between Italy and Bulgaria stated that "Italy is favored 10-11 to beat Bulgaria, which is rated at 10-3 to reach the final." Suppose this means that the odds that Italy wins are $11/10$ and the odds that Bulgaria wins are $3/10$. Find the probability that each team wins, and comment.

2.4. The odds ratio between treatment (A, B) and response (death, survival) equals 2.0.

a. Explain what is wrong with the interpretation, "The probability of death with treatment A is twice that with treatment B." Give the correct interpretation.

b. When is the quoted interpretation in **(a)** correct, in an approximate sense?

c. The odds of death equal 0.5 for treatment A. What is the probability of death for (i) treatment A, (ii) treatment B?

2.5. An estimated odds ratio for adult females between the presence of squamous cell carcinoma (yes, no) and smoking behavior (smoker, non-smoker) equals 11.7 when the smoker category consists of subjects whose smoking level s is $0 < s < 20$ cigarettes per day; it is 26.1 for smokers with $s \geq 20$ cigarettes per day (R. C. Brownson et al., *Epidemiology, 3*: 61–64, 1992). Show that the estimated odds ratio between carcinoma (yes, no) and the smoking levels $(s \geq 20, 0 < s < 20)$ equals $26.1/11.7 = 2.2$.

2.6. Table 2.11 was taken from the 1991 General Social Survey.

Table 2.11

Race	Belief in Afterlife	
	Yes	No or Undecided
White	621	239
Black	89	42

a. Identify each classification as a response or explanatory variable.

b. Describe the association. Interpret the direction and strength of association.

c. Obtain a 95% confidence interval for a population measure, and interpret.

2.7. A poll by Louis Harris and Associates of 1249 adult Americans in July 1994 indicated that 36% believe in ghosts and 37% believe in astrology. Can we compare the proportions using inferential methods for independent binomial samples? Explain.

2.8. Table 2.12 is based on records of accidents in 1988 compiled by the Department of Highway Safety and Motor Vehicles in Florida. Compute and interpret the sample odds ratio, relative risk, and difference of proportions, and explain why the odds ratio approximately equals the relative risk.

Table 2.12

Safety Equipment in Use	Injury	
	Fatal	Nonfatal
None	1601	162,527
Seat belt	510	412,368

Source: Department of Highway Safety and Motor Vehicles, State of Florida.

2.9. In an article about crime in the United States, *Newsweek* magazine (Jan. 10, 1994) quoted FBI statistics stating that of all blacks slain in 1992, 94% were slain by blacks, and of all whites slain in 1992, 83% were slain by whites. Let Y denote race of victim and X denote race of murderer.

   a. Which conditional distribution do these statistics refer to, Y given X, or X given Y?

   b. Calculate and interpret the odds ratio between X and Y.

   c. Given that a murderer was white, can you estimate the probability that the victim was white? What additional information would you need to do this? (*Hint:* Use Bayes Theorem.)

2.10. A 20-year study of British male physicians (R. Doll and R. Peto, *British Med. J., 2*: 1525–1536 (1976)) noted that the proportion who died from lung cancer was .00140 per year for cigarette smokers and .00010 per year for nonsmokers. The proportion who died from heart disease was .00669 for smokers and .00413 for nonsmokers.

   a. Describe the association of smoking with lung cancer and with heart disease, using the difference of proportions, the relative risk, and the odds ratio. Interpret.

   b. Which response is more strongly related to cigarette smoking in terms of the reduction in deaths that could occur with an absence of smoking?

2.11. Refer to Table 2.1.

   a. Construct a 90% confidence interval for the difference of proportions, and interpret.

   b. Construct a 90% confidence interval for the odds ratio, and interpret.

   c. Conduct a test of statistical independence. Interpret.

2.12. A large-sample confidence interval for the log of the relative risk is

$$\log\left(\frac{p_1}{p_2}\right) \pm z_{\alpha/2}\sqrt{\frac{1-p_1}{N_1 p_1} + \frac{1-p_2}{N_2 p_2}}.$$

Antilogs of the endpoints yield an interval for the true relative risk. For Table 2.1, construct a 90% confidence interval.

2.13. Refer to Table 2.3. Find the P-value for testing that the incidence of heart attacks is independent of aspirin intake, (a) using X^2, (b) using G^2. Interpret results.

2.14. Refer to Table 2.4. Do these data provide evidence of an association between myocardial infarction and smoking? Use an inferential procedure, and interpret.

2.15. Table 2.13 was taken from the 1991 General Social Survey.

Table 2.13

Race	Party Identification		
	Democrat	Independent	Republican
White	341	105	405
Black	103	15	11

a. Test the hypothesis of independence between party identification and race. Interpret.

b. Use adjusted residuals to describe the evidence.

c. Partition chi-squared into two components, and use the components to describe the evidence.

2.16. A recent article (D. J. Moritz and W. A. Satariano, *J. Clin. Epidemiol., 46*: 443–454 (1993)) investigated the relationship between stage of breast cancer at diagnosis (local or advanced) and a woman's living arrangement. Of 144 women living alone, 41.0% had an advanced case; of 209 living with spouse, 52.2% were advanced; of 89 living with others, 59.6% were advanced. The authors reported the P-value for the relationship as .02. Reconstruct the analysis they performed to obtain this P-value.

2.17. Give examples of contingency tables for which a chi-squared test of independence using X^2 or G^2 should *not* be used, because of (a) sample size, (b) measurement scale.

2.18. Table 2.14 classifies a sample of psychiatric patients by their diagnosis and by whether their treatment prescribed drugs.

Table 2.14

Diagnosis	Drugs	No Drugs
Schizophrenia	105	8
Affective disorder	12	2
Neurosis	18	19
Personality disorder	47	52
Special symptoms	0	13

Source: E. Helmes and G. C. Fekken, *J. Clin. Psychol., 42:* 569–576 (1986). © Clinical Psychology Publishing Co., Inc., Brandon, VT. Reproduced with permission of the publisher.

    a. Report the P-value for a test of independence, and interpret the result.

    b. Calculate adjusted residuals, and interpret.

    c. Partition chi-squared into three components to describe differences and similarities among the diagnoses, by comparing (i) the first two rows, (ii) the third and fourth rows, (iii) the last row to the first and second rows combined and the third and fourth rows combined.

2.19. Refer to Table 7.5 (Chapter 7). Combine data for the two genders, yielding a single 4 × 4 table.

    a. Use X^2 and G^2 to test independence. Interpret.

    b. Partition G^2 into three components for three 2 × 4 tables by (i) comparing the first two income levels on job satisfaction, (ii) comparing the last two income levels on job satisfaction, and (iii) comparing the first two income levels combined to the last two income levels combined. Interpret. Does job satisfaction appear to depend on whether one's income is below or above $15,000?

    c. Now test independence using the ordinality of income and job satisfaction. Interpret the result and explain why the result is so different from the X^2 and G^2 tests.

    d. Check the sensitivity to the choice of scores by redoing (**c**) for a different "reasonable" set of scores. Interpret.

2.20. A study on educational aspirations of high school students (S. Crysdale, *Int. J. Compar. Sociol., 16*: 19–36 (1975)) measured aspirations using the scale (some high school, high school graduate, some college, college graduate). For students whose family income was low, the counts in these categories were (9, 44, 13, 10); when family income was middle, the counts were (11, 52, 23, 22); when family income was high, the counts were (9, 41, 12, 27).

    a. Test independence of educational aspirations and family income using X^2 or G^2. Interpret, and explain the deficiency of this test for these data.

    b. Calculate the adjusted residuals. Do they suggest any association pattern?

    c. Conduct an alternative test that may be more powerful. Interpret results.

2.21. Table 2.15 refers to a study that assessed factors associated with women's attitudes toward mammography (Hosmer and Lemeshow, 1989, p. 220). The columns refer to their response to the question, "How likely is it that a mammogram could find a new case of breast cancer?" Analyze these data.

Table 2.15

Mammography Experience	Detection of Breast Cancer		
	Not Likely	Somewhat Likely	Very Likely
Never	13	77	144
Over one year ago	4	16	54
Within the past year	1	12	91

Source: Hosmer and Lemeshow (1989), p. 224. Reprinted with permission of John Wiley & Sons, Inc.

2.22. Refer to Table 8.12 (Chapter 8). Analyze these data using the methods of this chapter.

2.23. A study (B. Kristensen et al., *J. Intern. Med.*, *232*: 237–245 (1992)) considered the effect of prednisolone on severe hypercalcaemia in women with metastatic breast cancer. Of 30 patients, 15 were randomly selected to receive prednisolone, and the other 15 formed a control group. Seven of the 15 prednisolone-treated patients achieved normalization in their level of serum-ionized calcium. This happened for none of the 15 patients in the control group. Use Fisher's exact test to find a P-value for testing whether results were significantly better for treatment than control. Interpret.

2.24. Refer to the previous problem. Compute the sample odds ratio. Using software, obtain a 95% "exact" confidence interval for the true odds ratio. Interpret, and note the effect of the zero cell count.

2.25. Table 2.16 contains results of a study comparing radiation therapy with surgery in treating cancer of the larynx. Use Fisher's exact test to test $H_0 : \theta = 1$ against $H_a : \theta > 1$. Interpret results.

Table 2.16

	Cancer Controlled	Cancer Not Controlled
Surgery	21	2
Radiation therapy	15	3

Source: Reprinted from W. Mendenhall et al., *Int. J. Radiat. Oncol. Biol. Phys., 10:* 357–363 (1984), with kind permission of Elsevier Science Ltd.

2.26. Refer to the previous problem.

a. Obtain and interpret the one-sided mid-P value. Give advantages and disadvantages of this type of P-value compared to the ordinary one.

b. Obtain and interpret a two-sided exact P-value.

2.27. Suppose a researcher routinely conducts tests using a nominal probability of Type I error of .05, rejecting H_0 if the P-value satisfies $P \le .05$. Suppose an exact test using X^2 has null distribution $P(X^2 = 0) = .30$, $P(X^2 = 3) = .62$, and $P(X^2 = 9) = .08$.

a. Show that, with the usual P-value, the actual probability of Type I error equals 0.

b. Show that, with the mid P-value, the actual probability of Type I error equals .08.

c. Repeat **(a)** and **(b)** using the probabilities .30, .66, .04. Note that the test with mid P-value can be "conservative" or "liberal." The test with the ordinary P-value cannot be liberal.

2.28. Refer to Table 2.8.

    a. Construct the null distributions of the ordinary P-value and the mid P-value, for the one-sided alternative. Compute and compare their expected values.

    b. Repeat **(a)** for the test using X^2 for the two-sided alternative.

2.29. Consider the 3×3 table having entries, by row, of $(4, 2, 0/2, 2, 2/0, 2, 4)$.

    a. Using software, conduct an exact test of independence, using X^2. Interpret.

    b. Suppose the row and column classifications are ordinal. Using equally-spaced scores, conduct an ordinal exact test. Explain why results differ so much from **(a)**.

2.30. A diagnostic test is designed to detect whether subjects have a certain disease. A positive test outcome predicts that a subject has the disease. Given that the subject has the disease, the probability the diagnostic test is positive is called the *sensitivity*. Given that the subject does not have the disease, the probability the test is negative is called the *specificity*. Consider the 2×2 table having the true status as the row variable and the diagnosis as the column variable. If "positive" is the first level of each classification, then sensitivity is π_1 and specificity is $1 - \pi_2$. Let γ denote the probability that a subject has the disease.

    a. Given that the diagnosis is positive, use Bayes Theorem to show that the probability a subject truly has the disease is

$$\frac{\pi_1 \gamma}{\pi_1 \gamma + \pi_2(1 - \gamma)}.$$

    b. Suppose a diagnostic test for HIV+ status has sensitivity and specificity both equal to .95. If $\gamma = .005$, find the probability that a subject is HIV+, given that the diagnostic test is positive.

    c. To better understand the answer in **(b)**, find the four joint probabilities for the 2×2 table, and discuss their relative sizes.

2.31. For tests of independence, $\{\hat{\mu}_{ij} = n_{i+}n_{+j}/n\}$. Show that $\{\hat{\mu}_{ij}\}$ have the same row and column totals as the observed data. For 2×2 tables, show that their odds ratio equals 1.0. Hence, they satisfy the null hypothesis. (For $I \times J$ tables, the odds ratio equals 1.0 for every 2×2 subtable formed using a pair of rows and a pair of columns.)

2.32. The *Pearson residual* for a cell in a two-way table equals
$e_{ij} = (n_{ij} - \hat{\mu}_{ij})/\sqrt{\hat{\mu}_{ij}}$.

    a. Show that they provide a decomposition of the Pearson chi-squared statistic, through $X^2 = \sum e_{ij}^2$.

    b. Show that Pearson residuals are smaller than adjusted residuals and thus have smaller variance than standard normal variates.

    c. For 2×2 tables, show that (i) all four adjusted residuals have the same absolute value. (This is sensible, since $df = 1$ and the table contains only one piece of information about association.) Show that (ii) the four Pearson residuals may take different values; (iii) the square of each adjusted residual equals X^2. (*Note:* For 2×2 tables, X^2 simplifies to

$$X^2 = \frac{n(n_{11}n_{22} - n_{12}n_{21})^2}{(n_{1+}n_{2+}n_{+1}n_{+2})}.$$

Adjusted residuals are identical for Poisson and binomial sampling. The Pearson residual defined here refers to Poisson sampling, and a different Pearson residual applies for binomial sampling; see Section 5.3.3.)

2.33. Formula (2.4.3) has alternative expression $X^2 = n \sum (p_{ij} - p_{i+}p_{+j})^2 / p_{i+}p_{+j}$. For a particular set of $\{p_{ij}\}$, X^2 is directly proportional to n. Hence, X^2 can be large when n is large, regardless of whether the association is practically important. Explain why chi-squared tests, like other tests, simply indicate the degree of evidence against a hypothesis and do not give information about the strength of association. ("Like fire, the chi-square test is an excellent servant and a bad master," Sir Austin Bradford Hill, *Proc. R. Soc. Med., 58*: 295–300 (1965).)

2.34. Let Z denote a standard normal variate. Then Z^2 has a chi-squared distribution with $df = 1$. A chi-squared variate with degrees of freedom equal to df has representation $Z_1^2 + \cdots + Z_{df}^2$, where $Z_1, \ldots, Z_{df}$ are independent standard normal variates. Using this, show that if Y_1 and Y_2 are independent chi-squared variates with degrees of freedom df_1 and df_2, then $Y_1 + Y_2$ has a chi-squared distribution with $df = df_1 + df_2$.

C H A P T E R   3

Three-Way Contingency Tables

An important part of most research studies is the choice of control variables. In studying the effect of an explanatory variable X on a response variable Y, one should "control" covariates that can influence that relationship. That is, one should use some mechanism to hold such covariates constant while studying the effect of X on Y. Otherwise, an observed effect of X on Y may simply reflect effects of those covariates on both X and Y. This is particularly true for observational studies, where one does not have the luxury of randomly assigning subjects to different treatments.

To illustrate, suppose a study considers effects of passive smoking; that is, the effects on a nonsmoker of living with a smoker. To analyze whether passive smoking is associated with lung cancer, a cross-sectional study might compare lung cancer rates between nonsmokers whose spouses smoke and nonsmokers whose spouses do not smoke. In doing so, we should attempt to control for age, socioeconomic status, or other factors that might relate both to whether one's spouse smokes and to whether one has lung cancer. Unless we control such variables, results will have limited usefulness. Suppose spouses of nonsmokers tend to be younger than spouses of smokers, and suppose younger people are less likely to have lung cancer. Then, a lower proportion of lung cancer cases among spouses of nonsmokers may simply reflect their lower average age.

Including control variables in an analysis requires a multivariate rather than a bivariate analysis. This chapter generalizes the methods of Chapter 2 regarding two-way contingency tables to multi-way tables. The main topic is analyzing the association between two categorical variables X and Y, while controlling for effects of a possibly confounding variable Z. We do this by studying the X-Y relationship at fixed, constant levels of Z. For simplicity, the examples refer to three-way tables with binary responses. Later chapters treat more general cases as well as the use of models to perform statistical control.

Section 3.1 shows that the association between two variables may change dramatically under a control for another variable. Sections 3.2 and 3.3 present inferential methods for such associations; Section 3.2 presents large-sample methods and Section 3.3 discusses exact inference for small samples.

53

3.1 PARTIAL ASSOCIATION

We begin this chapter by discussing statistical control and the types of relationships one can encounter in performing such control with multivariate categorical data. We illustrate basic concepts for a response variable Y, an explanatory variable X, and a single control variable Z, all of which are categorical. A three-way contingency table displays counts for the combinations of levels of the three variables.

3.1.1 Partial Tables

Two-way cross-sectional slices of the three-way table cross classify X and Y at separate levels of the control variable Z. These cross sections are called *partial tables*. They display the X-Y relationship at fixed levels of Z, hence showing the effect of X on Y while controlling for Z. The partial tables remove the effect of Z by holding its value constant.

The two-way contingency table obtained by combining the partial tables is called the X-Y *marginal table*. Each cell count in the marginal table is a sum of counts from the same cell location in the partial tables. The marginal table, rather than controlling Z, ignores it. The marginal table contains no information about Z. It is simply a two-way table relating X and Y. Methods for two-way tables, discussed in Chapter 2, do not take into account effects of other variables.

The associations in partial tables are called *conditional associations*, because they refer to the effect of X on Y conditional on fixing Z at some level. Conditional associations in partial tables can be quite different from associations in marginal tables. In fact, it can be misleading to analyze only a marginal table of a multi-way contingency table, as the following example illustrates.

3.1.2 Death Penalty Example

Table 3.1 is a $2 \times 2 \times 2$ contingency table—two rows, two columns, and two layers—from an article that studied effects of racial characteristics on whether individuals convicted of homicide receive the death penalty. The 674 subjects classified in Ta-

Table 3.1 Death Penalty Verdict by Defendant's Race and Victims' Race

Victims' Race	Defendant's Race	Death Penalty		Percentage Yes
		Yes	No	
White	White	53	414	11.3
	Black	11	37	22.9
Black	White	0	16	0.0
	Black	4	139	2.8
Total	White	53	430	11.0
	Black	15	176	7.9

Source: M. L. Radelet and G. L. Pierce, *Florida Law Rev. 43*: 1–34 (1991).
Reprinted with permission of the *Florida Law Review*.

ble 3.1 were the defendants in indictments involving cases with multiple murders in Florida between 1976 and 1987. The variables in Table 3.1 are Y = "death penalty verdict," having categories (yes, no), and X = "race of defendant" and Z = "race of victims," each having categories (white, black). We study the effect of defendant's race on the death penalty verdict, treating victims' race as a control variable. Table 3.1 has a 2 × 2 partial table relating defendant's race and the death penalty verdict at each level of victims' race.

For each combination of defendant's race and victims' race, Table 3.1 lists and Figure 3.1 displays the percentage of defendants who received the death penalty. We use these to describe the conditional associations between defendant's race and the death penalty verdict, controlling for victims' race. When the victims were white, the death penalty was imposed 22.9% − 11.3% = 11.6% more often for black defendants than for white defendants. When the victim was black, the death penalty was imposed 2.8% more often for black defendants than for white defendants. Thus, *controlling* for victims' race by keeping it fixed, the percentage of "yes" death penalty verdicts was higher for black defendants than for white defendants.

The bottom portion of Table 3.1 displays the marginal table for defendant's race and the death penalty verdict. We obtain it by summing the cell counts in Table 3.1 over the two levels of victims' race, thus combining the two partial tables (e.g., 11 + 4 = 15). We see that, overall, 11.0% of white defendants and 7.9% of black defendants received the death penalty. *Ignoring* victims' race, the percentage of "yes"

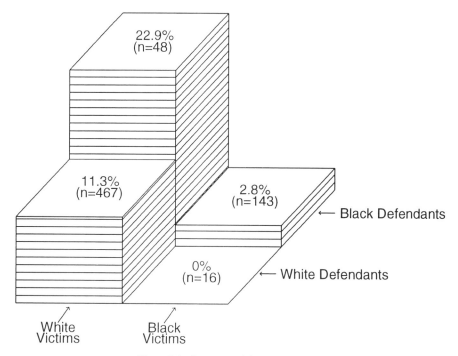

Figure 3.1   Percent receiving death penalty.

death penalty verdicts was lower for black defendants than for white defendants. The association reverses direction compared to the partial tables.

Why does the association between death penalty verdict and defendant's race differ so much when we ignore vs. control victims' race? This relates to the nature of the association between the control variable, victims' race, and each of the other variables. First, the association between victims' race and defendant's race is extremely strong. One can verify that the marginal table relating these variables has odds ratio $(467 \times 143)/(48 \times 16) = 87.0$; the odds that a white defendant had white victims are estimated to be 87.0 times the odds that a black defendant had white victims. Second, the percentages in Table 3.1 show that, regardless of defendant's race, the death penalty was considerably more likely when the victims were white than when the victims were black. So, whites are tending to kill whites, and killing whites is more likely to result in the death penalty. This suggests that the marginal association should show a greater tendency for white defendants to receive the death penalty than do the conditional associations. In fact, Table 3.1 shows this pattern.

Figure 3.2 may clarify why the conditional associations differ so from the marginal association. For each defendant's race, the figure plots the proportion receiving the death penalty at each level of victims' race. Each proportion is labeled by a letter

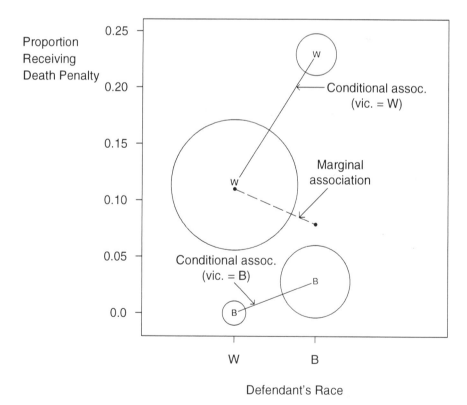

Figure 3.2 Proportion receiving death penalty by defendant's race, controlling and ignoring victim's race.

symbol giving the level of victims' race. Surrounding each observation is a circle having area proportional to the number of observations at that combination of defendant's race and victims' race. For instance, the W in the largest circle represents a proportion of .113 receiving the death penalty for cases with white defendants and white victims. That circle is largest, because the number of cases at that combination $(53 + 414 = 467)$ is larger than at the other three combinations. The next largest circle relates to cases in which blacks kill blacks.

We control for victims' race by comparing circles having the same victims' race letter at their centers. The line connecting the two W circles has a positive slope, as does the line connecting the two B circles. Controlling for victims' race, this reflects a higher chance of the death penalty for black defendants than white defendants. When we add results across victims' race to get a summary result for the marginal effect of defendant's race on the death penalty verdict, the larger circles having the greater number of cases have greater influence. Thus, the summary proportions for each defendant's race, marked on the figure by periods, fall closer to the center of the larger circles than the smaller circles. A line connecting the summary marginal proportions has negative slope, indicating that white defendants are more likely than black defendants to receive the death penalty.

The result that a marginal association can have different direction from the conditional associations is called *Simpson's paradox*. This result applies to quantitative as well as categorical variables.

3.1.3 Conditional and Marginal Odds Ratios

One can describe marginal and conditional associations using odds ratios. We illustrate for $2 \times 2 \times K$ tables, where K denotes the number of levels of a control variable, Z. Let $\{n_{ijk}\}$ denote observed frequencies and let $\{\mu_{ijk}\}$ denote their expected frequencies.

Within a fixed level k of Z,

$$\theta_{XY(k)} = \frac{\mu_{11k}\mu_{22k}}{\mu_{12k}\mu_{21k}} \tag{3.1.1}$$

describes conditional X-Y association. It is the ordinary odds ratio computed for the four expected frequencies in the kth partial table. We refer to the odds ratios for the K partial tables as the X-Y *conditional odds ratios*.

The conditional odds ratios can be quite different from marginal odds ratios, for which the third variable is ignored rather than controlled. The X-Y marginal table has expected frequencies $\{\mu_{ij+} = \sum_k \mu_{ijk}\}$ obtained by summing over the levels of Z. The X-Y marginal odds ratio is

$$\theta_{XY} = \frac{\mu_{11+}\mu_{22+}}{\mu_{12+}\mu_{21+}}.$$

Similar formulas with cell counts substituted for expected frequencies provide sample estimates of $\theta_{XY(k)}$ and θ_{XY}.

We illustrate by computing sample conditional and marginal odds ratios for the association between defendant's race and the death penalty. From Table 3.1, the estimated odds ratio in the first partial table, for which victims' race is white, equals

$$\hat{\theta}_{XY(1)} = \frac{53 \times 37}{414 \times 11} = 0.43.$$

The sample odds for white defendants receiving the death penalty were 43% of the sample odds for black defendants. In the second partial table, for which victim's race is black, the estimated odds ratio equals $\hat{\theta}_{XY(2)} = (0 \times 139)(16 \times 4) = 0.0$, since the death penalty was never given to white defendants having black victims.

Estimation of the marginal odds ratio for defendant's race and the death penalty uses the 2×2 marginal table in Table 3.1, collapsing over victims' race. The estimate equals $(53 \times 176)/(430 \times 15) = 1.45$. The sample odds of the death penalty were 45% higher for white defendants than for black defendants. Yet, we just observed that those odds were smaller for a white defendant than for a black defendant, within each level of victims' race. This reversal in the association when we control for victims' race illustrates Simpson's paradox. (Problems 5.16 and 6.3 consider further analyses of these data.)

3.1.4 Marginal versus Conditional Independence

Consider the true relationship between X and Y, controlling for Z. If X and Y are independent in each partial table, then X and Y are said to be *conditionally independent, given Z*. All conditional odds ratios between X and Y then equal 1. Conditional independence of X and Y, given Z, does not imply marginal independence of X and Y. That is, when odds ratios between X and Y equal 1 at each level of Z, the marginal odds ratio may differ from 1.

The expected frequencies in Table 3.2 show a hypothetical relationship among three variables: Y = response (success, failure), X = drug treatment (A, B), and Z = clinic (1, 2). The conditional association between X and Y at the two levels of

Table 3.2 Conditional Independence Does Not Imply Marginal Independence

Clinic	Treatment	Response	
		Success	Failure
1	A	18	12
	B	12	8
2	A	2	8
	B	8	32
Total	A	20	20
	B	20	40

Z is described by the odds ratios

$$\theta_{XY(1)} = \frac{18 \times 8}{12 \times 12} = 1.0$$

$$\theta_{XY(2)} = \frac{2 \times 32}{8 \times 8} = 1.0.$$

Given clinic, response and treatment are conditionally independent. The marginal table adds together the tables for the two clinics. The odds ratio for that marginal table equals $\theta_{XY} = (20 \times 40)/(20 \times 20) = 2.0$, so the variables are not marginally independent.

Why are the odds of a success twice as high for treatment A as treatment B when we ignore clinic? The conditional X-Z and Y-Z odds ratios give a clue. The odds ratio between Z and either X or Y, at each fixed level of the other variable, equals 6.0. For instance, the X-Z odds ratio at the first level of Y equals $(18)(8)/(12)(2) = 6.0$. The conditional odds (given response) of receiving treatment A are six times higher at clinic 1 than clinic 2, and the conditional odds (given treatment) of success are six times higher at clinic 1 than at clinic 2. Clinic 1 tends to use treatment A more often, and clinic 1 also tends to have more successes. For instance, if patients who attend clinic 1 tend to be in better health or tend to be younger than those who go to clinic 2, perhaps they have a better success rate than subjects in clinic 2 regardless of the treatment received.

It is misleading to study only the marginal table, concluding that successes are more likely with treatment A than with treatment B. Subjects within a particular clinic are likely to be more homogeneous than the overall sample, and response is independent of treatment in each clinic.

3.1.5 Homogeneous Association

There is *homogeneous X-Y association* in a $2 \times 2 \times K$ table when

$$\theta_{XY(1)} = \theta_{XY(2)} = \cdots = \theta_{XY(K)}.$$

The conditional odds ratio between X and Y is then identical at each level of Z. Thus, the effect of X on Y is the same at each level of Z, and a single number describes the X-Y conditional associations. Conditional independence of X and Y is the special case in which each conditional odds ratio equals 1.0.

Homogeneous X-Y association in an $I \times J \times K$ table means that any conditional odds ratio formed using two levels of X and two levels of Y is the same at each level of Z. When X-Y conditional odds ratios are identical at each level of Z, the same property holds for the other associations. For instance, the conditional odds ratio between two levels of X and two levels of Z is identical at each level of Y. Homogeneous association is a symmetric property, applying to any pair of the variables viewed across the levels of the third. When it occurs, there is said to be *no interaction* between two variables in their effects on the third variable.

When homogeneous association does not exist, the conditional odds ratio for any pair of variables changes across levels of the third variable. For X = smoking (yes, no), Y = lung cancer (yes, no), and Z = age (<45, 45–65, >65), suppose $\theta_{XY(1)} = 1.2$, $\theta_{XY(2)} = 2.8$, and $\theta_{XY(3)} = 6.2$. Then, smoking has a weak effect on lung cancer for young people, but the effect strengthens considerably with age.

The estimated conditional odds ratios for Table 3.1 are $\hat{\theta}_{XY(1)} = 0.43$ and $\hat{\theta}_{XY(2)} = 0.0$. The values are not close, but the second estimate is unstable because of the zero cell count. If we add $\frac{1}{2}$ to each cell count, we obtain 0.94 for the second estimate. Because the second estimate is so unstable and because further variation can occur from sampling variability, these data do not necessarily contradict homogeneous association. The next section shows how to check whether sample data are consistent with homogeneous association or conditional independence.

3.2 COCHRAN–MANTEL–HAENSZEL METHODS

This section introduces inferential analyses for three-way tables. We present a test of conditional independence and a test of homogeneous association for the K conditional odds ratios in $2 \times 2 \times K$ tables. We also show how to combine the sample odds ratios from the K partial tables into a single summary measure of partial association.

Analyses of conditional association are relevant in most applications having multivariate data. To illustrate, we analyze Table 3.3, which summarizes eight studies in China about smoking and lung cancer. The smokers and nonsmokers are the levels

Table 3.3 Chinese Smoking and Lung Cancer Study, with Information Relevant to Cochran–Haenszel Test

| City | Smoking | Lung Cancer | | Odds Ratio | μ_{11k} | $Var(n_{11k})$ |
		Yes	No			
Beijing	Smokers	126	100	2.20	113.0	16.9
	Nonsmokers	35	61			
Shanghai	Smokers	908	688	2.14	773.2	179.3
	Nonsmokers	497	807			
Shenyang	Smokers	913	747	2.18	799.3	149.3
	Nonsmokers	336	598			
Nanjing	Smokers	235	172	2.85	203.5	31.1
	Nonsmokers	58	121			
Harbin	Smokers	402	308	2.32	355.0	57.1
	Nonsmokers	121	215			
Zhengzhou	Smokers	182	156	1.59	169.0	28.3
	Nonsmokers	72	98			
Taiyuan	Smokers	60	99	2.37	53.0	9.0
	Nonsmokers	11	43			
Nanchang	Smokers	104	89	2.00	96.5	11.0
	Nonsmokers	21	36			

Source: Based on data in Z. Liu, Smoking and lung cancer in China, *Intern. J. Epidemiol.*, *21*: 197–201 (1992). Reprinted with permission of Oxford University Press.

of a group classification X, the two possible outcomes (yes, no) for lung cancer are the levels of a response variable Y, and different cities are levels of a control variable Z. Subjects may vary among cities on relevant characteristics such as socioeconomic status, which may cause heterogeneity among the cities in smoking rates and in the lung cancer rate. Thus, we investigate the association between X and Y while controlling for Z.

3.2.1 The Cochran–Mantel–Haenszel Test

For $2 \times 2 \times K$ tables, the null hypothesis that X and Y are conditionally independent, given Z, means that the conditional odds ratio $\theta_{XY(k)}$ between X and Y equals 1 in each partial table. The standard sampling models treat the cell counts as (1) independent Poisson variates, or (2) multinomial counts with fixed overall sample size, or (3) multinomial counts with fixed sample size for each partial table, with counts in different partial tables being independent, or (4) independent binomial samples within each partial table with row totals fixed. In partial table k, the row totals are $\{n_{1+k}, n_{2+k}\}$, and the column totals are $\{n_{+1k}, n_{+2k}\}$. Given both these totals, all these sampling schemes yield a hypergeometric distribution (Section 2.6.1) for the count n_{11k} in the cell in the first row and first column. That cell count determines all other counts in the partial table. The test statistic utilizes this cell in each partial table.

Under the null hypothesis, the mean and variance of n_{11k} are

$$\mu_{11k} = E(n_{11k}) = \frac{n_{1+k}n_{+1k}}{n_{++k}}$$

$$\text{Var}(n_{11k}) = \frac{n_{1+k}n_{2+k}n_{+1k}n_{+2k}}{n_{++k}^2(n_{++k} - 1)}.$$

When the true odds ratio $\theta_{XY(k)}$ exceeds 1.0 in partial table k, we expect to observe $(n_{11k} - \mu_{11k}) > 0$. The test statistic combines these differences across all K tables. When the odds ratio exceeds 1.0 in every partial table, the sum of such differences tends to be a relatively large positive number; when the odds ratio is less than 1.0 in each table, the sum of such differences tends to be a relatively large negative number.

The test statistic summarizes the information from the K partial tables using

$$CMH = \frac{\left[\sum_k (n_{11k} - \mu_{11k})\right]^2}{\sum_k \text{Var}(n_{11k})}. \tag{3.2.1}$$

This is called the *Cochran–Mantel–Haenszel* (*CMH*) statistic. It has a large-sample chi-squared distribution with $df = 1$.

The *CMH* statistic takes larger values when $(n_{11k} - \mu_{11k})$ is consistently positive or consistently negative for all tables, rather than positive for some and negative for others. This test is inappropriate when the association varies dramatically among the partial tables. It works best when the X-Y association is similar in each partial table.

The *CMH* statistic combines information across partial tables. When the true association is similar in each table, this test is more powerful than separate tests within each table. It is improper to combine results by adding the partial tables together to form a single 2×2 marginal table for the test. Simpson's paradox (Section 3.1.2) revealed the dangers of collapsing three-way tables.

3.2.2 Lung Cancer Meta Analysis Example

Table 3.3 summarizes eight case-control studies in China about smoking and lung cancer. Each study matched cases of lung cancer with controls not having lung cancer and then recorded whether each subject had ever been a smoker. In each partial table, we treat the counts in each column as a binomial sample, with column total fixed. We test the hypothesis of conditional independence between smoking and lung cancer, which states that the true odds ratio equals 1.0 for each city.

Table 3.3 reports the sample odds ratio for each table and the expected value and variance of the number of lung cancer cases who were smokers (the count in the first row and first column), under this hypothesis. In each table, the sample odds ratio shows a moderate positive association, so it makes sense to combine results through the *CMH* statistic. We obtain $\sum_k n_{11k} = 2930$, $\sum_k \mu_{11k} = 2562.5$, and $\sum_k \text{Var}(n_{11k}) = 482.1$, for which $CMH = (2930.0 - 2562.5)^2/482.1 = 280.1$, with $df = 1$. There is extremely strong evidence against conditional independence ($P < .0001$). This is not surprising, given the large sample size for the combined studies ($n = 8419$).

A statistical analysis that combines information from several studies is called a *meta analysis*. The meta analysis of Table 3.3 provides stronger evidence of an association than any single partial table gives by itself.

3.2.3 Estimation of Common Odds Ratio

It is more informative to estimate the strength of association than simply to test a hypothesis about it. When the association seems stable across partial tables, we can estimate an assumed common value of the K true odds ratios.

In a $2 \times 2 \times K$ table, suppose that $\theta_{XY(1)} = \cdots = \theta_{XY(K)}$. The *Mantel–Haenszel estimator* of that common value equals

$$\hat{\theta}_{MH} = \frac{\sum_k (n_{11k} n_{22k}/n_{++k})}{\sum_k (n_{12k} n_{21k}/n_{++k})}. \tag{3.2.2}$$

The standard error for $\log(\hat{\theta}_{MH})$ has a complex formula (Agresti (1990), p. 236), so we shall not report it here. Some software (such as SAS-PROC FREQ) computes a standard error, and one can also obtain an estimate and standard error using logit models (Section 5.4.4).

For the Chinese smoking studies summarized in Table 3.3, the Mantel–Haenszel odds ratio estimate equals

$$\hat{\theta}_{MH} = \frac{(126)(61)/(322) + \cdots + (104)(36)/250}{(35)(100)/(322) + \cdots + (21)(89)/(250)} = 2.17.$$

The estimated standard error of $\log(\hat{\theta}_{MH}) = \log(2.17) = 0.777$ equals 0.046. An approximate 95% confidence interval for the common log odds ratio is $0.777 \pm 1.96 \times 0.046$, or $(0.686, 0.868)$, corresponding to $(\exp(.686), \exp(.868)) = (1.98, 2.38)$ for the odds ratio. The odds of lung cancer for smokers equal about twice the odds for non-smokers. Such odds ratios are typically much larger in Western society. They may be lower in China partly because, until recent years, pipes with long stems were more common than cigarettes.

If the true odds ratios are not identical but do not vary drastically, $\hat{\theta}_{MH}$ still provides a useful summary of the K conditional associations. Similarly, the CMH test is a powerful summary of evidence against the hypothesis of conditional independence, as long as the sample associations fall primarily in a single direction.

3.2.4 Testing Homogeneity of Odds Ratios*

One can test the hypothesis that the odds ratio between X and Y is the same at each level of Z; that is, $H_0 : \theta_{XY(1)} = \cdots = \theta_{XY(K)}$. This is a test of homogeneous association for $2 \times 2 \times K$ tables.

Let $\{\hat{\mu}_{11k}, \hat{\mu}_{12k}, \hat{\mu}_{21k}, \hat{\mu}_{22k}\}$ denote estimated expected frequencies in the kth partial table that have the same marginal totals as the observed data, yet have odds ratio equal to the Mantel–Haenszel estimate $\hat{\theta}_{MH}$ of a common odds ratio. The test statistic, called the *Breslow–Day statistic*, has the Pearson form

$$\sum \frac{(n_{ijk} - \hat{\mu}_{ijk})^2}{\hat{\mu}_{ijk}}, \tag{3.2.3}$$

where the sum is taken over all cells in the table. The closer the cell counts fall to the values having a common odds ratio, the smaller the statistic and the less the evidence against H_0.

Calculation of the $\{\hat{\mu}_{ijk}\}$ satisfying a common odds ratio is complex and not discussed here since standard software (such as PROC FREQ in SAS) reports this statistic. The Breslow–Day statistic is an approximate chi-squared statistic with $df = K - 1$. The sample size should be relatively large in each partial table, with $\{\hat{\mu}_{ijk} \geq 5\}$ in at least about 80% of the cells.

For the meta analysis of smoking and lung cancer (Table 3.3), software reports a Breslow–Day statistic equal to 5.2, based on $df = 7$, for which $P = .64$. This evidence does not contradict the hypothesis of equal odds ratios. We are justified in summarizing the conditional association by a single odds ratio for all eight partial tables.

3.2.5 Some Caveats*

To ensure that the distribution of the Breslow–Day statistic (3.2.3) converges to chi-squared as the sample size increases, R. Tarone showed (*Biometrika*, 72: 91–95 (1985)) that one must adjust it by subtracting

$$\frac{\left[\sum_k (n_{11k} - \hat{\mu}_{11k})\right]^2}{\sum_k \left[\frac{1}{\hat{\mu}_{11k}} + \frac{1}{\hat{\mu}_{12k}} + \frac{1}{\hat{\mu}_{21k}} + \frac{1}{\hat{\mu}_{22k}}\right]^{-1}}.$$

This adjustment is usually minor. For Table 3.3, the correction is less than 0.01, making no difference to the conclusion.

Each partial table in Table 3.3 refers to a case-control study. Because each subject having lung cancer was matched with one or more controls not having lung cancer, the two columns are not truly independent binomial samples. In practice, whether a case had ever been a smoker is probably essentially independent of whether a control had ever been a smoker, so the different columns in each table should be similar to independent binomial samples. If for each case-control pair we had information about whether each subject had been a smoker, we could form a 2×2 table relating whether the control had ever been a smoker (yes, no) to whether the case had ever been a smoker (yes, no). Then specialized methods are available for comparing cases and controls on the proportions who ever smoked that take into account potential dependence. Chapter 9 discusses some of these. Positive correlations between proportions yield smaller P-values than we get by treating the samples as independent.

Section 6.5.1 discusses an alternative test of homogeneity of odds ratios, based on models. Section 7.3 presents generalizations of the *CHM* test for $I \times J \times K$ tables.

3.3   EXACT INFERENCE ABOUT CONDITIONAL ASSOCIATIONS*

The chi-squared tests presented in the previous section, like chi-squared tests of independence for $I \times J$ tables, are large-sample tests. The true sampling distributions converge to chi-squared as the sample size n increases. It is difficult to provide general guidelines about how large n must be. The tests' adequacy depends more on the two-way marginal totals than on counts in the separate partial tables. For the *CMH* statistic, for instance, cell counts in the partial tables can be small (which often happens when K is large), but the *X-Y* marginal totals should be relatively large.

In practice, small sample sizes are not problematic, since one can conduct *exact* inference about conditional associations. For instance, exact tests of conditional independence generalize Fisher's exact test for 2×2 tables.

3.3.1   Exact Test of Conditional Independence for $2 \times 2 \times K$ Tables

For $2 \times 2 \times K$ tables, conditional on the marginal totals in each partial table, the Cochran–Mantel–Haenszel test of conditional independence depends on the cell counts through $\sum_k n_{11k}$. Exact tests use $\sum_k n_{11k}$ in the way they use n_{11} in Fisher's exact test for 2×2 tables. Hypergeometric distributions in each partial table determine probabilities for $\{n_{11k}, k = 1, \ldots, K\}$. These determine the distribution of their sum.

The null hypothesis of conditional independence states that all conditional odds ratios $\{\theta_{XY(k)}\}$ equal 1. A "positive" conditional association corresponds to the one-sided alternative to this hypothesis, $\theta_{XY(k)} > 1$. The P-value then equals the right-tail probability that $\sum_k n_{11k}$ is at least as large as observed, for the fixed marginal totals. For the one-sided alternative $\theta_{XY(k)} < 1$, the P-value equals the left-tail probability that $\sum_k n_{11k}$ is no greater than observed. Two-sided alternatives can use a two-tail probability of those outcomes that are no more likely than the observed one.

Exact tests of conditional independence are computationally highly intensive. They require software for practical implementation. We used StatXact in the following example.

3.3.2 Promotion Discrimination Example

Table 3.4 refers to U.S. government computer specialists of similar seniority considered for promotion from classification level GS-13 to level GS-14. The table cross classifies promotion decision by employee's race, considered for three separate months. We test conditional independence of promotion decision and race. The table contains several small counts. The overall sample size is not small ($n = 74$), but one marginal count (collapsing over month of decision) equals zero, so we might be wary of using the *CMH* test.

We first use the one-sided alternative of an odds ratio less than 1. This corresponds to potential discrimination against black employees, their probability of promotion being lower than for white employees. Fixing the row and column marginal totals in each partial table, the test uses n_{11k}, the first cell count in each. For the margins of the partial tables in Table 3.4, n_{111} can range between 0 and 4, n_{112} can range between 0 and 4, and n_{113} can range between 0 and 2. The total $\sum_k n_{11k}$ can take values between 0 and 10. The sample data represent the most extreme possible result in each of the three cases. The observed value of $\sum_k n_{11k}$ equals 0, and the P-value is the null probability of this outcome, which software reveals to equal .026. A two-sided P-value, based on summing the probabilities of all tables having probabilities no greater than the observed table, equals .056. There is some evidence that promotion is related to race.

Table 3.4 Promotion Decisions by Race and by Month

Race	July Promotions		August Promotions		September Promotions	
	Yes	No	Yes	No	Yes	No
Black	0	7	0	7	0	8
White	4	16	4	13	2	13

Source: J. Gastwirth, *Statistical Reasoning in Law and Public Policy* (San Diego: Academic Press, 1988, p. 266).

3.3.3 Exact Confidence Interval for Common Odds Ratio

As discussed in Section 2.6.3, for small samples the discreteness implies that exact tests are conservative. When H_0 is true, for instance, the P-value may fall below .05 less than 5% of the time. One can alleviate conservativeness by using the mid-P value. It sums half the probability of the observed result with the probability of "more extreme" tables.

One can also construct "exact" confidence intervals for an assumed common value θ of $\{\theta_{XY(k)}\}$. Because of discreteness, these are also conservative. For a 95% "exact" confidence interval, the true confidence level is at least as large as .95, but is unknown. A more useful 95% confidence interval is the one containing the values θ_0 having mid P-values exceeding .05 in tests of $H_0 : \theta = \theta_0$. Though mid P-based confidence intervals do not guarantee that the true confidence level is at least as large as the nominal one, they are narrower and usually more closely match the nominal level than either the exact or large-sample intervals. Both approaches are computationally complex and require software.

We illustrate with Table 3.4. Because the sample result is the most extreme possible, the Mantel–Haenszel estimator (3.2.2) of an assumed common odds ratio equals $\hat{\theta}_{MH} = 0.0$. StatXact reports an "exact" 95% confidence interval for a common odds ratio of $(0, 1.01)$. We can be at least 95% confident that the true odds ratio falls in this interval. A 95% confidence interval based on correspondence with tests using the mid P-value is $(0, 0.78)$. We can be approximately 95% confident that the odds of promotion for blacks are no more than 78% of the odds for whites.

3.3.4 Exact Test of Homogeneity of Odds Ratios

The Breslow–Day test of homogeneity of odds ratios (Section 3.2.4) is also a large-sample test. It applies when the sample size is relatively large in each partial table. When the total sample size is small, or when the total sample size is large but the number K of partial tables is large and individual tables have small sample sizes, this test is not valid. An exact test of homogeneity of odds ratios, sometimes called *Zelen's exact test*, handles such cases. The exact distribution is calculated using the set of all $2 \times 2 \times K$ tables that have the same two-way marginal totals as the observed table. The P-value is the sum of probabilities of all $2 \times 2 \times K$ tables that are no more likely than the observed table.

For Table 3.4, the values $\{\hat{\mu}_{ijk}\}$ that yield the Mantel–Haenszel estimate $\hat{\theta}_{MH} = 0.0$ of the common odds ratio are identical to the observed counts in each partial table. The Breslow–Day statistic contains terms of the form $0/0$, and it is undefined. In addition, no table other than the observed table has all the two-way marginal totals of that table. Therefore, Zelen's exact test is degenerate, giving $P = 1.0$. For this pattern of data, one cannot obtain any meaningful information about whether the true odds ratios differ.

A disadvantage of exact inference is that the small-sample conditional distribution is often highly discrete. In some cases, such as this one, it concentrates at only a single point. One can do only so much with small sample sizes!

PROBLEMS

3.1. In murder trials in 20 Florida counties during 1976 and 1977, the death penalty was given in 19 out of 151 cases in which a white killed a white, in 0 out of 9 cases in which a white killed a black, in 11 out of 63 cases in which a black killed a white, and in 6 out of 103 cases in which a black killed a black (M. Radelet, *Amer. Sociol. Rev., 46*: 918–927 (1981)).

    a. Exhibit the data as a three-way contingency table.

    b. Construct the partial tables needed to study the conditional association between defendant's race and the death penalty verdict. Compute and interpret the sample conditional odds ratios, adding 0.5 to each cell to reduce the impact of the 0 cell count.

    c. Compute and interpret the sample marginal odds ratio between defendant's race and the death penalty verdict. Do these data exhibit Simpson's paradox? Explain.

3.2. For all trials in Florida involving homicides between 1976 and 1987, M. Radelet and G. Pierce (*Florida Law Review, 43*: 1–34 (1991)) reported the following results: The death penalty was given in 227 out of 4645 cases in which a white killed a white, in 92 out of 731 cases in which a black killed a white, in 9 out of 264 cases in which a white killed a black, and in 36 out of 4428 cases in which a black killed a black. Compute and interpret the sample conditional odds ratios between defendant's race and the death penalty verdict. Do the conditional associations seem to be homogeneous? Explain.

3.3. Smith and Jones are baseball players. Smith had a higher batting average than Jones in 1994 and 1995. Is it possible that for the combined data for these two years, Jones had the higher batting average? Explain, and illustrate using data.

3.4. Give a "real world" example of three variables X, Y, and Z, for which you expect X and Y to be marginally associated but conditionally independent, controlling for Z.

3.5. Based on 1987 murder rates in the United States, the Associated Press reported that the probability a newborn child has of eventually being a murder victim is 0.0263 for nonwhite males, 0.0049 for white males, 0.0072 for nonwhite females, and 0.0023 for white females.

    a. Find the conditional odds ratios between race and whether a murder victim, given gender. Interpret, and discuss whether these variables exhibit homogeneous association.

    b. If half the newborns are of each gender, for each race, find the marginal odds ratio between race and whether a murder victim.

3.6. Using graphs or tables to illustrate, explain what is meant by "no interaction" in modeling a response Y and explanatory variables X and Z, when (a) all variables are continuous (multiple regression), (b) Y and X are continuous, Z is categorical (analysis of covariance), (c) Y is continuous, X and Z are categorical (two-way ANOVA), (d) all variables are categorical ("no interaction" = "homogenous association").

3.7. For three-way contingency tables, when any pair of variables is conditionally independent, explain why there is homogenous association. When there is not homogeneous association, explain why no pair of variables can be conditionally independent.

3.8. Table 3.5 refers to the effect of passive smoking on lung cancer. It summarizes results of case-control studies from three countries among nonsmoking women married to smokers. Test the hypothesis that having lung cancer is independent of passive smoking, controlling for country. Report the P-value, and interpret. (*Note:* Weak associations in observational studies are suspect. With relatively small changes in the data, perhaps representing effects of misclassification or other bias, the association could disappear. See, for instance, R. L. Tweedie et al., Garbage in, garbage out, *Chance, 7*: no. 2, 20–27 (1994)).

Table 3.5

Country	Spouse Smoked	Cases	Controls
Japan	No	21	82
	Yes	73	188
Great Britain	No	5	16
	Yes	19	38
United States	No	71	249
	Yes	137	363

Source: Blot and Fraumeni, *J. Nat. Cancer Inst., 77*: 993–1000 (1986).

3.9. Refer to the previous problem. Assume that the true odds ratio between passive smoking and lung cancer is the same for each study. Estimate its value, and use software to find a 95% confidence interval. Interpret. Analyze whether the odds ratios truly are identical.

3.10. Table 3.6 shows results of a three-center clinical trial designed to compare a drug to placebo for treating severe migraine headaches. At each center, subjects were randomly assigned to treatments.

Table 3.6

Center	Group	Response Success	Failure
1	Drug	6	4
	Placebo	2	8
2	Drug	4	3
	Placebo	1	5
3	Drug	5	3
	Placebo	3	6

a. Describe the associations in the partial tables. Are results similar among centers?

    b. Find the P-value for testing that response is conditionally independent of whether a subject receives drug or placebo. Interpret.

    c. Calculate and interpret an estimate of the average conditional association between group and response.

    d. Test whether the true odds ratio is the same in each center. Interpret.

3.11. Refer to Table 3.1. Treating this as a sample, analyze the data.

3.12. Refer to Problem 3.2. Test whether the odds ratios are the same at each level of victims' race. Interpret.

3.13. Refer to Table 3.7, which classifies police officers by rank, race, and promotion decisions made in 1988.

Table 3.7

	Sergeant Promotions		SA Sergeant Promotions		Master Sgt Promotions		SA Master Sgt Promotions	
Race	Yes	No	Yes	No	Yes	No	Yes	No
Black	4	103	1	35	2	18	1	4
White	43	548	4	109	43	155	18	94

Source: Dixon vs. Margolis, 56 FEP Cases 401 (N.D. Illinois, 1991), as quoted by J. Gastwirth and C. Mehta in The usefulness of exact statistical methods in equal employment litigation, in *Computational Statistics*, Vol. 2, Y. Dodge and J. Whittaker, eds. (Berlin: Springer-Verlag, 1992), pp. 91–95.

    a. Conduct an exact test of conditional independence of promotion and race, given rank. Interpret, and compare results to the large-sample test.

    b. Conduct an exact test of whether the odds ratio is identical for the four ranks. Interpret, and compare results to the large-sample test.

    c. Construct and interpret a confidence interval for an assumed common odds ratio.

3.14. Refer to Problem 3.10.

    a. Use an exact test to conduct this analysis. Compare results to the large-sample test.

    b. Conduct an exact test that the odds ratio is identical for all three centers. Compare results to the large-sample test.

    c. Construct and interpret a confidence interval for an assumed common odds ratio.

3.15. Table 3.8 refers to ratings of agricultural extension agents in North Carolina. In each of five districts, agents were classified by their race and by whether they qualified for a merit pay increase. Analyze these data.

Table 3.8

District	Blacks, Merit Pay		Whites, Merit Pay	
	Yes	No	Yes	No
NC	24	9	47	12
NE	10	3	45	8
NW	5	4	57	9
SE	16	7	54	10
SW	7	4	59	12

Source: J. Gastwirth, *Statistical Reasoning in Law and Public Policy*, Vol. 1, (San Diego: Academic Press, 1988), p. 268.

CHAPTER 4

Generalized Linear Models

Chapters 2 and 3 presented methods for analyzing associations in two-way and three-way contingency tables. Those methods help us investigate effects of explanatory variables on categorical response variables. We next study ways of using *models* as the basis of such analyses.

A good-fitting model has several benefits. The structural form of the model describes the patterns of association and interaction. Inferences for model parameters help us evaluate which explanatory variables affect the response, while controlling effects of possible confounding variables. The sizes of the estimated model parameters determine the strength and importance of the effects. Finally, the model's predicted values smooth the data and provide improved estimates of the mean of the response distribution.

The inferences presented in Chapters 2 and 3 also result from analyzing effects in certain models. But models can handle more complicated situations than these, such as analyzing simultaneously the effects of several explanatory variables. In addition, the model-building paradigm focuses on estimating parameters that describe the effects, which is more informative than mere significance testing. Modeling categorical response variables is the primary theme of the rest of this book. The explanatory variables in the models can be continuous or categorical or both types.

Nearly all models presented in this book are special cases of *generalized linear models*. This is a broad class of models that includes ordinary regression and ANOVA models for continuous response variables as well as models for categorical response variables. This chapter introduces generalized linear models for categorical and other discrete response data. We use the acronym GLM as shorthand for *generalized linear model*.

The first section discusses three components that are common to all GLMs. Section 4.2 introduces models for binary response variables, appropriate for binomial data; an important special case is the *logistic regression* model, which Chapter 5 presents in detail. Section 4.3 introduces GLMs for count-type response variables modeled as Poisson data; an important special case is the *loglinear model*, which Chapter 6 presents in detail. Section 4.4 discusses checks of the adequacy of model fit for GLMs, illustrating for Poisson loglinear models. Section 4.5 presents further details about the fitting and checking of GLMs.

4.1 COMPONENTS OF A GENERALIZED LINEAR MODEL

All generalized linear models have three components: The *random component* identifies the response variable Y and assumes a probability distribution for it. The
systematic component specifies the explanatory variables used as predictors in the
model. The *link* describes the functional relationship between the systematic component and the expected value (mean) of the random component. The GLM relates
a function of that mean to the explanatory variables through a prediction equation
having linear form.

4.1.1 Random Component

For a sample of size N, denote the observations on the response variable Y by
$(Y_1, \ldots, Y_N)$. The GLMs discussed in this text treat $Y_1, \ldots, Y_N$ as independent. The
random component of a GLM consists of identifying the response variable Y and
selecting a probability distribution for $(Y_1, \ldots, Y_N)$.

    In many applications, the potential outcomes for each observation Y_i are binary,
such as "success" or "failure"; or, more generally, each Y_i might be the number
of "successes" out of a certain fixed number of trials. We then assume a *binomial*
distribution for the random component. In some other applications, each response
observation is a nonnegative count, such as a cell count in a contingency table.
We might then assume a *Poisson* distribution for the random component. If each
observation is continuous, such as a subject's weight in a dietary study, we might
assume a *normal* random component.

4.1.2 Systematic Component

Denote the expected value of Y, the mean of its probability distribution, by $\mu = E(Y)$.
In a GLM, the value of μ varies according to levels of explanatory variables.

    The *systematic component* of a GLM specifies the explanatory variables. These
enter linearly as predictors on the right hand side of the model equation. That is, the
systematic component specifies the variables that play the roles of $\{x_j\}$ in the formula

$$\alpha + \beta_1 x_1 + \cdots + \beta_k x_k.$$

This linear combination of the explanatory variables is called the *linear predictor*.
Some $\{x_j\}$ may be based on others in the model; for instance, perhaps $x_3 = x_1 x_2$,
to allow interaction between x_1 and x_2 in their effects on Y, or perhaps $x_3 = x_1^2$, to
allow a curvilinear effect of x_1.

4.1.3 Link

The third component of a GLM is the *link* between the random and systematic
components. It specifies how $\mu = E(Y)$ relates to the explanatory variables in the
linear predictor. One can model the mean μ directly, or model a monotone function

$g(\mu)$ of the mean. The model formula states that

$$g(\mu) = \alpha + \beta_1 x_1 + \cdots + \beta_k x_k.$$

The function $g(.)$ is called the *link function.*

The simplest possible link function has the form $g(\mu) = \mu$. This models the mean directly and is called the *identity link.* It specifies a linear model for the mean response,

$$\mu = \alpha + \beta_1 x_1 + \cdots + \beta_k x_k. \tag{4.1.1}$$

This is the form of ordinary regression models for continuous responses.

Other links permit the mean to be nonlinearly related to the predictors. For instance, the link function $g(\mu) = \log(\mu)$ models the log of the mean. The log function applies to positive numbers, so this "log link" is appropriate when μ cannot be negative, such as with count data. A GLM that uses the log link is called a *loglinear model.* It has form

$$\log(\mu) = \alpha + \beta_1 x_1 + \cdots + \beta_k x_k. \tag{4.1.2}$$

The link function $g(\mu) = \log[\mu/(1 - \mu)]$ models the log of an odds. This is called the *logit link.* It is appropriate when μ is between 0 and 1, such as a probability. A GLM that uses the logit link is called a *logit model.*

Each potential probability distribution for the random component has one special function of the mean that is called its *natural parameter.* For the normal distribution, it is the mean itself. For the Poisson, the natural parameter is the log of the mean. For the binomial, the natural parameter is the logit of the success probability. The link function that uses the natural parameter as $g(\mu)$ in the GLM is called the *canonical link.* For instance, the model formula (4.1.1) is the form of the GLM with canonical link for a normally distributed response. The GLM with canonical link for Poisson response data has form (4.1.2). Though other links are possible, in practice the canonical links are most common.

4.1.4 Normal GLM

Ordinary regression and ANOVA models for continuous variates are special cases of GLMs. One assumes a normal distribution for the random component and models the mean directly, using the identity link $g(\mu) = \mu$. A GLM generalizes ordinary regression models in two ways: First, it allows the random component to have a distribution other than the normal. Second, it allows modeling some function of the mean. Both generalizations are important for categorical data.

A traditional way of analyzing nonnormal data attempts to transform the response so it is approximately normal, with constant variance. Then, ordinary regression methods using least squares are applicable. In practice, this is usually not possible. A transform that produces constant variance may not produce normality, or else simple linear models for the explanatory variables may fit poorly on that scale. With the theory and methodology of GLMs, it is unnecessary to transform data so that normal-theory methods apply. This is because the GLM fitting process utilizes maximum likelihood methods for our choice of random component, and we are not restricted to normality for that choice. In addition, in GLMs the choice of link is separate from the choice of random component. If a link produces additivity of effects (i.e., if a linear model holds for that link), it is not necessary that it also stabilize variance or produce normality.

We have introduced GLMs in order to unify a wide variety of statistical methods. Regression, ANOVA, and models for categorical data are special cases of one super model. In fact, the same fitting method yields ML estimates of parameters for all GLMs. This method is the basis of software for fitting GLMs, such as GLIM and SAS (PROC GENMOD).

The next two sections illustrate the three GLM components by introducing the two most important GLMs for categorical response variables: logistic regression models for binary data with binomial random component and loglinear models for count data with Poisson random component.

4.2 GENERALIZED LINEAR MODELS FOR BINARY DATA

Many categorical response variables have only two categories: for instance, a vote in an election (Democrat, Republican), a choice of automobile (domestic, foreign import), or a diagnosis regarding whether a woman has breast cancer (present, absent). Denote a binary response by Y and the two possible outcomes by 1 and 0, or by the generic terminology "success" and "failure."

A binary response is sometimes called a *Bernoulli variable*. Its distribution is specified by probabilities $P(Y = 1) = \pi$ of success and $P(Y = 0) = (1 - \pi)$ of failure. This distribution has mean $E(Y) = \pi$ and variance $\text{Var}(Y) = \pi(1 - \pi)$. For n independent observations on a binary response with parameter π, the number of successes has the binomial distribution specified by the indices n and π.

This section introduces GLMs for binary response data. We assume that the random component in the model has a binomial distribution. Though GLMs can have multiple explanatory variables, for simplicity we introduce them using a single X. The value of π can vary as the value x of X changes, and we replace the π notation by $\pi(x)$ to reflect its dependence on that value.

4.2.1 Linear Probability Model

One approach to modeling the effect of X uses the form of ordinary regression, by which the expected value of Y is a linear function of X. The model

$$\pi(x) = \alpha + \beta x \tag{4.2.1}$$

is called a *linear probability model*, because the probability of success changes linearly in x. The parameter β represents the change in the probability per unit change in x. This model is a GLM with binomial random component and identity link function.

Unfortunately, this model has a major structural defect. Probabilities fall between 0 and 1, whereas linear functions take values over the entire real line. This model predicts $\pi(x) < 0$ and $\pi(x) > 1$ for sufficiently large or small x values. The model can be valid over a finite range of x values. However, most applications, particularly those having several predictors, require a more complex model form.

Though (4.2.1) looks like an ordinary regression model, least squares estimators of the model parameters are not optimal. The variance of the binary outcome for each subject, $\text{Var}(Y) = \pi(x)[1 - \pi(x)]$, is not constant for all x, but rather depends on x through its influence on $\pi(x)$. Because of the nonconstant variance, maximum likelihood (ML) estimators can have smaller standard errors than least squares estimators.

The ML estimates for this model, like most GLMs, do not have closed-form; that is, no formula exists for them. Computer software for GLMs calculates the ML estimates by numerically determining the maximum of the likelihood function. Section 4.5.1 describes an iterative method for doing this. When predicted values fall outside the (0, 1) range at some observed x values, the ML fitting procedure fails, but one can still compute the least squares estimates. (With GLM software, one does this by identifying the binary response as normal rather than binomial.)

4.2.2   Snoring and Heart Disease Example

To illustrate the linear probability model, we refer to Table 4.1, based on an epidemiological survey of 2484 subjects to investigate snoring as a possible risk factor for heart disease. Those surveyed were classified according to their spouses' report of how much they snored. The model states that the probability of heart disease $\pi(x)$ is linearly related to the level of snoring x. We treat the rows of the table as independent binomial samples with that probability as the parameter. We use scores (0, 2, 4, 5) for the snoring categories, treating the last two levels as closer than the other adjacent pairs.

Software for GLMs reports the ML fit of the model as $\hat{\pi} = 0.0172 + 0.0198x$. For instance, for nonsnorers ($x = 0$), the estimated proportion of subjects having heart disease is $\hat{\pi} = 0.0172 + 0.0198(0) = .0172$. We refer to the estimated values of $E(Y)$

Table 4.1   Relationship Between Snoring and Heart Disease

	Heart Disease		Proportion	Linear	Logit	Probit
Snoring	Yes	No	Yes	Fit[a]	Fit	Fit
Never	24	1355	.017	.017	.021	.020
Occasional	35	603	.055	.057	.044	.046
Nearly every night	21	192	.099	.096	.093	.095
Every night	30	224	.118	.116	.132	.131

[a]Model fits refer to proportion of yes responses.

Source: P. G. Norton and E .V. Dunn, *Brit. Med. J., 291*: 630–632 (1985), published by BMJ Publishing Group. See also *Small Data Sets*, D. J. Hand et al., ed. (London: Chapman and Hall, 1994).

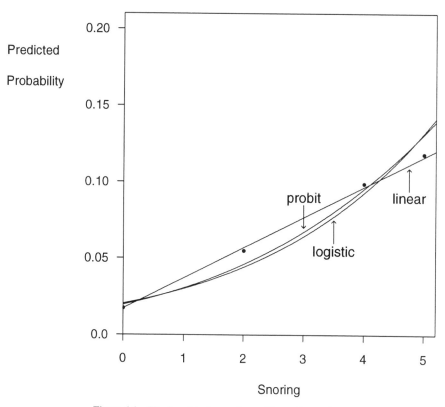

Figure 4.1   Fit of models for snoring and heart disease data.

for a GLM as *fitted values*. Table 4.1 shows the sample proportions and the fitted values for the linear probability model. Figure 4.1 graphs the sample and fitted values. The table and graph suggest that the model fits these data well. (Section 5.4 discusses formal goodness-of-fit analyses for binary-response GLMs.) The model interpretation is simple. The estimated probability of heart disease is about .02 (namely, .0172) for nonsnorers, it increases 2(0.0198) = .04 for occasional snorers, another .04 for those who snore nearly every night, and another .02 for those who always snore. This rather surprising effect is significant, as the standard error of the slope estimate of 0.0198 equals 0.0028.

Suppose we had chosen scores for snoring level having different relative spacings from the scores $\{0, 2, 4, 5\}$. Examples are $\{0, 2, 4, 4.5\}$ or $\{0, 1, 2, 3\}$. Then the fitted values for the four snoring categories would change somewhat. They would not change if the relative spacings between scores were the same, such as $\{0, 4, 8, 10\}$ or $\{1, 3, 5, 6\}$. For these data, any increasing scores yield the conclusion that the chance of heart disease increases as snoring level increases. Incidentally, if we entered the data as 2484 binary observations of 0 or 1 and fitted the model using ordinary least squares rather than ML, we would obtain $\hat{\pi} = 0.0169 + 0.0200x$. In practice, when the model fit is good, least squares and ML estimates are usually similar.

4.2.3 Logistic Regression Model

Relationships between $\pi(x)$ and x are usually nonlinear rather than linear. A fixed change in X may have less impact when π is near 0 or 1 than when π is near the middle of its range. In the purchase of an automobile, for instance, consider the choice between buying new or used. Let $\pi(x)$ denote the probability of selecting new, when annual family income $= x$. An increase of \$50,000 in annual family income would likely have less effect when $x = \$1,000,000$ (for which π is near 1) than when $x = \$50,000$.

In practice, nonlinear relationships between $\pi(x)$ and x are often monotonic, with $\pi(x)$ increasing continuously as x increases, or $\pi(x)$ decreasing continuously as x increases. The S-shaped curves displayed in Figure 4.2 are often realistic shapes for the relationship. The most important function having this shape has the model form

$$\log\left(\frac{\pi(x)}{1 - \pi(x)}\right) = \alpha + \beta x. \qquad (4.2.2)$$

This is called the *logistic regression function*. Interpretations for the model formula are not obvious, and we devote much of Section 5.1 to making sense of it.

Logistic regression models are special cases of GLMs. The random component for the (success, failure) determinations is binomial. The link function is the *logit*

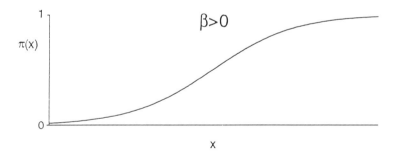

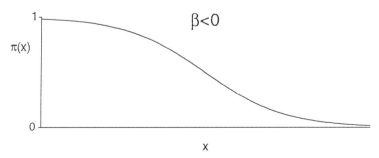

Figure 4.2   Logistic regression functions.

transformation $\log[\pi/(1 - \pi)]$ of π, symbolized by logit(π). Logistic regression models are often called *logit models*. The logit is the natural parameter of the binomial distribution, so the logit link is its canonical link. Whereas π is restricted to the range $(0, 1)$, the logit can be any real number. The real numbers are also the potential range for linear predictors (such as $\alpha + \beta x$) that form the systematic component of a GLM, so this model does not have the structural problem that the linear probability model has.

The parameter β in (4.2.2) determines the rate of increase or decrease of the curve. When $\beta > 0$, $\pi(x)$ increases as x increases, as in Figure 4.2a. When $\beta < 0$, $\pi(x)$ decreases as x increases, as in Figure 4.2b. The magnitude of β determines how fast the curve increases or decreases. As $|\beta|$ increases, the curve has a steeper rate of change. When $\beta = 0$, the curve flattens to a horizontal straight line.

For the snoring and heart disease data in Table 4.1, software reports the ML fit for the logistic regression model of

$$\text{logit}[\hat{\pi}(x)] = -3.87 + 0.40x.$$

The positive value of $\hat{\beta} = 0.40$ reflects the increased chance of heart disease at higher levels of snoring. Chapter 5 presents several ways of interpreting such equations. For instance, it shows how to calculate the predicted probabilities for the model fit. Table 4.1 also reports these fitted values, and Figure 4.1 displays the fit. The fit is close to linear over this rather narrow range of predicted probabilities, and results are similar to those for the linear probability model.

4.2.4   Alternative Binary Links*

For the logistic regression curves pictured in Figure 4.2, the probability of a success increases continuously or decreases continuously as x increases. Let X denote a random variable, and let x denote a potential value for X. The cumulative distribution function (cdf) $F(x)$ for X is defined as

$$F(x) = P(X \le x), \qquad -\infty < x < \infty.$$

Such a function, plotted as a function of x, has appearance like that in Figure 4.2a. As x increases, $F(x)$ increases gradually from 0 to 1, since $P(X \le x)$ increases as x increases. This suggests a class of models for binary responses whereby the dependence of $\pi(x)$ on x has form

$$\pi(x) = F(x), \qquad\qquad (4.2.3)$$

where F is a cdf for some distribution.

The logistic regression curve has this form. When $\beta > 0$, $F(x)$ is the cdf of a two-parameter *logistic distribution*. When $\beta < 0$, the formula for $1 - \pi(x)$ has the logistic cdf appearance. Each choice of α and of $\beta > 0$ corresponds to a different logistic distribution. The logistic cdf corresponds to a probability distribution with a symmetric, bell shape. In fact, it looks similar to a normal distribution but with slightly thicker tails.

Model form (4.2.3) occurs naturally when a *tolerance distribution* applies to subjects' responses. For instance, in a toxicology study, suppose that researchers spray an insectide at various dosage levels on batches of mosquitoes. For each mosquito, the response is whether it dies. Each mosquito may have a certain tolerance to the insecticide, such that it dies if the dosage level exceeds its tolerance and survives if the dosage level is less than its tolerance. Tolerances would vary among mosquitoes. If a *cdf F* describes the distribution of tolerances, then the model for the probability $\pi(x)$ of death at dosage level x necessarily has form (4.2.3). For instance, if the tolerances vary among mosquitoes according to a logistic distribution, then the logistic regression model applies.

4.2.5 Probit Models*

When F is the *cdf* of a normal distribution, model type (4.2.3) is called the *probit model*. The link function for the model is then called the *probit link*. The probit model has alternative expression

$$\text{probit}[\pi(x)] = \alpha + \beta x.$$

The probit link applied to a probability $\pi(x)$ transforms it to the standard normal z-score at which the left-tail probability equals $\pi(x)$. For instance, probit(.05) $=$ -1.645, probit(.50) $= 0$, probit(.95) $= 1.645$, and probit(.975) $= 1.96$. The probit model is a GLM with binomial random component and probit link.

We illustrate using the snoring and heart disease data. The ML fit of the probit model, using scores $\{0, 2, 4, 5\}$ for snoring level, is

$$\text{probit}[\hat{\pi}(x)] = -2.061 + 0.188x.$$

At snoring level $x = 0$, the fitted probit equals $-2.061 + 0.188(0) = -2.06$. The fitted probability $\hat{\pi}(0)$ is the left-tail probability for the standard normal distribution at -2.06, which equals .020. At snoring level $x = 5$, the fitted probit equals $-2.061 + 0.188(5) = -1.12$, which corresponds to a fitted probability of .131.

The fitted values, shown in Table 4.1 and Figure 4.1, are similar to those obtained with the linear probability and logistic regression models. For practical purposes, probit and logistic regression curves look the same. It is rare, and requires enormous sample sizes, to find data for which a logistic regression model fits well but the probit model fits poorly, or conversely. Parameter estimates differ for the two models, since their links have different scales. When both models fit well, slope estimates in logistic regression models are roughly about 1.6–2.0 times those in probit models.

The probit transform maps $\pi(x)$ so that the regression curve for $\pi(x)$ (or $1 - \pi(x)$, when $\beta < 0$) has the appearance of the normal *cdf* with mean $\mu = -\alpha/\beta$ and standard deviation $\sigma = 1/|\beta|$. For the snoring and heart disease data, the probit fit corresponds to a normal *cdf* having mean of $-\hat{\alpha}/\hat{\beta} = 2.061/0.188 = 11.0$ and standard deviation of $1/|\hat{\beta}| = 1/0.188 = 5.3$. The predicted probability of heart disease equals $\frac{1}{2}$ at snoring level $x = 11.0$; that is, $x = 11.0$ has a fitted probit of $-2.061 + 0.188(11) = 0$, which is the z-score corresponding to a left-tail probability of $\frac{1}{2}$. The fitted probit value of -2.06 at $x = 0$ means that 0 is 2.06 standard deviations below the mean of a normal distribution with mean 11.0 and standard deviation 5.3. Since snoring level is restricted to the range 0–5 for these data, well below 11, the fitted probabilities over this range are quite small.

The probit model was introduced in 1934 for models in toxicology. The logistic regression model was not studied until about a decade later, but it is now much more popular than the probit. Partly this is because one can also interpret the logistic regression effects using odds ratios. Thus, one can fit those models to data from case-control studies, because one can estimate odds ratios for such data (Sections 2.3.4, 5.1.4, 9.2.3).

4.3 GENERALIZED LINEAR MODELS FOR COUNT DATA: POISSON REGRESSION

Many discrete response variables have counts as possible outcomes. For instance, for a sample of cities worldwide, each observation might be the number of automobile thefts in 1995. Or, for a sample of silicon wafers used in manufacturing computer chips, each observation might be the number of imperfections on a wafer. This section introduces GLMs for count data. These GLMs assume a Poisson distribution for the random component. Like counts, Poisson variates can take any nonnegative integer value.

Section 1.2 introduced the Poisson distribution as a sampling model for counts. Chapter 6 presents Poisson GLMs for counts in contingency tables. The response data are cell counts obtained by cross-classifying subjects on two or more categorical response variables. This section introduces Poisson regression-type models using an alternative application: modeling count or rate data for a single response variable.

4.3.1 Poisson Regression

The Poisson distribution has a positive mean. Though one can model the Poisson mean in GLMs using the identity link, it is more common to model the log of the mean. Like the linear predictor $\alpha + \beta x$, the log of the mean can take any real value. The log mean is the natural parameter for the Poisson distribution, and the log link is the canonical link for a GLM with Poisson random component. A *Poisson loglinear model* is a GLM that assumes a Poisson distribution for Y and uses the log link.

Let μ denote the expected value for a Poisson variate Y, and let X denote an explanatory variable. The Poisson loglinear model has form

$$\log \mu = \alpha + \beta x. \qquad (4.3.1)$$

For this model, the mean satisfies the exponential relationship

$$\mu = \exp(\alpha + \beta x) = e^{\alpha}(e^{\beta})^x. \qquad (4.3.2)$$

A one-unit increase in X has a multiplicative impact of e^β on μ. The mean of Y at $x + 1$ equals the mean of Y at x multiplied by e^β. If $\beta = 0$, then $e^\beta = e^0 = 1$ and the multiplicative factor is 1; that is, the mean of Y does not change as X changes. If $\beta > 0$, then $e^\beta > 1$, and the mean of Y increases as X increases. If $\beta < 0$, the mean decreases as X increases.

4.3.2 Horseshoe Crabs and Satellites

To illustrate Poisson regression, we analyze some data in Table 4.2 from a study of nesting horseshoe crabs (J. Brockmann, to appear in *Ethology* (1996)). Each female horseshoe crab in the study had a male crab attached to her in her nest. The study investigated factors that affect whether the female crab had any other males, called *satellites*, residing nearby her. Explanatory variables thought possibly to affect this included the female crab's color, spine condition, weight, and carapace width. The response outcome for each female crab is her number of satellites. For now, we use width alone as a predictor of the response. Table 4.2 lists this variable in centimeters. (Other analyses of these data occur in Chapter 5.)

Figure 4.3 plots the response counts against width, with numbered symbols indicating the number of observations at each point. The substantial variability in counts makes it difficult to discern a clear pattern. To obtain a clearer picture of overall trend, we grouped the female crabs into a set of width categories, (≤ 23.25, 23.25–24.25, 24.25–25.25, 25.25–26.25, 26.25–27.25, 27.25–28.25, 28.25–29.25, > 29.25), and calculated the sample mean number of satellites for female crabs in each category. Figure 4.4 on page 85 plots these sample means against the sample mean width for crabs in each category. Our choice of width categories was somewhat arbitrary. The sample mean width equals 26.3 and the standard deviation equals 2.1. We chose six width categories, each of about half a standard deviation in width, with two outlying categories to capture the lowest and highest measurements; we used 26.25 rather than 26.3 for the midpoint of the eight classes so that no observation would fall exactly on the boundary between two categories.

Some software has more sophisticated ways of portraying the overall trend in the data, based on smoothing the data without grouping the width values or assuming a particular functional form for the relationship. Figure 4.4 also shows such a smoothed curve. The sample means and the smoothed curve both show a strong increasing trend. (The means tend to fall above the curve, since the response counts in a category tend to be skewed to the right; the smoothed curve is less susceptible to outlying observations.) The trend seems approximately linear, and we next discuss models for which the mean or the log of the mean is linear in width.

Let μ denote the expected number of satellites for a female crab, and let x denote her width. From GLM software, the ML fit of the Poisson loglinear model (4.3.1) is

$$\log \hat{\mu} = \hat{\alpha} + \hat{\beta} x = -3.305 + 0.164x.$$

The effect $\hat{\beta} = 0.164$ of width has an asymptotic (large-sample) standard error of $ASE = 0.020$. Since $\hat{\beta} > 0$, width has a positive estimated effect on the number of satellites.

Table 4.2 Number of Crab Satellites by Female's Color, Spine Condition, Width, and Weight

C	S	W	Wt	Sa	C	S	W	Wt	Sa	C	S	W	Wt	Sa	C	S	W	Wt	Sa
2	3	28.3	3.05	8	3	3	22.5	1.55	0	1	1	26.0	2.30	9	3	3	24.8	2.10	0
3	3	26.0	2.60	4	2	3	23.8	2.10	0	3	2	24.7	1.90	0	2	1	23.7	1.95	0
3	3	25.6	2.15	0	3	3	24.3	2.15	0	2	3	25.8	2.65	0	2	3	28.2	3.05	11
4	2	21.0	1.85	0	2	1	26.0	2.30	14	1	1	27.1	2.95	8	2	3	25.2	2.00	1
2	3	29.0	3.00	1	4	3	24.7	2.20	0	2	3	27.4	2.70	5	2	2	23.2	1.95	4
1	2	25.0	2.30	3	2	1	22.5	1.60	1	3	3	26.7	2.60	2	4	3	25.8	2.00	3
4	3	26.2	1.30	0	2	3	28.7	3.15	3	2	1	26.8	2.70	5	4	3	27.5	2.60	0
2	3	24.9	2.10	0	1	1	29.3	3.20	4	1	3	25.8	2.60	0	2	2	25.7	2.00	0
2	1	25.7	2.00	8	2	1	26.7	2.70	5	4	3	23.7	1.85	0	2	3	26.8	2.65	0
2	3	27.5	3.15	6	4	3	23.4	1.90	0	2	3	27.9	2.80	6	3	3	27.5	3.10	3
1	1	26.1	2.80	5	1	1	27.7	2.50	6	2	1	30.0	3.30	5	3	1	28.5	3.25	9
3	3	28.9	2.80	4	2	3	28.2	2.60	6	2	3	25.0	2.10	4	2	3	28.5	3.00	3
2	1	30.3	3.60	3	4	3	24.7	2.10	5	2	3	27.7	2.90	5	1	1	27.4	2.70	6
2	3	22.9	1.60	4	2	1	25.7	2.00	5	2	3	28.3	3.00	15	2	3	27.2	2.70	6
3	3	26.2	2.30	3	2	1	27.8	2.75	0	4	3	25.5	2.25	0	3	3	27.1	2.55	3
3	3	24.5	2.05	5	3	1	27.0	2.45	3	2	3	26.0	2.15	5	2	3	28.0	2.80	1
3	3	30.0	3.05	8	2	3	29.0	3.20	10	2	3	26.2	2.40	0	2	1	26.5	1.30	0
2	3	26.2	2.40	3	3	3	25.6	2.80	7	3	3	23.0	1.65	1	3	3	23.0	1.80	0
2	3	25.4	2.25	6	3	3	24.2	1.90	0	2	2	22.9	1.60	0	3	2	26.0	2.20	3
2	3	25.4	2.25	4	3	3	25.7	1.20	0	2	3	25.1	2.10	5	3	2	24.5	2.25	0
4	3	27.5	2.90	0	3	3	23.1	1.65	0	3	1	25.9	2.55	4	2	3	25.8	2.30	0
4	3	27.0	2.25	3	2	3	28.5	3.05	0	4	1	25.5	2.75	0	4	3	23.5	1.90	0

C	S	W	Wt	Sa	C	S	W	Wt	Sa	C	S	W	Wt	Sa	C	S	W	Wt	Sa
2	2	24.0	1.70	0	2	1	29.7	3.85	5	2	1	26.8	2.55	0	3	3	26.7	2.45	0
2	1	28.7	3.20	0	3	3	23.1	1.55	0	2	1	29.0	2.80	1	2	3	25.5	2.25	0
3	3	26.5	1.97	1	3	3	24.5	2.20	1	3	3	28.5	3.00	1	2	3	28.2	2.87	1
2	3	24.5	1.60	1	2	3	27.5	2.55	1	2	2	24.7	2.55	4	2	1	25.2	2.00	1
3	3	27.3	2.90	1	2	3	26.3	2.40	3	2	3	29.0	3.10	1	3	3	25.3	1.90	2
2	3	26.5	2.30	4	2	3	27.8	3.25	2	2	3	27.0	2.50	6	4	3	25.7	2.10	0
2	3	25.0	2.10	2	2	3	31.9	3.33	5	4	3	23.7	1.80	0	3	3	29.3	3.23	12
2	3	22.0	1.40	0	2	3	25.0	2.40	0	3	3	27.0	2.50	6	2	3	23.8	1.80	6
3	3	30.2	3.28	2	2	3	26.2	2.22	3	2	3	24.2	1.65	2	2	3	27.4	2.90	3
1	1	25.4	2.30	0	3	3	28.4	3.20	6	4	3	22.5	1.47	4	2	3	26.2	2.02	2
2	2	24.9	2.30	6	1	2	24.5	1.95	7	2	3	25.1	1.80	0	2	1	28.0	2.90	4
2	1	25.8	2.25	10	2	3	27.9	3.05	6	2	3	24.9	2.20	0	2	1	28.4	3.10	5
4	3	27.2	2.40	5	2	2	25.0	2.25	3	2	3	27.5	2.63	6	2	1	33.5	5.20	7
3	3	30.5	3.32	3	3	3	29.0	2.92	4	2	1	24.3	2.00	0	2	3	25.8	2.40	0
2	3	25.0	2.10	8	2	1	31.7	3.73	4	2	3	29.5	3.02	4	3	3	24.0	1.90	10
4	3	30.0	3.00	9	2	3	27.6	2.85	0	2	3	26.2	2.30	0	2	1	23.1	2.00	0
2	3	22.9	1.60	0	4	3	24.5	1.90	0	2	3	24.7	1.95	4	2	3	28.3	3.20	0
2	3	23.9	1.85	2	3	3	23.8	1.80	8	3	2	29.8	3.50	4	2	3	26.5	2.35	4
2	3	26.0	2.28	3	2	3	28.2	3.05	0	4	3	25.7	2.15	0	2	3	26.5	2.75	7
2	3	25.8	2.20	0	3	3	24.1	1.80	0	3	3	26.2	2.17	2	3	3	26.1	2.75	3
2	3	29.0	3.28	4	1	1	28.0	2.62	0	4	3	27.0	2.63	0	2	2	24.5	2.00	0
1	1	26.5	2.35	0															

Note: C = Color (1 = light medium, 2 = medium, 3 = dark medium, 4 = dark); S = spine condition (1 = both good, 2 = one worn or broken, 3 = both worn or broken); W = carapace width (cm); Wt = weight (kg); Sa = number of satellites.
Source: Data provided by Dr. Jane Brockmann, Zoology Dept., University of Florida. The data are available as part of Table A.6 in the file accessible on the World Wide Web at http://lib.stat.cmu.edu/datasets/agresti.

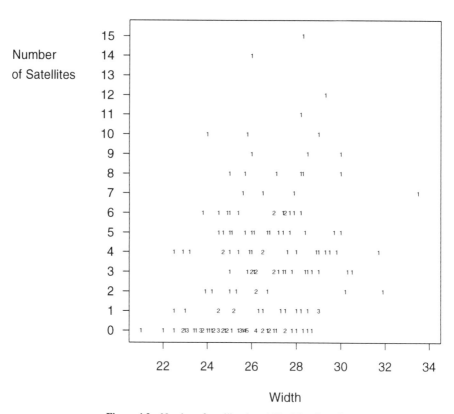

Figure 4.3   Number of satellites by width of female crab.

The model fit yields an estimated mean number of satellites $\hat{\mu}$, a *fitted value*, at any width level. For instance, from (4.3.2), the fitted value at the mean width of $x = 26.3$ is

$$\hat{\mu} = \exp(\hat{\alpha} + \hat{\beta}x) = \exp[-3.305 + 0.164(26.3)] = 2.74.$$

For this model, $\exp(\hat{\beta}) = \exp(0.164) = 1.18$ represents the multiplicative effect on the fitted value for each 1-unit increase in x. For instance, the fitted value at $x = 27.3 = 26.3 + 1$ is $\exp[-3.305 + 0.164(27.3)] = 3.23$, which equals $(1.18)(2.74)$. A 1-cm increase in width yields an 18% increase in the estimated mean number of satellites.

Figure 4.3 shows that one crab had somewhat greater width than the others, 33.5 cm. An observation having explanatory variable values much different from the rest of the sample can have an undue influence on the model fit. To check the effect of this observation, we deleted it and refitted the model for the remaining 172 crabs. The ML estimates then equal $\hat{\alpha} = -3.461$ and $\hat{\beta} = 0.170$ (*ASE* = 0.022). The estimates differ somewhat, but again provide strong evidence of a positive width effect.

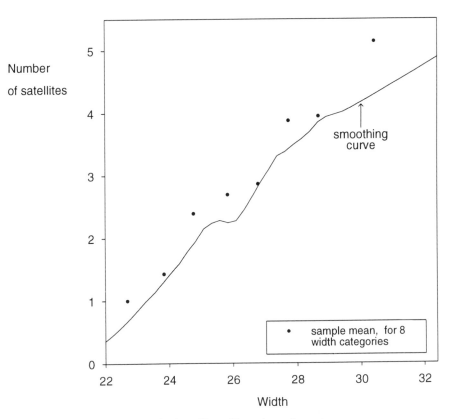

Figure 4.4   Smoothings of horseshoe crab counts.

Figure 4.4 reveals that the number of satellites may grow approximately linearly with width. This suggests the Poisson regression model with identity link,

$$\mu = \alpha + \beta x.$$

Its ML estimates equal $\alpha = -11.53$ and $\beta = 0.55$ ($ASE = 0.059$). The effect of X on μ in this model is additive, rather than multiplicative. A 1-cm increase in width has a predicted increase of $\hat{\beta} = 0.55$ in the expected number of satellites. For instance, the fitted value at the mean width of $x = 26.3$ is $\hat{\mu} = -11.53 + 0.55(26.3) = 2.93$; at $x = 27.3$, it is $2.93 + 0.55 = 3.48$. The fitted values are positive at all widths observed in the sample, and the model provides a simple description of the effect of width on the number of satellites: On the average, an extra satellite results from roughly a 2-cm increase in width.

Figure 4.5 plots the fitted number of satellites against width, for the models with log link and with identity link. Though they diverge somewhat for relatively small and relatively large widths, they provide similar predictions over the portion of the width range in which most observations occur. Sections 4.4.2 and 4.4.3 study whether either model provides an adequate fit to these data.

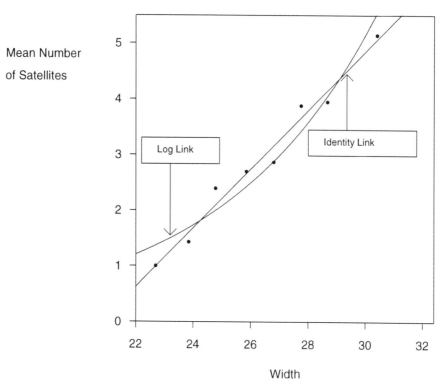

Figure 4.5   Estimated mean number of satellites for log and identity links.

4.3.3   Poisson Regression for Rate Data

When events of a certain type occur over time, space, or some other index of size, it is often relevant to model the *rate* at which events occur. For instance, in modeling numbers of auto thefts in 1995 for a sample of cities, we could form a rate for each city by dividing the number of thefts by the city's population size. The model might describe how the rate depends on explanatory variables such as the city's unemployment rate, median income, and percentage of residents having completed high school.

When a response count Y has index (such as population size) equal to t, the sample rate of outcomes is Y/t. The expected value of the rate is μ/t. A loglinear model for the expected rate has form

$$\log(\mu/t) = \alpha + \beta x. \tag{4.3.3}$$

This model has equivalent representation

$$\log \mu - \log t = \alpha + \beta x.$$

The adjustment term, $-\log t$, to the log link of the mean on the left-hand side of the equation is called an *offset*. Standard GLM software can fit models having offsets.

For model (4.3.3), the expected number of outcomes satisfies

$$\mu = t \exp(\alpha + \beta x). \qquad (4.3.4)$$

The mean is proportional to the index t, with proportionality constant depending on the value of the explanatory variable. For a fixed value of x, doubling the population size t also doubles the expected number of auto thefts μ.

4.3.4 Examples of Rate Models

To illustrate models for rates, we first use data (from an article by W. A. Ray et al., *Amer. J. Epidemiol.*, *132*: 873–884 (1992)) dealing with motor vehicle accident rates for elderly drivers. The sample consisted of 16,262 Medicaid enrollees aged 65–84 years, with data on each subject for a period of somewhere between 0 and 4 years. The total observation time for women in the sample was 17.30 thousand years. During this period, they had 175 accidents in which an injury occurred. The total observation time for the men was 21.40 thousand years, during which they had 320 injurious accidents. The sample rates of injurious accidents are $320/21.40 = 14.95$ crashes per thousand years of driving for males, and $175/17.30 = 10.12$ for females.

Let μ denote the expected number of injurious accidents, for an observation period of t thousand years. To model the effect of gender on the accident rate, we use model (4.3.3) with $x = 0$ for females and $x = 1$ for males. The explanatory variable x is a *dummy variable* for gender. The log of the accident rate equals α for females, and it equals $\alpha + \beta$ for males. The rates are identical if $\beta = 0$.

The estimate of α in model (4.3.3) is simply the sample log(rate) for females, namely $\log(10.12) = 2.31$; the estimate of $\alpha + \beta$ is the sample log(rate) for males, namely $\log(14.95) = 2.70$. The estimated difference is $\hat{\beta} = 0.39$. The estimated accident rate for men was $\exp(\hat{\beta}) = \exp(0.39) = 1.48$ times the rate for women. That is, $14.95/10.12 = 1.48$, the sample rate being 48% higher for men.

To test whether the true rates are the same, we test $H_0 : \beta = 0$. GLM software reports an *ASE* for $\hat{\beta} = 0.39$ of 0.09, so there is strong evidence that the accident rate was higher for males (i.e., that $\beta > 0$). Of course, the accident rates do not take into account possibly different yearly levels of driving for the two groups.

We next reconsider the horseshoe crab data. In this sample, the 173 female crabs had 66 distinct width values. Let Y now denote the total number of satellites for the t crabs having a particular width, and let $\mu = E(Y)$. Then μ/t is the expected number of satellites per crab at that width, and the rate model (4.3.3) applies. Fitting this model to the 66 counts of total satellites and the numbers of female crabs at the 66 distinct widths necessarily gives the same ML estimates of the parameters as fitting model (4.3.1) to the 173 individual observations.

When all counts in a data set have the same index value t, or when counts do not refer to an index such as time or group size, the model does not need an offset term. One can then analyze effects of explanatory variables on the response counts using model (4.3.3) with $t = 1$, which is simply model (4.3.1).

MODEL INFERENCE AND MODEL CHECKING

Having introduced the two primary GLMs for categorical data, we now turn our attention to statistical inference and model checking for GLMs. We illustrate for GLMs having a Poisson random component. Chapter 5 presents similar methods for GLMs having a binomial random component.

For most GLMs for categorical responses, calculation of ML parameter estimates is computationally complex. GLM software uses a numerical method described in Section 4.5.1. The ML estimates are approximately normally distributed for large samples. Thus, a confidence interval for a model parameter β equals $\hat{\beta} \pm z_{\alpha/2}ASE$, where ASE is the asymptotic standard error of $\hat{\beta}$.

4.4.1   The Wald, Likelihood-Ratio, Score Test Triad

There are three methods for performing significance tests of hypotheses $H_0 : \beta = 0$ about parameters in GLMs. The simplest uses the large-sample normality of ML estimates. The test statistic

$$z = \hat{\beta}/ASE$$

has an approximate standard normal distribution when $\beta = 0$. One refers z to the standard normal table to get one-sided or two-sided P-values. Equivalently, for the two-sided alternative, z^2 has a chi-squared distribution with $df = 1$; the P-value is then the right-tail chi-squared probability above the observed value. This type of statistic, which divides a parameter estimate by its standard error and then squares it, is called a *Wald statistic*.

A second method uses the likelihood function through the ratio of two maximizations of it: (1) the maximum over the possible parameter values that assume the null hypothesis, (2) the maximum over the larger set of possible parameter values for the full model, permitting the null or the alternative hypothesis to be true. Let ℓ_1 denote the maximized value of the likelihood function for the full model, and let ℓ_0 denote the maximized value for the simpler model, representing the null hypothesis. For instance, when the linear predictor is $\alpha + \beta x$ and the null hypothesis is $H_0 : \beta = 0$, ℓ_1 is the likelihood function calculated at the (α, β) combination for which the data would have been most likely; ℓ_0 is the likelihood function calculated at the α value for which the data would have been most likely, when $\beta = 0$. Then ℓ_1 is always at least as large as ℓ_0, since ℓ_0 refers to maximizing over a restricted set of the parameter values that yield ℓ_1.

The *likelihood-ratio* test statistic equals

$$-2\log(\ell_0/\ell_1) = -2\left[\log(\ell_0) - \log(\ell_1)\right] = -2(L_0 - L_1),$$

where L_0 and L_1 denote the maximized log-likelihood functions. This transformation of ℓ_0 and ℓ_1 yields a chi-squared statistic. Under $H_0 : \beta = 0$, it also has a large-sample chi-squared distribution with $df = 1$. Most software for GLMs reports the maximized log-likelihood values and the likelihood-ratio statistic $-2(L_0 - L_1)$.

Some software for GLMs also reports a third chi-squared test, using a *score* statistic, sometimes also called the *efficient score* statistic. We defer discussion of this test to Section 4.5.2. The Wald, likelihood-ratio, and score tests are the three major types of statistical tests for GLMs. For very large samples, they have similar behavior. For sample sizes used in practice, the likelihood-ratio test is usually more reliable than the Wald test. A marked divergence in the values of the three statistics indicates that the distribution of $\hat{\beta}$ may be far from normality. In that case, small-sample methods are more appropriate than large-sample methods. For the most well-known GLM, regression for normal data using the identity link, the three tests provide identical results.

We illustrate the Wald and likelihood-ratio tests for the Poisson loglinear model fitted to the horseshoe crab data of Section 4.3.2. The ML estimate of the width effect equals $\hat{\beta} = 0.164$, with $ASE = 0.020$. The Wald test of $H_0 : \beta = 0$ treats $z = \hat{\beta}/ASE = 0.164/0.020 = 8.2$ as standard normal, or $z^2 = 67.4$ as chi-squared with $df = 1$. This provides strong evidence of a positive effect of width on the presence of satellites ($P < .0001$). We obtain similar strong evidence from a likelihood-ratio test comparing this model to the simpler one having $\beta = 0$. That chi-squared statistic equals $-2(L_0 - L_1) = 64.9$ with $df = 1$ ($P < .0001$).

4.4.2 Poisson Model Checking

A GLM provides accurate description and inference for a data set only if it fits that data set well. Summary goodness-of-fit statistics and residuals help us to investigate the adequacy of a GLM fit. We illustrate for GLMs with Poisson random component.

We first test the null hypothesis that the chosen Poisson GLM holds. At the ith of N settings of explanatory variables, denote the observed count by y_i and the fitted value by $\hat{\mu}_i$. The Pearson and the likelihood-ratio goodness-of-fit statistics equal

$$X^2 = \sum \frac{(y_i - \hat{\mu}_i)^2}{\hat{\mu}_i}, \qquad G^2 = 2\sum y_i \log\left(\frac{y_i}{\hat{\mu}_i}\right).$$

When the fitted values $\{\hat{\mu}_i\}$ are relatively large (mostly exceeding 5) and the number of settings N is fixed, these test statistics have approximate chi-squared distributions. The df equal the number of response counts minus the number of model parameters. This df value is called the *residual df* for the model. The P-value is the right-tail probability; large test statistics and small P-values suggest a poor model fit.

We illustrate with the Poisson loglinear model for the number of satellites for a female horseshoe crab, using her width as the predictor (Section 4.3.2). We use the satellite totals for all female crabs at a given width (i.e., version (4.3.3) of the model), to increase the counts and fitted values relative to those for individual female crabs. The $N = 66$ distinct width levels each have a total count y_i for the number of satellites and a fitted value $\hat{\mu}_i$. The goodness-of-fit statistics comparing these equal $X^2 = 174.3$ and $G^2 = 190.0$, with $df = 64$. However, the large-sample chi-squared theory for these statistics is violated in two ways. First, many observed total counts are small, several equaling zero. Second, the chi-squared approximation improves as the Poisson fitted values increase for a *fixed* number N of settings of explanatory variables. In this case, additional width values would occur when more female crabs are sampled, so the number N of width settings would increase rather than remain fixed. Thus, X^2 and G^2 are unreliable here as measures of lack of fit.

To achieve potentially a better chi-squared approximation, one can perform this analysis for a fixed grouping of the data into width categories. We grouped width into the eight categories shown in Table 4.3. For female crabs within each width category, we sum the observed counts of satellites and sum the fitted values, yielding a new set of eight pairs $(y_i, \hat{\mu}_i)$ having much larger sizes than the original 66 pairs. Substituting these, we get $X^2 = 6.5$ and $G^2 = 6.9$, with $df =$ (number of response counts − number of parameters) $= 8 - 2 = 6$. These statistics indicate a good fit $(P > .3)$. The method of grouping is arbitrary, but for most purposes about six to ten categories are adequate. The fitted values should satisfy $\hat{\mu}_i \geq 5$ for each category.

A simpler approximation results from refitting the model to the grouped counts. This requires assigning scores to the width categories. One could simply average the widths of all crabs in a category, which yields the width scores (22.69, 23.84, 24.77, 25.84, 26.79, 27.74, 28.67, 30.41). This model fit, which is cruder since it treats all crabs in a category as if they have the same width, has $\hat{\alpha} = -3.535$ and $\hat{\beta} = 0.173$ $(ASE = 0.021)$. At the lowest width score of $x = 22.69$, for instance, the estimated expected number of satellites per female crab is $\exp[-3.535 + 0.173(22.69)] = 1.47$. Since 14 female crabs had that score, this category has a fitted value of $\hat{\mu}_1 = 14(1.47) = 20.5$ satellites; by contrast, the observed count of satellites in the first width category is $y_1 = 14$. A comparison of observed and fitted values for the 8 categories yields $X^2 = 6.2$ and $G^2 = 6.5$ with $df = 6$ $(P > .3)$. This approach also shows no evidence of model lack of fit.

4.4.3   Model Residuals

For any GLM, goodness-of-fit statistics only broadly summarize how well models fit data. We obtain further insight by comparing observed and fitted counts individually.

For observation i, the residual difference $y_i - \hat{\mu}_i$ between an observed and fitted count has limited usefulness. For Poisson sampling, for instance, the standard deviation of a count is $\sqrt{\mu_i}$, so larger differences tend to occur when μ_i is larger. The

Table 4.3   Fit and Residuals for Poisson Loglinear Model

Width	Number Cases	Number Satellites	Fitted Count	Pearson Residual	Adjusted Residual
< 23.25	14	14	20.5	−1.44	−1.63
23.25–24.25	14	20	25.1	−1.01	−1.11
24.25–25.25	28	67	58.9	1.06	1.23
25.25–26.25	39	105	98.6	0.64	0.75
26.25–27.25	22	63	65.5	−0.31	−0.34
27.25–28.25	24	93	84.3	0.95	1.06
28.25–29.25	18	71	74.2	−0.37	−0.42
> 29.25	14	72	77.9	−0.67	−1.00

Pearson residual is a standardization of this difference, defined by

$$\text{Pearson residual} = \frac{(\text{observed} - \text{fitted})}{\sqrt{\hat{V}\text{ar}(\text{observed})}}. \qquad (4.4.1)$$

For Poisson GLMs, this simplifies for count i to

$$e_i = \frac{y_i - \hat{\mu}_i}{\sqrt{\hat{\mu}_i}}, \qquad (4.4.2)$$

which standardizes by dividing the difference by the estimated Poisson standard deviation. These residuals relate to the Pearson goodness-of-fit statistic by $\sum e_i^2 = X^2$.

Table 4.3 shows Pearson residuals for the Poisson loglinear model fitted to the grouped crab data of Table 4.3, using the average width scores. For the observed count of $y_1 = 14$ and the fitted value of $\hat{\mu}_1 = 20.5$ at $x = 22.69$, for instance, the Pearson residual equals $(14 - 20.5)/\sqrt{20.5} = -1.44$. Counts having larger residuals made greater contributions to the overall X^2 value for testing model fit.

Pearson residual values fluctuate around zero, following approximately a normal distribution when μ_i is large. When the model holds, these residuals are less variable than standard normal, however, because the numerator must use the fitted value $\hat{\mu}_i$ rather than the true mean μ_i. Since the sample data determine the fitted value, $y_i - \hat{\mu}_i$ tends to be smaller than $y_i - \mu_i$. The Pearson residual divided by its estimated standard error is called an *adjusted residual*. It does have an approximate standard normal distribution when μ_i is large. Thus, with adjusted residuals it is easier to tell when a deviation $y_i - \hat{\mu}_i$ is "large." Adjusted residuals larger than about 2 in absolute value are worthy of attention, though one expects some values of this size by chance alone when the number of categories is large. Adjusted residuals are preferable to Pearson residuals.

Section 2.4.5 introduced adjusted residuals for tests of independence in two-way contingency tables. For Poisson GLMs, the general form of the adjusted residual is

$$\frac{(y_i - \hat{\mu}_i)}{\sqrt{\hat{\mu}_i(1 - h_i)}} = \frac{e_i}{\sqrt{(1 - h_i)}},$$

where h_i is called the *leverage* of observation i. The formula for the leverage is complex, but roughly speaking, the greater the leverage, the greater potential that observation has for influencing the fit of the model.

Most GLM software (such as PROC GENMOD in SAS) reports adjusted residuals for GLMs. Table 4.3 shows adjusted residuals for the Poisson loglinear model fitted to the grouped crab data. Though slightly larger than the Pearson residuals, they show the same basic pattern. The poorest fit occurs in the first category, the number of satellites being somewhat less than predicted by the model. None of the adjusted residuals is large enough, however, to indicate potential problems with model fit.

The Poisson model using identity link also fits the grouped data well. For this link, both X^2 and G^2 equal 3.0 based on $df = 6$, showing an even closer fit to the data than the log link provides. This link fails, however, when other explanatory variables are added to the model, because it yields some negative predicted counts.

Other diagnostic tools from regression modeling are also helpful in assessing fits of GLMs. For instance, to assess the influence of an observation on the overall fit, one can refit the model with that observation deleted. This is equivalent to fitting a slightly more complex model having a parameter applying to that observation alone. One studies the change in goodness of fit and the changes in parameter estimates and fitted values. Section 5.3.4 presents diagnostics of this type.

4.4.4 Overdispersion in Poisson Regression*

Count data often show greater variability in the response counts than one would expect if the response distribution truly were Poisson. For instance, for the grouped horseshoe crab data, Table 4.4 shows the sample mean and variance for the response counts of number of satellites for the female crabs in each width category. The variances are much larger than the means, whereas Poisson distributions have identical mean and variance. The phenomenon of a GLM having greater variability than predicted by the random component of the model is called *overdispersion*.

A common cause of overdispersion is heterogeneity among subjects. For instance, suppose width, weight, color, and spine condition all affect a female crab's number of satellites. Suppose the number of satellites has a Poisson distribution at each fixed combination of those four variables. Now, suppose our model uses width alone as a predictor. Crabs having a certain fixed width are then a mixture of crabs of various weights, colors, and spine conditions; thus, the population of crabs having that fixed width is a mixture of several Poisson populations, each having its own mean for the response. This heterogeneity in the crabs having a given width yields an overall response distribution at that width that shows greater variation than the Poisson predicts. If the variance equals the mean when all relevant variables are controlled, it exceeds the mean when only one is controlled.

Overdispersion is not an issue in ordinary regression models for a normally-distributed random component, because the normal distribution has a separate pa-

Table 4.4   Sample Mean and Variance of Number of Satellites

Width	Number Cases	Number Satellites	Sample Mean	Sample Variance
< 23.25	14	14	1.00	2.77
23.25–24.25	14	20	1.43	8.88
24.25–25.25	28	67	2.39	6.54
25.25–26.25	39	105	2.69	11.38
26.25–27.25	22	63	2.86	6.88
27.25–28.25	24	93	3.87	8.81
28.25–29.25	18	71	3.94	16.88
> 29.25	14	72	5.14	8.29

rameter from the mean (i.e., the variance, σ^2) to describe variability. For binomial and Poisson distributions, however, the variance is a function of the mean. Overdispersion is common in the modeling of binomial and Poisson counts.

Many methods deal with overdispersion for discrete data, but it is beyond our scope to discuss them (Good references are Collett (1992), Ch. 6, and Morgan (1992), Ch. 6). We simply mention one elementary way. When the model holds, df is the approximate expected value of X^2. When the form of the model relating the expected count to explanatory variables holds, but the actual response distribution is more variable than Poisson, then this is not true. If the response variance is *proportional* to the mean, rather than equal (as the Poisson requires), then X^2/df estimates the proportionality constant. One obtains adjusted *ASE* values by multiplying the values that GLM software reports by the scaling factor $\sqrt{X^2/df}$. The regular ML parameter estimates still apply, and one performs inference in the usual way with these adjusted *ASE* values.

The ungrouped data provide a better estimate of the scaling factor than the grouped data. For the goodness of fit of the Poisson loglinear model to the counts at the 66 distinct width levels, $X^2 = 174.3$ based on $df = 64$, for which $X^2/df = 2.7$. In fact, on the average, the variance in Table 4.4 is nearly three times the mean. The overdispersion adjustment for standard errors equals $\sqrt{174.3/64} = 1.65$. Increasing the standard errors by 65% helps adjust for the apparent overdispersion. The ML fit of the model to the ungrouped data had width effect equal to $\hat{\beta} = 0.164$, with $ASE = 0.020$. Increasing the ASE by 65% yields $ASE = 1.65(0.020) = 0.033$. To test $H_0 : \beta = 0$, the adjusted Wald statistic equals $(0.164/0.033)^2 = 24.8$ with $df = 1$, which yields $P < .0001$. Even with this increase in ASE, strong evidence remains of a positive effect of width on the number of satellites.

An overdispersion-adjusted 95% confidence interval for β is $0.164 \pm 1.96(0.033)$, or $(.099, .229)$. The corresponding interval for e^β, the multiplicative effect of a one-unit increase in x, is $(e^{.099}, e^{.229}) = (1.10, 1.26)$. We conclude that the mean number of satellites at width $x + 1$ is between 10% and 26% higher than the mean number of satellites at width x.

4.5   FITTING GENERALIZED LINEAR MODELS*

We finish this chapter by discussing model-fitting for generalized linear models. We first describe an algorithm used to calculate the ML estimates of model parameters. We then provide further details about how basic inference utilizes the likelihood function. Finally, we discuss a measure, called the *deviance*, that summarizes the fit of GLMs.

4.5.1   The Newton–Raphson Algorithm

For most GLMs, the equations that determine the ML parameter estimates are non-linear, and the estimates do not have a closed-form expression. Software calculates the estimates using an iterative algorithm for solving nonlinear equations. The algorithm requires an initial guess for the parameter values that maximize the likelihood function. Successive approximations produced by the algorithm tend to fall closer to the ML estimates. A popular algorithm for doing this, called *Fisher scoring*, was first proposed by R. A. Fisher for fitting probit models. For binomial logistic regression and Poisson loglinear models, Fisher scoring simplifies to a general-purpose method called the *Newton–Raphson* algorithm.

The Newton–Raphson algorithm approximates the log-likelihood function in a neighborhood of the initial guess by a simpler polynomial function that has the shape of a concave (mound-shaped) parabola. It has the same slope and curvature at the initial guess as does the log-likelihood function. It is simple to determine the location of the maximum of this approximating polynomial. That location comprises the second guess for the ML estimates. One then approximates the log-likelihood function in a neighborhood of the second guess by another concave parabolic function, and the third guess is the location of its maximum. The successive approximations converge rapidly to the ML estimates, often within a few cycles. Software for GLMs does not require the user to provide the initial guess.

Each cycle in the Newton–Raphson method represents a type of weighted least squares fitting. This is a generalization of ordinary least squares that accounts for nonconstant variance of Y in GLMs. Observations taken where the variability is smaller receive greater weight in determining the parameter estimates. The weights change somewhat from cycle to cycle, as we update the approximation for the ML estimates and thus for variance estimates. ML estimation for GLMs is sometimes called *iteratively reweighted least squares*.

The Newton–Raphson method utilizes a matrix, called the *information matrix*, that provides *ASE* values for the parameter estimates. That matrix is based on the curvature of the log-likelihood function at the ML estimate. The greater the curvature, the greater the information about the parameter values. The standard errors are the square roots of the diagonal elements for the inverse of the information matrix. The greater the curvature of the log likelihood, the smaller the standard errors. This is reasonable, since large curvature implies that the log likelihood drops quickly as β moves away from $\hat{\beta}$; hence, the data would have been much more likely to occur if β took value $\hat{\beta}$ than if it took some value not close to $\hat{\beta}$. Software for GLMs routinely calculates the information matrix and the associated standard errors.

4.5.2 Inference Using the Likelihood Function

Section 4.4.1 introduced three methods for testing $H_0 : \beta = 0$ for a GLM model parameter: Wald tests, likelihood-ratio tests, and efficient score tests. Figure 4.6, showing a generic plot of a log-likelihood L as a function of a parameter β, illustrates the three tests. The Wald test is based on the behavior of the log-likelihood function at the ML estimate $\hat{\beta}$, having chi-squared form $(\hat{\beta}/ASE)^2$. The *ASE* of $\hat{\beta}$ depends on the curvature of the log-likelihood function at the point where it is maximized.

The efficient score test is based on the behavior of the log-likelihood function at the null value for β of 0. It utilizes the size of the derivative (slope) of the log-likelihood function, evaluated at the null hypothesis value of the parameter. The log-likelihood

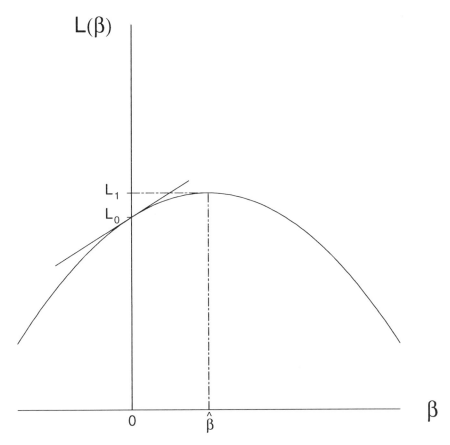

Figure 4.6 Information used in Wald, likelihood-ratio, and score tests.

function for some GLMs, including Poisson loglinear models and binomial logistic models, has concave shape. The ML estimate $\hat{\beta}$ is then the point at which the derivative equals 0. The derivative at $\beta = 0$ tends to be larger in absolute value when $\hat{\beta}$ is further from that null value. Figure 4.6 illustrates the slope information used in the efficient score test. The score statistic is the square of the ratio of this derivative to its *ASE*. It also has an approximate chi-squared distribution with $df = 1$.

We shall not present the general formula here for score statistics, but many test statistics in this text are this type. Examples include the Pearson statistic for testing independence and the *CMH* statistic for testing conditional independence. One can usually compute the score statistic without fitting the model. Another advantage is that it exists even when the ML estimate $\hat{\beta}$ is infinite. In that case, one cannot compute the Wald statistic.

The likelihood-ratio test combines information about the log-likelihood function both at $\hat{\beta}$ and at the null value for β of 0. It compares the log-likelihood values L_1 at $\hat{\beta}$ and L_0 at $\beta = 0$, using the chi-squared statistic $-2(L_0 - L_1)$. In Figure 4.6, this statistic is twice the vertical distance between values of the log-likelihood function at

$\hat{\beta}$ and at $\beta = 0$. In a sense, this statistic uses the most information of the three types of test statistic and is usually the most reliable.

4.5.3 The Deviance

For normal-response models, comparison of a nested set of regression models utilizes a decomposition of a sum of squares representing the variability in the data. This "analysis of variance" generalizes to an "analysis of deviance" for GLMs. Let L_M denote the maximized log-likelihood value for the model of interest. Let L_S denote the maximized log-likelihood value for the most complex model, which has a separate parameter at each explanatory setting; that model is said to be *saturated*. The *deviance* of a model is defined to be

$$\text{Deviance} = -2[L_M - L_S].$$

The deviance is the likelihood-ratio statistic for comparing model M to the saturated model; it is the statistic for testing the hypothesis that all parameters that are in the saturated model but not in model M equal zero. For models in this book, the deviance has the same form as the G^2 likelihood-ratio goodness-of-fit statistic for model M. For instance, the deviance for a Poisson regression model is the G^2 statistic for testing its fit by comparing observed counts to fitted values.

For many GLMs, the deviance has approximately a chi-squared distribution. An example is Poisson regression with a fixed number of explanatory levels and relatively large counts. Another example is binary regression models with a fixed number of explanatory levels and relatively large counts of successes and failures. For such models, one can use the deviance to test the model fit. In either case, the residual df equals the number of responses (Poisson counts or binomial "success" totals) minus the number of nonredundant model parameters.

Components of the deviance, called *deviance residuals*, provide diagnostic measures of lack of fit for individual observations. They are alternatives to Pearson and adjusted residuals. Like the Pearson residuals, deviance residuals are approximately normally distributed but (when the model holds) somewhat less variable than standard normal variates.

For two models, suppose that one (M_0) is a special case of the other (M_1). Given that the more complex model holds, the likelihood-ratio statistic for testing that the simpler model holds is

$$-2[L_0 - L_1] = -2[L_0 - L_S] - \{-2[L_1 - L_S]\} = \text{Deviance}_0 - \text{Deviance}_1.$$

One can compare the models by comparing their deviances. For large samples, this is an approximate chi-squared statistic, with df equal to the difference between the residual df values for the separate models. This df value equals the number of additional nonredundant parameters that are in M_1 but not in M_0. This likelihood-ratio test is a GLM analog of the F test for comparing regression models for normally distributed responses.

Table 4.5 Types of Models for Statistical Analysis

Random Component	Link	Systematic Component	Model	Chapter
Normal	Identity	Continuous	Regression	
Normal	Identity	Categorical	Anal. of variance	
Normal	Identity	Mixed	Anal. of covariance	
Binomial	Logit	Mixed	Logistic regression	5, 7
Poisson	Log	Mixed	Loglinear	6, 7
Multinomial	Generalized logit	Mixed	Multinomial response	8

When df for comparing two GLMs is small, such as when M_1 has one more parameter than M_0, chi-squared approximations work well for this test even when they are poor for the overall goodness-of-fit test. For instance, one can compare two similar Poisson regression models even when the explanatory variables are continuous or when fitted values are small.

4.5.4 Advantages of GLMs

The development of GLM theory in the past two decades has unified important models for continuous and categorical response variables. For theoretical reasons, the random component in the definition of a GLM must have what is called an *exponential family distribution*. This restriction is not serious, since this family contains most important distributions, including the Poisson, binomial, normal, and gamma. Table 4.5 lists several popular GLMs for practical application.

A nice feature of GLMs is that the model-fitting algorithm, Fisher scoring, is the same for any GLM. This holds regardless of the choice of distribution for the random component or the choice of link function. So, GLM software such as GLIM and PROC GENMOD in SAS can fit a very wide variety of useful models.

PROBLEMS

4.1. Describe the purpose of the link function of a GLM. Define the identity link, and explain why it is not often used with binomial or Poisson data.

4.2. Refer to Table 4.1. Refit the linear probability model or the logistic regression model using the scores (i) (0, 2, 4, 6), (ii) (0, 1, 2, 3), (iii). (1, 2, 3, 4). Compare the model parameter estimates under the three choices. Compare the fitted values. What can you conclude about the effect of transformations of scores (called *linear*) that preserve relative sizes of spacings between scores?

4.3. Refer to Table 4.2. Let $Y = 1$ if a crab has at least one satellite, and $Y = 0$ otherwise. Using weight as the predictor, fit the linear probability model.

  a. Use ordinary least squares. Interpret the parameter estimates. Find the predicted probability at the highest observed weight of 5.20 kg. Comment.

b. Attempt to fit the model using ML, treating Y as binomial. (The failure is due to a fitted probability falling outside the $(0, 1)$ range. In (a), least squares corresponds to ML for a normal random component, for which fitted values outside this range are permissible.)

c. Fit the logistic regression model. Show that the predicted logit at a weight of 5.20 kg equals 5.74. Show that $\hat{\pi} = 0.9968$ at that point by checking that $\log[\hat{\pi}/(1 - \hat{\pi})] = 5.74$.

d. Fit the probit model. Interpret the parameter estimates using characteristics of the normal *cdf* that describes the response curve. Find the predicted probability at weight = 5.20 kg.

4.4. Refer to Table 2.7 on alcohol consumption and infant malformation.

a. Using scores $\{0, .5, 1.5, 4, 7\}$, fit a linear probability model. Interpret, and compare the sample proportions to the fitted probabilities.

b. Fit a logit or probit model, and interpret.

4.5. Table 4.6 refers to a sample of subjects randomly selected for an Italian study on the relation between income and whether one possesses a travel credit card (such as American Express or Diners Club). At each level of annual income in millions of lira, the table indicates the number of subjects sampled and the number of them possessing at least one travel credit card. Analyze these data.

Table 4.6

Income	Number Cases	Credit Cards	Income	Number Cases	Credit Cards
24	1	0	48	1	0
27	1	0	49	1	0
28	5	2	50	10	2
29	3	0	52	1	0
30	9	1	59	1	0
31	5	1	60	5	2
32	8	0	65	6	6
33	1	0	68	3	3
34	7	1	70	5	3
35	1	1	79	1	0
38	3	1	80	1	0
39	2	0	84	1	0
40	5	0	94	1	0
41	2	0	120	6	6
42	2	0	130	1	1
45	1	1			

Source: Categorical Data Analysis, Quaderni del Corso Estivo di Statistica e Calcolo delle Probabilità, n. 4., Istituto di Metodi Quantitativi, Università Luigi Bocconi, a cura di R. Piccarreta.

4.6. An experiment analyzes imperfection rates for two processes used to fabricate silicon wafers for computer chips. For treatment A applied to 10 wafers, the numbers of imperfections are 8, 7, 6, 6, 3, 4, 7, 2, 3, 4. Treatment B applied to 10 wafers has 9, 9, 8, 14, 8, 13, 11, 5, 7, 6 imperfections. Treat the counts as independent Poisson variates having means μ_A and μ_B.

  a. Fit the model $\log \mu = \alpha + \beta x$, where $x = 1$ for treatment B and $x = 0$ for treatment A. Show that $\beta = \log \mu_B - \log \mu_A$, and interpret its estimate.

  b. Test $H_0 : \mu_A = \mu_B$ by conducting the Wald or likelihood-ratio test of $H_0 : \beta = 0$. Interpret.

  c. Construct a 95% confidence interval for μ_B / μ_A. (*Hint*: First construct one for $\beta = \log \mu_B - \log_A = \log(\mu_B / \mu_A)$.)

4.7. Refer to the previous problem. Conduct the test of $H_0 : \mu_A = \mu_B$ by using the fact that if X is Poisson with mean μ_1 and Y is an independent Poisson variate with mean μ_2, then X given $X + Y$ is binomial with index $n = X + Y$ and parameter $\pi = \mu_1 / (\mu_1 + \mu_2)$. (*Hint*: Let X and Y be the total numbers of imperfections for the two treatments; then the test of equal Poisson means reduces to a binomial test of $\pi = \frac{1}{2}$ when 140 trials have 50 successes.)

4.8. Refer to Problem 4.6. The wafers are also classified by thickness of silicon coating ($z = 0$, low; $z = 1$, high). The first five imperfection counts reported for each treatment refer to $z = 0$ and the last five refer to $z = 1$. Analyze these data.

4.9. Refer to Table 4.2.

  a. Using weight as the predictor and the number of satellites as the response, fit a Poisson loglinear model. Estimate the mean number of satellites for female crabs of average weight, 2.44 kg.

  b. Use $\hat{\beta}$ to describe the effect of weight. Construct a confidence interval for the effect.

  c. Conduct a Wald test of the hypothesis that the mean number of satellites is independent of weight. Interpret.

  d. Conduct a likelihood-ratio test of independence. Interpret.

4.10. Refer to the previous problem.

  a. Test goodness of fit by grouping levels of weight. Use residuals to describe lack of fit.

  b. Is there evidence of overdispersion? If necessary, adjust the standard errors for the parameter estimates, and interpret.

4.11. Refer to Table 4.2.

  a. Fit a Poisson loglinear model using both weight and color to predict the number of satellites. Assigning dummy variables, treat color as a nominal factor. Interpret the parameter estimates.

  b. Estimate the mean number of satellites for female crabs of average weight (2.44 kg) that are (i) medium light, (ii) dark.

  c. Test whether color is needed in the model. (*Hint*: From Section 4.5.3, the likelihood-ratio statistic for comparing two models is the difference in their G^2 goodness-of-fit statistics.)

   d. Note that the estimated color effects are monotone across the four categories. Fit a simpler model that treats color as quantitative by assuming a linear color effect. Interpret the color effect in this model, and repeat the analyses of (**b**) and (**c**). Compare the fit to the model in (**a**). Interpret.

   e. Add width to the model. Note the effect of the strong positive correlation between width and weight. Are both width and weight needed?

4.12. In Section 4.3.2, refer to the Poisson regression model with identity link for the crab data. Explain why the fit differs from the least squares fit. (*Hint*: The least squares fit is the same as the ML fit of the GLM assuming normal rather than Poisson random component. What do the two approaches assume about the variance of Y?)

4.13. Refer to the injurious accident data in Section 4.3.4.

   a. Test the hypothesis of equal rates for men and women using a likelihood-ratio test. Compare results to the Wald test.

   b. White drivers had 348 injurious accidents in 29.4 thousand years of driving, and black drivers had 147 injurious accidents in 9.2 thousand years. Test whether the rates of injurious accidents differ by race.

   c. In the same study, accident rates were much higher for subjects residing in urban areas than in rural areas. Explain why, controlling for residence, the difference in rates between blacks and whites is likely to be smaller than that observed in (**a**).

   d. The entire cohort of elderly drivers had 495 injurious accidents in 38.7 thousand years of driving. Find a 95% confidence interval for the true rate. (*Hint*: Compute a confidence interval first for the log rate by obtaining the estimate and standard error for the intercept term in a Poisson loglinear model that has no other predictor and uses $\log(38.7)$ as an offset.)

Table 4.7

Team	Attendance	Arrests	Team	Attendance	Arrests
Aston Villa	404	308	Shrewsbury	108	68
Bradford City	286	197	Swindon Town	210	67
Leeds United	443	184	Sheffield Utd.	224	60
Bournemouth	169	149	Stoke City	211	57
West Brom	222	132	Barnsley	168	55
Huddersfield	150	126	Millwall	185	44
Middlesbro	321	110	Hull City	158	38
Birmingham	189	101	Manchester City	429	35
Ipswich Town	258	99	Plymouth	226	29
Leicester City	223	81	Reading	150	20
Blackburn	211	79	Oldham	148	19
Crystal Palace	215	78			

Source: *The Independent* (London), Dec. 21, 1988.

4.14. Table 4.7 lists total attendance (in thousands) and the total number of arrests, in the 1987–1988 season for soccer teams in the Second Division of the British football league. (Thanks to Dr. P. M. E. Altham for showing me these data.)

    a. Let Y denote the number of arrests for a team with total attendance t. Explain why the model $E(Y) = \mu t$ might be plausible. Show that it has alternative form $\log[E(Y)/t] = \alpha$, where $\alpha = \log(\mu)$. Assuming Poisson sampling, fit the model, and interpret the estimate of μ.

    b. Plot arrests against attendance, and overlay the prediction equation. Use residuals to identify teams that had a much larger or smaller than expected number of arrests.

4.15. Table 4.8 shows the number of train miles (in millions) and the number of collisions involving British Rail passenger trains between 1970 and 1984. Is it plausible that the collision counts are independent Poisson variates with constant rate over the 14-year period? Respond by testing the goodness of fit of a GLM for collision rates that contains only an intercept term.

Table 4.8

Year	Collisions	Miles	Year	Collisions	Miles
1970	3	281	1977	4	264
1971	6	276	1978	1	267
1972	4	268	1979	7	265
1973	7	269	1980	3	267
1974	6	281	1981	5	260
1975	2	271	1982	6	231
1976	2	265	1983	1	249

Source: British Dept. of Transport; see also C. Chatfield, *Problem Solving: A Statistician's Guide* (London: Chapman and Hall, 1988), p. 86.

4.16. Table 4.9, based on a study with British doctors conducted by R. Doll and A. B. Hill, was analyzed by N. R. Breslow in *A Celebration of Statistics*, A. C. Atkinson and S. E. Fienberg, eds. (Berlin: Springer-Verlag, 1985).

Table 4.9

Age	Person-Years		Coronary Deaths	
	Nonsmokers	Smokers	Nonsmokers	Smokers
35–44	18,793	52,407	2	32
45–54	10,673	43,248	12	104
55–64	5,710	28,612	28	206
65–74	2,585	12,663	28	186
75–84	1,462	5,317	31	102

Source: R. Doll and A. B. Hill, *Natl. Cancer Inst. Monogr.*, *19*: 205–268 (1966).

a. For each age, compute the sample coronary death rates per 1000 person-years, for nonsmokers and smokers. To compare them, take their ratio and describe its dependence on age.

b. Fit a main-effects model for the log rates having four parameters for age and one for smoking. In discussing lack of fit, show that this model assumes a constant ratio of nonsmokers' to smokers' coronary death rates over levels of age.

c. Based on (**a**), explain why it is sensible to add a quantitative interaction of age and smoking. Assign scores to the levels of age, and add a term based on the product of age and smoking. For this model, show that the log of the ratio of coronary death rates changes linearly with age. Fit the model, and interpret.

4.17. For rate data, the Poisson GLM with identity link is

$$\frac{\mu}{t} = \alpha + \beta x.$$

a. Since the model has form $\mu = \alpha t + \beta t x$, argue that it is equivalent to a Poisson GLM for the response totals at the various levels of x, using identity link with t and tx as explanatory variables and no intercept or offset terms.

b. Fit this model to the grouped data in Table 4.3, using average width scores. Compare results, including interpretations, goodness of fit, and residual analyses, to those obtained with the log link.

CHAPTER 5

Logistic Regression

Let's now take a closer look at the statistical modeling of binary response variables, for which the response measurement for each subject is a "success" or "failure." Binary data are perhaps the most common form of categorical data, and the methods of this chapter are of fundamental importance. The most popular model for binary data is *logistic regression*. Section 4.2.3 introduced this model as a generalized linear model (GLM) for a binomial random component. This chapter studies the application of logistic regression in greater detail.

The first section discusses interpretation of the logistic regression model. Section 5.2 presents statistical inference for the model parameters, and Section 5.3 presents ways of checking the model fit. Section 5.4 shows how to handle qualitative predictors in the model, and Section 5.5 discusses the extension of the model for multiple explanatory variables. Section 5.6 discusses determination of the sample size needed to obtain adequate inferential power. Finally, Section 5.7 presents small-sample, exact inference for logistic regression.

5.1 INTERPRETING THE LOGISTIC REGRESSION MODEL

For a binary response Y and a quantitative explanatory variable X, let $\pi(x)$ denote the "success" probability when X takes value x. This probability is the parameter for the binomial distribution. The logistic regression model has linear form for the logit of this probability,

$$\text{logit}[\pi(x)] = \log\left(\frac{\pi(x)}{1 - \pi(x)}\right) = \alpha + \beta x. \tag{5.1.1}$$

The formula implies that $\pi(x)$ increases or decreases as an S-shaped function of x (recall Figure 4.2).

An alternative formula for logistic regression refers directly to the success probability. This formula uses the exponential function $\exp(x) = e^x$, in the form

$$\pi(x) = \frac{\exp(\alpha + \beta x)}{1 + \exp(\alpha + \beta x)}. \tag{5.1.2}$$

This section shows ways of interpreting these model formulas.

103

5.1.1 Linear Approximation Interpretations

The parameter β determines the rate of increase or decrease of the S-shaped curve. The sign of β indicates whether the curve ascends or descends, and the rate of change increases as $|\beta|$ increases. When the model holds with $\beta = 0$, the right-hand side of (5.1.2) simplifies to a constant. Then, $\pi(x)$ is identical at all x, so the curve becomes a horizontal straight line. The binary response Y is then independent of X.

Figure 5.1 shows the S-shaped appearance of the logistic regression model for $\pi(x)$, as fitted for the example in the following subsection. Since it has a curved rather than linear appearance, the function (5.1.2) implies that the rate of change in $\pi(x)$ per unit change in x varies. A straight line drawn tangent to the curve at a particular x value, such as shown in Figure 5.1, describes the rate of change at that point. For logistic regression parameter β, that line has slope equal to $\beta\pi(x)[1 - \pi(x)]$. For instance, the line tangent to the curve at x for which $\pi(x) = .5$ has slope $\beta(.5)(.5) = .25\beta$; by contrast, when $\pi(x) = .9$ or $.1$, it has slope $.09\beta$. The slope approaches 0 as the probability approaches 1.0 or 0.

The steepest slope of the curve occurs at x for which $\pi(x) = .5$; that x value is $x = -\alpha/\beta$. (One can check that $\pi(x) = .5$ at this point by substituting $-\alpha/\beta$ for x in (5.1.2), or by substituting $\pi(x) = .5$ in (5.1.1) and solving for x.) This x value

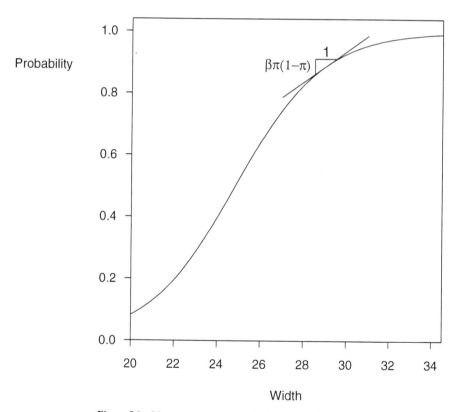

Figure 5.1   Linear approximation to logistic regression curve.

is sometimes called the *median effective level* and is denoted EL_{50}. It represents the level at which each outcome has a 50% chance.

5.1.2 Horseshoe Crabs Revisited

Maximum likelihood (ML) computations for fitting logistic regression models are complex, but are easy to perform using statistical software. To illustrate the model, we reanalyze the horseshoe crab data introduced in Section 4.3.2. We use the binary response of whether a female crab has any satellites present; that is, $Y = 1$ if a female crab has at least one satellite, and $Y = 0$ if she has no satellite. We first use the female crab's width as the sole predictor. Section 5.5 discusses models having additional predictors.

Figure 5.2 shows a plot of the data. It consists of a set of points at the level $Y = 1$ and a second set of points at the level $Y = 0$. The numbered symbols indicate the number of observations at each point. It appears that $Y = 1$ tends to occur relatively more often at higher x values. Since Y takes only values 0 and 1, however, it is difficult to determine whether a logistic regression model is reasonable

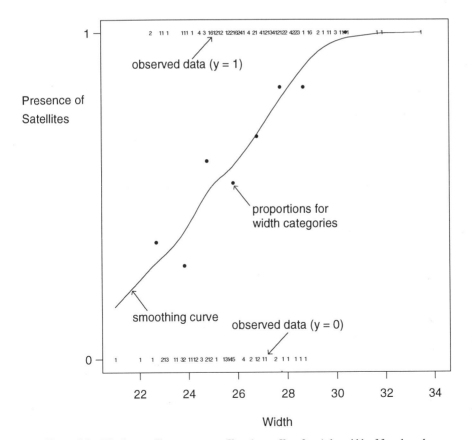

Figure 5.2 Whether satellites are present ($Y = 1$, yes; $Y = 0$, no), by width of female crab.

Table 5.1   Relation Between Width of Female Crab and Existence of Satellites, and Predicted Values for Logistic Regression Model

Width	Number Cases	Number Having Satellites	Sample Proportion	Predicted Probability	Predicted Number Crabs with Satellites
< 23.25	14	5	.36	.26	3.64
23.25–24.25	14	4	.29	.38	5.31
24.25–25.25	28	17	.61	.49	13.78
25.25–26.25	39	21	.54	.62	24.23
26.25–27.25	22	15	.68	.72	15.94
27.25–28.25	24	20	.83	.81	19.38
28.25–29.25	18	15	.83	.87	15.65
> 29.25	14	14	1.00	.93	13.08

by plotting Y against x. Better information results from grouping the width values into categories and calculating a sample proportion of crabs having satellites for each category. This reveals whether the true proportions follow approximately the trend required by this model. Consider the grouping used to investigate adequacy of Poisson regression models in Section 4.4.2, shown again in Table 5.1. In each of the eight width categories, we computed the sample proportion of crabs having satellites and the mean width for the crabs in that category. Figure 5.2 also contains eight dots representing the sample proportions of female crabs having satellites plotted against the mean widths for the eight categories.

Alternatively, some software can smooth the data, revealing a general trend without assuming a particular functional form for the relationship. Smoothing methods based on *generalized additive models* do this by providing even more general structural form than GLMs. For instance, they find possibly complex functions of the explanatory variables that serve as the "best" predictors of a certain type. Figure 5.2 also shows a curve based on smoothing the data using this method. The eight plotted sample proportions and this smoothing curve both show a roughly increasing trend, so we proceed with fitting models that imply such trends.

For the ungrouped data from Table 4.2, let $\pi(x)$ denote the probability that a female horseshoe crab of width x has a satellite. The simplest model to interpret is the linear probability model, $\pi(x) = \alpha + \beta x$. For these data, some predicted values for this GLM fall outside the legitimate range for a binomial parameter, so ML fitting fails. Ordinary least squares fitting yields $\hat{\pi}(x) = -1.766 + 0.092x$. The predicted probability of a satellite increases by .092 for each 1-cm increase in width. This model provides a simple interpretation and realistic predictions over most of the width range, but it is inadequate for extreme values. For instance, at the maximum width in this sample of 33.5, its predicted probability equals $-1.766 + 0.092(33.5) = 1.3$.

The ML parameter estimates for the logistic regression model are $\hat{\alpha} = -12.351$ and $\hat{\beta} = 0.497$. The predicted probability of a satellite is the sample analog of (5.1.2),

$$\hat{\pi} = \frac{\exp(-12.351 + 0.497x)}{1 + \exp(-12.351 + 0.497x)}.$$

Since $\hat{\beta} > 0$, the predicted probability $\hat{\pi}$ is higher at larger width values. At the minimum width in this sample of 21.0 cm, the predicted probability is $\hat{\pi} = \exp(-12.351 + 0.497(21.0))/[1 + \exp(-12.351 + 0.497(21.0))] = .129$; at the maximum width of 33.5 cm, the predicted probability equals $\exp(-12.351 + 0.497(33.5))/[1 + \exp(-12.351+0.497(33.5))] = .987$. The median effective level is the width at which the predicted probability equals .5, which is $x = EL_{50} = -\hat{\alpha}/\hat{\beta} = 12.351/0.497 = 24.8$. Figure 5.1 plots the predicted probabilities as a function of width.

At the sample mean width of 26.3 cm, the predicted probability of a satellite equals .674. The incremental rate of change in the fitted probability at that point is $\hat{\beta}\hat{\pi}(1 - \hat{\pi}) = 0.497(0.674)(0.326) = .11$. For female crabs near the mean width, the estimated probability of a satellite increases at the rate of .11 per cm increase in width. The predicted rate of change is greatest at the x value (24.8) at which $\hat{\pi} = .5$; there, the predicted probability increases at the rate of $(0.497)(0.5)(0.5) = .12$ per cm increase in width. Unlike the linear probability model, the logistic regression model permits the rate of change to vary as x varies.

To further describe the fit, for each category of width Table 5.1 reports the predicted number of crabs having satellites (i.e., the fitted values). To get these, one adds the predicted probabilities for all crabs in the category; for instance, the predicted probabilities for the 14 crabs with widths below 23.25 cm sum to 3.64. The average predicted probability for female crabs in a given width category equals the fitted value divided by the number of female crabs in that category. For the first width category, $3.64/14 = .26$ is the average predicted probability. Table 5.1 reports the fitted values and the average predicted probabilities in grouped fashion. An eyeball comparison of these to the sample counts of crabs having satellites and the sample proportions suggests that the model fits decently. Section 5.3 presents objective criteria for making this comparison.

5.1.3   Odds Ratio Interpretation

Another interpretion of the logistic regression model uses the *odds* and the *odds ratio*. For model (5.1.1), the odds of response 1 (i.e., the odds of a "success") are

$$\frac{\pi(x)}{1 - \pi(x)} = \exp(\alpha + \beta x) = e^{\alpha}(e^{\beta})^x. \tag{5.1.3}$$

This exponential relationship provides an interpretation for β: The odds increase multiplicatively by e^{β} for every one-unit increase in x. That is, the odds at level $x + 1$ equal the odds at x multiplied by e^{β}. When $\beta = 0$, $e^{\beta} = 1$, and the odds do not change as x changes.

For the female horseshoe crabs, the estimated odds of a satellite multiply by $\exp(\hat{\beta}) = \exp(0.497) = 1.64$ for each centimeter increase in width; that is, there is a 64% increase. To illustrate, the mean width value of $x = 26.3$ has a predicted probability of a satellite equal to .674, and odds $= .674/.326 = 2.07$. At $x = 27.3 = 26.3 + 1.0$, one can check that the predicted probability equals .773, and odds $= .773/.227 = 3.40$. But this is a 64% increase; that is, $3.40 = 2.07(1.64)$.

The logarithm of the odds, which is the logit transform of $\pi(x)$, has the linear relationship (5.1.1). This is the logit expression of the model, originally introduced in Section 4.2.3. It states that the logit increases by β units for every 1-unit change in x. Most of us do not think naturally on a logit scale, so this interpretation has limited use.

5.1.4   Logistic Regression with Case-Control Studies*

Another property of the logistic regression model relates to situations in which the explanatory variable X rather than the response variable Y is random. Most commonly, this occurs with retrospective sampling designs, such as case-control studies (Section 2.3.4). For samples of subjects having $Y = 1$ ("cases") and having $Y = 0$ ("controls"), the value of X is observed. Evidence exists of an association between X and Y if the distribution of X values differs between cases and controls.

Many biomedical studies, particularly epidemiological studies, use case-control designs. For retrospective data, Section 2.3.4 noted that one can estimate odds ratios. Logistic regression parameters refer to odds and odds ratios. Thus, one can fit such models to retrospective data, and one can estimate effects in case-control studies. This is not true of other models for binary responses, since the odds ratio is not their natural measure for describing effects. This provides an important advantage of the logit link over links such as the probit and is a major reason why the logit model has surpassed the others in popularity. Section 9.2.3 discusses the application of logistic regression to case-control studies that match each case with a single control.

Regardless of the sampling mechanism, the logistic regression model may or may not describe a relationship well. In one special case, it does necessarily hold. Suppose that the distribution of X for all subjects having $Y = 1$ is normal $N(\mu_1, \sigma)$, and suppose the distribution of X for all subjects having $Y = 0$ is normal $N(\mu_0, \sigma)$; that is, with different mean but the same standard deviation. Then one can show that $\pi(x)$ satisfies the logistic regression curve, with β having the same sign as $\mu_1 - \mu_0$. When a population consists of a mixture of two types of subjects, one set with $Y = 1$ having a bell-shaped distribution on X and the other set with $Y = 0$ having another bell-shaped distribution with similar spread, then the logistic regression function (5.1.2) approximates well the curve for $\pi(x)$. If the distributions are bell-shaped but with highly different spreads, then a model containing also a quadratic term (i.e., both x and x^2) often fits well. In that case, the relationship is nonmonotone, with $\pi(x)$ increasing and then decreasing, or the reverse (Problem 5.5).

5.2   INFERENCE FOR LOGISTIC REGRESSION

We have studied how the fit of a logistic regression model helps us describe the effects of a predictor on a binary response variable. We next present statistical inference for the model parameters, to help judge the significance and size of the effects. Widely available software reports the parameter estimates and their standard errors, as well as other information about the model fit.

5.2.1 Confidence Intervals for Effects

A large-sample confidence interval for the parameter β in the logistic regression model, $\text{logit}[\pi(x)] = \alpha + \beta x$, is

$$\hat{\beta} \pm z_{\alpha/2}(ASE).$$

Exponentiating the endpoints of this interval yields one for e^{β}, the multiplicative effect on the odds of a 1-unit increase in X.

To illustrate, we continue our logistic regression analysis of the horseshoe crab data. The estimated effect of width in the fitted equation for the probability of a satellite is $\hat{\beta} = 0.497$, with $ASE = 0.102$. A 95% confidence interval for β is $0.497 \pm 1.96(0.102)$, or $(0.298, 0.697)$. The confidence interval for the effect on the odds per centimeter increase in width equals $(e^{.298}, e^{.697}) = (1.35, 2.01)$. We infer that each centimeter increase in width has at least a 35% increase and at most a doubling in the odds that a female crab has a satellite.

Section 5.1.1 noted that simpler interpretations result from linear approximations to the logistic regression curve. The term $\beta\pi(1 - \pi)$ approximates the change in the probability per unit change in x. For instance, at $\pi = .5$, the estimated rate of change is $0.25\hat{\beta} = .124$. A 95% confidence interval for 0.25β equals 0.25 times the endpoints of the interval for β, or $[0.25(.298), 0.25(.697)] = (.074, .174)$. If the logistic regression model holds, for values of x near the width at which $\pi = .5$, the rate of increase in the probability of a satellite per centimeter increase in width falls between about .07 and .17.

5.2.2 Significance Testing

We next discuss significance tests for the effect of X on the binary response. For the logistic regression model, the null hypothesis $H_0 : \beta = 0$ states that the probability of success is independent of X.

For large samples, the test statistic

$$z = \frac{\hat{\beta}}{ASE}$$

has a standard normal distribution when $\beta = 0$. One can refer z to the standard normal table to get a one-sided or two-sided P-value in the usual manner. Equivalently, for the two-sided alternative $\beta \neq 0$, $(\hat{\beta}/ASE)^2$ is a Wald statistic (Section 4.4.1) having a large-sample chi-squared distribution with $df = 1$.

Though the Wald test works well for very large samples, the likelihood-ratio test (Section 4.4.1) is more powerful and reliable for sample sizes used in practice. The test statistic compares the maximum L_0 of the log-likelihood function when $\beta = 0$ (i.e., when $\pi(x)$ is forced to be identical at all x values) to the maximum L_1 of the log-likelihood function for unrestricted β. The test statistic, $-2(L_0 - L_1)$, also has a large-sample chi-squared distribution with $df = 1$. Most software for logistic regression reports the maximized log-likelihoods L_0 and L_1 and the likelihood-ratio statistic derived from those maxima.

For the horseshoe crab data, the statistic $z = \hat{\beta}/ASE = 0.497/0.102 = 4.9$ shows strong evidence of a positive effect of width on the presence of satellites ($P < .0001$). The equivalent Wald chi-squared statistic, $z^2 = 23.9$, has $df = 1$. The maximized log likelihoods equal $L_0 = -112.88$ under $H_0 : \beta = 0$ and $L_1 = -97.23$ for the full model. The likelihood-ratio statistic equals $-2(L_0 - L_1) = 31.3$, with $df = 1$. This provides even stronger evidence than the Wald statistic of a width effect.

5.2.3 Distribution of Probability Estimates*

The estimated probability that $Y = 1$ at a fixed setting x of X equals

$$\hat{\pi}(x) = \frac{\exp(\hat{\alpha} + \hat{\beta}x)}{1 + \exp(\hat{\alpha} + \hat{\beta}x)}. \tag{5.2.1}$$

Most software for logistic regression can report such estimates as well as confidence intervals for the true probabilities.

We illustrate by estimating the probability of a satellite for female crabs of width $x = 26.5$, which is near the mean width. The logistic regression fit yields the estimate, $\hat{\pi}(26.5) = \exp(-12.351 + 0.497(26.5))/[1 + \exp(-12.351 + 0.497(26.5))] = .695$. From software, a 95% confidence interval for the true probability is $(.61, .77)$.

One can construct confidence intervals for probabilities using the covariance matrix of the model parameter estimates. The term $\hat{\alpha} + \hat{\beta}x$ in the exponents of the prediction equation (5.2.1) is the estimated linear predictor in the logit transform (5.1.1) of $\pi(x)$. This estimated logit has large-sample ASE given by the estimated square root of

$$\text{Var}(\hat{\alpha} + \hat{\beta}x) = \text{Var}(\hat{\alpha}) + x^2\,\text{Var}(\hat{\beta}) + 2x\,\text{Cov}(\hat{\alpha}, \hat{\beta}).$$

A 95% confidence interval for the true logit is $(\hat{\alpha} + \hat{\beta}x) \pm 1.96ASE$. Substituting the endpoints of this interval for $\alpha + \beta x$ in the exponents of (5.2.1) gives a corresponding interval for the probability.

For instance, at $x = 26.5$, the predicted logit is $-12.351 + 0.497(26.5) = 0.825$. Software reports $\hat{V}\text{ar}(\hat{\alpha}) = 6.910$, $\hat{V}\text{ar}(\hat{\beta}) = 0.01035$, $\hat{C}\text{ov}(\hat{\alpha}, \hat{\beta}) = -0.2668$, from which the estimated variance of this predicted logit equals $(6.910) + (26.5)^2(0.01035) + 2(26.5)(-0.2668) = 0.038$. The 95% confidence interval for the true logit equals $0.825 \pm (1.96)\sqrt{0.038}$, or $(0.44, 1.21)$. From (5.2.1), this translates to the interval

$$\left(\frac{\exp(0.44)}{1 + \exp(0.44)}, \frac{\exp(1.21)}{1 + \exp(1.21)} \right) = (.61, .77)$$

for the probability of satellites at width 26.5 cm.

One could ignore the model fit and simply use sample proportions to estimate such probabilities. Six crabs in the sample had width 26.5, and four of them had satellites.

The sample proportion estimate at $x = 26.5$ is $p = \frac{4}{6} = .67$, similar to the model-based estimate. From inverting small-sample tests using the binomial distribution, a 95% confidence interval based on these six observations alone equals $(.22, .96)$.

When the logistic regression model truly holds, the model-based estimator of a probability is considerably better than the sample proportion. The model has only two parameters to estimate, whereas the nonmodel-based approach has a separate parameter for every distinct value of X. For instance, at $x = 26.5$, software reports an $ASE = 0.04$ for the model-based estimate $.695$, whereas the estimated standard error is $\sqrt{p(1-p)/n} = \sqrt{(.67)(.33)/6} = 0.19$ for the sample proportion of $.67$ based on only 6 observations. The 95% confidence intervals are $(.61, .77)$ versus $(.22, .96)$. Instead of using only 6 observations, the model uses the information that all 173 observations provide in estimating the two model parameters. The result is a much more precise estimate.

Reality is a bit more complicated. In practice, any model will not exactly represent the true relationship between $\pi(x)$ and x. Thus, as the sample size increases, the model-based estimator may not converge exactly to the true value of the probability. This does not imply, however, that the sample proportion is actually a better estimator in practice. If the model approximates the true probabilities decently, its estimator still tends to be much closer than the sample proportion to the true value. The model smooths the sample data, somewhat dampening the observed variability. The resulting estimators tend to be better unless each sample proportion is based on an extremely large sample.

In summary, if the logistic regression model approximates well the true dependence of $\pi(x)$ on x, point and interval estimates of $\pi(x)$ based on it are quite useful. The next section shows how to investigate whether the model does in fact provide an adequate fit to the sample data.

5.3   MODEL CHECKING

So far, we have used logistic regression for description and inference about the effects of predictors on binary responses. There is no guarantee, however, that a particular model of this form is appropriate or that it provides a good fit to the data. This section discusses ways of checking the model fit.

Fitted logistic regression models provide predicted probabilities that $Y = 1$. At each setting of the explanatory variables, one can multiply the predicted probability by the number of subjects to obtain a fitted count. Similarly, one can obtain the fitted count for $Y = 0$ at each setting. The test of the null hypothesis that the model holds compares the fitted and observed counts using a Pearson X^2 or likelihood-ratio G^2 test statistic.

For a fixed number of settings, when most fitted counts equal at least about 5, X^2 and G^2 have approximate chi-squared distributions. The degrees of freedom, called the *residual df* for the model, equal the number of sample logits (i.e., the number of settings of explanatory variables) minus the number of model parameters. As usual, large X^2 or G^2 values provide evidence of lack of fit, and the P-value is the right-tail probability above the observed value. When the fit is poor, residuals and other diagnostic measures describe the influence of individual observations on the model fit and highlight reasons for the inadequacy.

5.3.1 Goodness of Fit for Models with Continuous Predictors

We first present a goodness-of-fit analysis for the model using $x = $ width to predict the probability $\pi(x)$ that a female crab has a satellite,

$$\text{logit}[\pi(x)] = \alpha + \beta x. \tag{5.3.1}$$

Width takes 66 distinct values for the 173 crabs, with few observations at most widths. One could regard the data as a 66×2 contingency table, in which the two cells in each row give the counts of the number of crabs with satellites and the number of crabs without satellites, at that width. The cell counts in this table are small, as are the fitted counts.

The large-sample theory for X^2 and G^2 applies for a fixed number of cells when the fitted counts are large. This theory is violated for the 66×2 table in two ways. First, most fitted counts are very small. Second, when more data are collected, additional width values would occur, so the contingency table would contain more cells rather than a fixed number. Because of this, X^2 and G^2 for logistic regression models fitted with continuous or nearly-continuous predictors do not have approximate chi-squared distributions. These indices of fit are more properly applied when the explanatory variables are categorical, and relatively few fitted counts are small.

To check the adequacy of logistic regression for these data, we compare the observed and fitted values in the grouped form of Table 5.1, shown again in Table 5.2, which is an 8×2 table. In each width category, the fitted value for response "yes" is the sum of the predicted probabilities $\hat{\pi}(x)$ for all crabs having width in that category; the fitted value for response "no" is the sum of $1 - \hat{\pi}(x)$ for those crabs. The fitted values displayed in this form are much larger than in the original 66×2 table, and chi-squared statistics for testing the model have better validity. Substituting the 16

Table 5.2   Grouping of Observed and Fitted Values for Fit of Logistic Regression Model to Horseshoe Crab Data

Width	Number Yes	Number No	Fitted Yes	Fitted No
< 23.25	5	9	3.64	10.36
23.25-24.25	4	10	5.31	8.69
24.25-25.25	17	11	13.78	14.22
25.25-26.25	21	18	24.23	14.77
26.25-27.25	15	7	15.94	6.06
27.25-28.25	20	4	19.38	4.62
28.25-29.25	15	3	15.65	2.35
> 29.25	14	0	13.08	0.92

grouped observed counts and fitted values into the standard chi-squared statistics,

$$X^2 = \sum \frac{(\text{observed} - \text{fitted})^2}{\text{fitted}} = 5.3,$$

and

$$G^2 = 2 \sum (\text{observed}) \log \left(\frac{\text{observed}}{\text{fitted}} \right) = 6.2.$$

Table 5.2 has 8 sample logits, one for each width setting; the logistic regression model (5.3.1) has two parameters, so $df = 8 - 2 = 6$. Neither X^2 nor G^2 shows evidence of lack of fit ($P > .4$). Thus, we can feel more comfortable about the use of the model in Sections 5.1 and 5.2 for the original ungrouped data.

A simpler but more approximate method for obtaining goodness-of-fit statistics fits the logistic regression model directly to the observed counts in the 8×2 table. To treat width as quantitative, we assign scores to its categories, such as the mean widths $\{22.69, 23.84, 24.77, 25.84, 26.79, 27.74, 28.67, 30.41\}$ for the crabs in each category. The logit prediction equation then equals $\text{logit}[\hat{\pi}(x)] = -11.51 + 0.465x$, which yields a set of predicted probabilities and fitted values. For this fit, $X^2 = 5.0$ and $G^2 = 6.0$, based on $df = 6$. Results are similar to the statistics using fitted values based on predicted probabilities at the individual width values.

When explanatory variables are continuous, it is difficult to analyze lack of fit without some type of grouping. As the number of explanatory variables increases, however, simultaneous grouping of values for each variable can produce a contingency table with a large number of cells, many of which have small counts. An alternative way of grouping forms observed and fitted values based on a partitioning of predicted probabilities. One can regard the grouping in Table 5.2 as having been done in this way. For the model fitted, the 14 crabs in the first width category are the ones with the smallest predicted probabilities of a satellite; the 14 crabs in the second width category have higher predicted probabilities than the crabs in the first category, but smaller predicted probabilities than crabs in the other categories, and so forth.

Regardless of how many predictors are in the model, one can partition observed and fitted values according to the predicted probabilities. One common approach forms the groups in the partition so they have approximately equal size. To form 10 groups, for instance, one pair of observed and fitted counts refers to the $n/10$ observations having the highest predicted probabilities, another pair refers to the $n/10$ observations having the second decile of predicted probabilities, and so forth. In practice, it is usually not possible to form groups of exactly equal size because sets of observations have the same predicted probability, and all observations having the same predicted probability are kept in the same group. For each group, the fitted value for an outcome is the sum of the predicted probabilities for that outcome for all observations in that group.

This construction is the basis of a test due to Hosmer and Lemeshow (1989, p. 140). Their Pearson-like statistic does not actually have a chi-squared distribution, but simulations have shown that its distribution is roughly approximated by chi-squared with $df = g - 2$, where g denotes the number of groups. We applied their test with $g = 10$ approximately equally sized groups for the logistic regression model fitted to the ungrouped data (Table 4.2). The Hosmer–Lemeshow statistic equals 3.5, based on $df = 8$, indicating a decent fit.

One can also detect lack of fit by using a likelihood-ratio test to compare the working model to more complex ones, as we will illustrate in Section 5.5. For instance, we might consider more complex models containing nonlinear effects (such as quadratic terms) for quantitative predictors or interaction terms. If we do not find a more complex model that provides a better fit, this provides some assurance that our fitted model is reasonable. This approach is more useful from a scientific perspective. A large goodness-of-fit statistic simply indicates there is *some* lack of fit, but provides no insight about its nature. Comparing a model to a more complex model, on the other hand, indicates whether lack of fit exists of a particular type.

5.3.2 Goodness of Fit and Likelihood-Ratio Model Comparison Tests

Sections 4.4.1 and 4.5.2 introduced the likelihood-ratio statistic $-2(L_0 - L_1)$ for testing whether certain parameters in a model equal zero. The test compares the maximized log likelihood (L_1) for the model to the maximized log likelihood (L_0) for the simpler model that deletes those parameters. Denote the fitted model by M_1 and the simpler model for which those parameters equal zero by M_0.

The goodness-of-fit statistic G^2 for testing the fit of a logistic regression model M is the special case of the likelihood-ratio statistic in which $M_0 = M$ and M_1 is the most complex model possible. That complex model has a separate parameter for each logit, and provides a perfect fit to the sample logits. It is called the *saturated model*. In testing whether M fits, we test whether *all* parameters that are in the saturated model but not in M equal zero. Denote this statistic for testing the fit of M by $G^2(M)$. In GLM terminology, it is called the *deviance* of the model (Section 4.5.3). Let L_S denote the maximized log likelihood for the saturated model. Then, for instance, the deviances for models M_0 and M_1 are $G^2(M_0) = -2(L_0 - L_S)$ and $G^2(M_1) = -2(L_1 - L_S)$.

Denote the likelihood-ratio statistic for testing M_0, given that M_1 holds, by $G^2(M_0 \mid M_1)$. This statistic for comparing these models equals

$$G^2(M_0 \mid M_1) = -2(L_0 - L_1) = -2(L_0 - L_S) - [-2(L_1 - L_S)] = G^2(M_0) - G^2(M_1),$$

the difference in G^2 goodness-of-fit statistics for the two models. That is, the likelihood-ratio statistic for comparing two models is simply the difference in the deviances of those models. This statistic is large when M_0 fits poorly compared to M_1. It is a large-sample chi-squared statistic, with df equal to the difference between the residual df values for the two models.

We illustrate this comparison for two models fitted to the grouped crab data. Denote the logistic regression model (5.3.1) with width as the sole predictor by M_1 and the simpler model having only an intercept parameter as M_0. That simpler model posits independence of width and having a satellite, and the G^2 goodness-of-fit statistic for testing it is simply the G^2 statistic (2.4.3) for testing independence in a two-way contingency table. For the observed counts in the 8×2 Table 5.2, it equals $G^2(M_0) = 34.0$, based on $df = 7$. Since the fit of the model with width as a predictor has $G_2(M_1) = 6.0$ with $df = 6$, the comparison statistic for the two models is $G^2(M_0 \mid M_1) = G^2(M_0) - G^2(M_1) = 34.0 - 6.0 = 28.0$, based on $df = 7 - 6 = 1$. In fact, this equals the likelihood-ratio statistic $-2(L_0 - L_1)$ for testing that $\beta = 0$ in the logistic regression model fitted to the grouped data of Table 5.2.

5.3.3 Residuals for Logit Models

Goodness-of-fit statistics such as G^2 and X^2 are summary indicators of the overall quality of fit. Additional diagnostic analyses are necessary to describe the nature of any lack of fit. Residuals comparing observed and fitted counts are useful for this purpose.

Let y_i denote the number of "successes" for n_i trials at the ith setting of the explanatory variables. Let $\hat{\pi}_i$ denote the predicted probability of success for the model fit. Then $n_i \hat{\pi}_i$ is the fitted number of successes. For a GLM with binomial random component, the Pearson residual (4.4.1) for the fit at setting i is

$$e_i = \frac{y_i - n_i \hat{\pi}_i}{\sqrt{[n_i \hat{\pi}_i (1 - \hat{\pi}_i)]}}. \tag{5.3.2}$$

Each residual divides the difference between an observed count and its fitted value by the estimated binomial standard deviation of the observed count.

The Pearson statistic for testing the model fit satisfies

$$X^2 = \sum e_i^2. \tag{5.3.3}$$

Each squared Pearson residual is a component of X^2. When the binomial index n_i is large, the Pearson residual e_i has an approximate normal distribution. When the model holds, it has an approximate expected value of zero but a smaller variance than a standard normal variate. If the number of model parameters is small compared to the number of sample logits, Pearson residuals are treated like standard normal deviates, with absolute values larger than 2 indicating possible lack of fit.

Table 5.3 shows Pearson residuals for two logistic regression models fitted to the grouped crab data—model (5.3.1) with width as the predictor and the model having only an intercept term. The latter model, which is (5.3.1) with $\beta = 0$, treats the response as independent of width. Some residuals for that model are large, and they show an increasing trend. This trend disappears for the model with a linear effect of width.

Graphical displays are also useful for showing lack of fit. One can compare observed and fitted proportions by plotting them against each other, or by plotting both of them against explanatory variables. For Table 5.2, Figure 5.3 plots both the observed proportions and the predicted probabilities of a satellite against width. The

Table 5.3   Residuals for Logistic Regression Models Fitted to Grouped Crab Data

Width	Number Cases	Number Yes	Fitted[a] Yes	Pearson[a] Residual	Fitted Yes	Pearson Residual	Adjusted Residual
< 23.25	14	5	8.98	−2.22	3.85	0.69	0.85
23.25–24.25	14	4	8.98	−2.78	5.50	−0.82	−0.93
24.25–25.25	28	17	17.96	−0.38	13.97	1.14	1.35
25.25–26.25	39	21	25.02	−1.34	24.21	−1.06	−1.24
26.25–27.25	22	15	14.12	0.39	15.80	−0.38	−0.42
27.25–28.25	24	20	15.40	1.96	19.16	0.43	0.49
28.25–29.25	18	15	11.55	1.70	15.46	−0.31	−0.36
> 29.25	14	14	8.98	2.80	13.05	1.01	1.14

[a]Independence model, other fitted values and residuals refer to model (5.3.1) with width predictor.

fit seems decent, which is not surprising since the formal goodness-of-fit tests showed no evidence of lack of fit.

We have noted that X^2 and G^2 are invalid when fitted values are very small. Similarly, residuals have limited meaning in that case. When explanatory variables are continuous, often $n_i = 1$ at many settings. Then, y_i can equal only 0 or 1, and

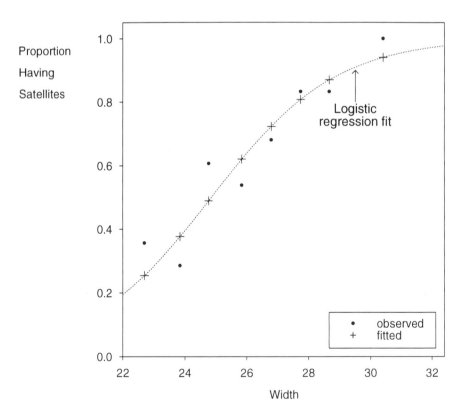

Figure 5.3   Observed and fitted proportions of satellites by width of female crab.

e_i can assume only two values. One must then be cautious about regarding either outcome as "extreme," and a single residual is usually uninformative.

5.3.4 Diagnostic Measures of Influence*

As in ordinary regression modeling, some observations may have much, perhaps too much, influence in determining parameter estimates. The fit could be quite different if they were deleted. An observation is more likely to have a large influence when it takes an extreme value on one or more of the explanatory variables. It may be informative to report the fit of the model after deleting one or two observations, if the fit with them seems misleading.

Several measures describe various aspects of influence. Many of them relate to the effect on certain characteristics of removing the observation from the data set. These measures are algebraically related to an observation's *leverage*, its element from the diagonal of the so-called *hat matrix*. (Roughly speaking, the hat matrix is a matrix that, when applied to the sample logits, yields the predicted logit values for the model.) The greater an observation's leverage, the greater its potential influence. Formulas for leverages and the diagnostic measures of influence are complex, so we do not reproduce them here. Most software for logistic regression produces these diagnostics. Influence measures for each observation include the following:

1. For each parameter in the model, the change in the parameter estimate when the observation is deleted. This change, divided by its standard error, is called *Dfbeta*.

2. A measure of the change in a joint confidence interval for the parameters produced by deleting the observation. This confidence interval displacement diagnostic is denoted by c.

3. The change in X^2 or G^2 goodness-of-fit statistics when the observation is deleted.

For each measure, the larger the value, the greater the observation's influence. We illustrate them using the logistic regression model with width as a predictor for the grouped crab data. Table 5.4 contains the *Dfbeta* measure for the coefficient of width, the confidence interval diagnostic c, the change in X^2, and the change in G^2. None of their values reveal any highly influential observations. By contrast, Table 5.4 also contains the changes in X^2 and G^2 for deleting observations from the fit of the independence model ((5.3.1) with $\beta = 0$). At the low and high ends of the width values, several of these changes are large. The severe influence of these values partly reflects the poor fit of the model that does not permit width to have an effect on the response.

One can also use leverage values to construct an adjustment to the Pearson residual e_i that is slightly larger in absolute value and *does* have an approximate standard normal distribution when the model holds. For observation i with leverage h_i, the

Table 5.4 Diagnostic Measures for Logistic Regression Models Fitted to Grouped Crab Data

Width	Dfbeta	c	Pearson Diff.	Likelihood-Ratio Diff.	Pearson[a] Diff.	Likelihood-Ratio[a] Diff.
< 23.25	−0.54	0.38	0.73	0.70	5.36	5.09
23.25–24.25	0.37	0.25	0.87	0.89	8.39	8.00
24.25–25.25	−0.43	0.71	1.82	1.83	0.17	0.17
25.25–26.25	−0.02	0.58	1.55	1.52	2.33	2.27
26.25–27.25	−0.09	0.04	0.17	0.17	0.18	0.18
27.25–28.25	0.21	0.08	0.24	0.25	4.45	4.95
28.25–29.25	−0.17	0.04	0.13	0.13	3.21	3.58
> 29.25	0.55	0.34	1.29	2.24	8.51	13.11

[a]Independence model, other values refer to model (5.3.1) with width predictor.

adjusted residual has form

$$\frac{e_i}{\sqrt{1 - h_i}} = \frac{y_i - n_i\hat{\pi}_i}{\sqrt{[n_i\hat{\pi}_i(1 - \hat{\pi}_i)(1 - h_i)]}}.$$

These serve the same purpose as the adjusted residuals defined in Section 2.4.5 for detecting patterns of dependence in two-way contingency tables and in Section 4.4.3 for describing lack of fit in Poisson regression models.

Table 5.3 contains adjusted residuals for the logistic regression model with width predictor fitted to the grouped crab data. Though slightly larger than the Pearson residuals, they show a similar pattern and do not suggest any lack of fit.

5.4 LOGIT MODELS FOR QUALITATIVE PREDICTORS

Logistic regression, like ordinary regression, extends to models incorporating multiple explanatory variables. Moreover, some or all of those explanatory variables can be qualitative, rather than quantitative. This section shows the use of dummy variables for including qualitative predictors, often called *factors*, and Section 5.5 presents the general form of multiple logistic regression models.

5.4.1 Dummy Variables in Logit Models

Suppose that a binary response Y has two binary predictors, X and Z. Denote the two levels for each variable by $(0, 1)$. For the $2 \times 2 \times 2$ contingency table, the model for the probability π that $Y = 1$,

$$\text{logit}(\pi) = \alpha + \beta_1 x + \beta_2 z, \tag{5.4.1}$$

has separate main effects for the two predictors. It assumes an absence of interaction, the effect of one factor being the same at each level of the other factor.

The variables x and z in this model are *dummy variables* that indicate categories for the predictors. At a fixed level z of Z, the effect on the logit of changing from $x = 0$ to $x = 1$ is

$$= [\alpha + \beta_1(1) + \beta_2 z] - [\alpha + \beta_1(0) + \beta_2 z] = \beta_1.$$

This difference between two logits equals the difference of log odds, which equals the log of the odds ratio between X and Y, at a fixed level of Z. Thus, $\exp(\beta_1)$ describes the conditional odds ratio between X and Y. Controlling for Z, the odds of "success" at $x = 1$ equal $\exp(\beta_1)$ times the odds of success at $x = 0$. This conditional odds ratio is the same at each level z of Z. The lack of an interaction term in this model implies a common value of the odds ratio for the partial tables at the two levels of Z. The model satisfies homogeneous association (Sections 3.1.5, 3.2.3, 3.2.4).

Conditional independence exists between X and Y, controlling for Z, if $\beta_1 = 0$, in which case the common odds ratio equals 1. The simpler model

$$\text{logit}(\pi) = \alpha + \beta_2 z \qquad (5.4.2)$$

then applies to the three-way table. One can test whether $\beta_1 = 0$ using a Wald statistic or a likelihood-ratio statistic comparing the two models.

5.4.2 AZT and AIDS Example

We illustrate models with qualitative predictors using Table 5.5, based on a study described in the *New York Times* (Feb. 15, 1991) on the effects of AZT in slowing the development of AIDS symptoms. In the study, 338 veterans whose immune systems were beginning to falter after infection with the AIDS virus were randomly assigned either to receive AZT immediately or to wait until their T cells showed severe immune weakness. Table 5.5 is a $2 \times 2 \times 2$ cross classification of the veterans' race, whether they received AZT immediately, and whether they developed AIDS symptoms during the three-year study.

In model (5.4.1), we identify X with AZT treatment ($x = 1$ for those who took AZT immediately and $x = 0$ otherwise) and Z with race ($z = 1$ for whites and $z = 0$ for blacks), for predicting the probability that AIDS symptoms developed. The ML estimate of the effect of AZT is $\beta_1 = -0.720$ (*ASE* $= 0.279$). The estimated

Table 5.5 Development of AIDS Symptoms by AZT Use and Race

		Symptoms	
Race	AZT Use	Yes	No
White	Yes	14	93
	No	32	81
Black	Yes	11	52
	No	12	43

odds ratio between immediate AZT use and development of AIDS symptoms equals $\exp(-0.720) = 0.49$. For each race, the estimated odds of developing symptoms are half as high for those who took AZT immediately.

The hypothesis of conditional independence of AZT treatment and the development of AIDS symptoms, controlling for race, is $H_0 : \beta_1 = 0$. The likelihood-ratio statistic $-2(L_0 - L_1)$ based on comparing models (5.4.2) and (5.4.1) equals 6.9, based on $df = 1$, showing evidence of association ($P = .01$). The Wald statistic $(\hat{\beta}_1/ASE)^2 = (-0.720/0.279)^2 = 6.6$ provides similar results ($P = .01$).

We next analyze the goodness of fit of model (5.4.1). For its fit, white veterans with immediate AZT use had predicted probability .150 of developing AIDS symptoms during the study. Since 107 white veterans took AZT, the fitted number developing symptoms is $107(.150) = 16.0$, and the fitted number not developing symptoms is $107(.850) = 91.0$. Similarly, one can obtain fitted values for all eight cells in Table 5.5. Substituting these and the cell counts into the usual goodness-of-fit statistics, we obtain $G^2 = 1.4$ and $X^2 = 1.4$. The model has four sample logits, one for each binomial response distribution at the four combinations of AZT use and race. The model has three parameters, so the residual $df = 4 - 3 = 1$. The small values for G^2 and X^2 suggest that the model fits decently ($P > .2$). Further analysis suggests that an even simpler model may be adequate, since the effect of race is not significant.

5.4.3 ANOVA-Type Representation of Factors

A factor having two levels requires only a single dummy variable. A factor having I levels requires $I - 1$ dummy variables, as shown in Section 5.5.1.

An alternative representation of factors in logistic regression models resembles the way ANOVA models ordinarily express them. The model formula

$$\text{logit}(\pi) = \alpha + \beta_i^X + \beta_k^Z \tag{5.4.3}$$

represents the effects of X through parameters $\{\beta_i^X\}$ and the effects of Z through parameters $\{\beta_k^Z\}$. (The X and Z superscripts are simply labels, and do not represent powers.) Model form (5.4.3) applies for any number of levels for X and Z. Each factor has as many parameters as it has levels, but one is redundant. For instance, if X has I levels, it has $I - 1$ nonredundant parameters; β_i^X denotes the effect on the logit of being classified in level i of X. Conditional independence between X and Y, given Z, corresponds to $\beta_1^X = \beta_2^X = \cdots = \beta_I^X$.

One can account for redundancies in parameters in (5.4.3) by setting the parameter for the last category equal to zero. When X and Z have two categories, as in Table 5.5, the parameterization in model (5.4.3) then corresponds to that in model (5.4.1) with $\beta_1^X = \beta_1$ and $\beta_2^X = 0$, and with $\beta_1^Z = \beta_2$ and $\beta_2^Z = 0$. For model (5.4.3) fitted to Table 5.5, Table 5.6 shows parameter estimates for three ways of defining parameters: (1) the approach just described that sets the last parameter (e.g., β_2^X) equal to 0, (2) an analogous approach for which the first parameter equals 0, and (3) an approach whereby each factor's parameters sum to zero. For the second approach, model (5.4.3) corresponds to model (5.4.1) with the dummy variable $x = 0$ in

Table 5.6   Parameter Estimates for Logit Model Fitted to Table 5.5

Parameter	Last = zero	Definition of Parameters First = zero	Sum = zero
Intercept	−1.074	−1.738	−1.406
AZT—yes	−0.720	0.000	−0.360
AZT—no	0.000	0.720	0.360
Race—W	0.055	0.000	0.028
Race—B	0.000	−0.055	−0.028

category 1 and $x = 1$ in category 2 of AZT use. For the third approach, when a factor has two levels, one estimate is the negative of the other (e.g., $\hat{\beta}_1^X = -\hat{\beta}_2^X$). This results from "effect coding" for a dummy variable, such as $x = 1$ in category 1 and $x = -1$ in category 2.

For any of the three coding schemes, the differences $\beta_1^X - \beta_2^X$ and $\beta_1^Z - \beta_2^Z$ are identical and represent the conditional log odds ratios of X and Z with the response, given the other variable. For instance, $\exp(\hat{\beta}_1^X - \hat{\beta}_2^X) = \exp(-0.720) = 0.49$ refers to the estimated common odds ratio between immediate AZT use and development of symptoms, for each race. The estimate of a parameter for a single category of a factor is irrelevant; different ways of handling parameter redundancies result in different values for that estimate. An estimate makes sense only by comparison with one for another category. Exponentiating a difference between estimates for two categories determines the odds ratio relating to the effect of classification in one category rather than the other.

Similarly, different parameter coding schemes yield the same estimated probabilities. The sum of the intercept estimate and the estimates for given factor levels is identical for each scheme. For instance, from Table 5.6, the intercept estimate plus the estimate for immediate AZT use plus the estimate for being white is −1.738 for each scheme, leading to a predicted probability that white veterans with immediate AZT use develop AIDS symptoms equal to $\exp(-1.738)/[1 + \exp(-1.738)] = .15$.

5.4.4   Logit Models for $2 \times 2 \times K$ Contingency Tables

An important special case of logit models with qualitative predictors occurs when X is a binary classification of two groups, and Z is a control variable with K levels. For instance, X might refer to two experimental treatments and Z might refer to several locations for conducting the experiment. Model (5.4.3) then refers to a $2 \times 2 \times K$ contingency table that cross classifies X, Y, and Z. In this model, conditional independence exists between X and Y, controlling for Z, if $\beta_1^X = \beta_2^X$, in which case the common X-Y odds ratio $\exp(\beta_1^X - \beta_2^X)$ for the K partial tables equals 1. One can then absorb the common value of β_i^X into the α term, yielding the simpler model

$$\text{logit}(\pi) = \alpha + \beta_k^Z \qquad (5.4.4)$$

for the three-way table. When $K = 2$, this is equivalent to model (5.4.2).

Given that model (5.4.3) holds, one can test conditional independence of X and Y by the likelihood-ratio statistic $-2(L_0 - L_1)$, comparing that model to the simpler model (5.4.4) not having the X main effect, as illustrated above in Section 5.4.2. Section 3.2.1 presented the Cochran–Mantel–Haenszel (*CMH*) test of this same hypothesis for $2 \times 2 \times K$ tables. That test performs well when the association between X and Y is similar in each partial table. In fact, the *CMH* statistic is the efficient score statistic (Section 4.5.2) for testing X-Y conditional independence in model (5.4.3), which assumes a common odds ratio for the partial tables. The likelihood-ratio test is an alternative to the *CMH* procedure for testing X-Y conditional independence in $2 \times 2 \times K$ tables. For model (5.4.3), the ML estimate $\exp(\hat{\beta}_1^X - \hat{\beta}_2^X)$ of the common X-Y odds ratio for the K partial tables is an alternative to the Mantel–Haenszel estimate (3.2.2).

For Table 5.5, we noted that the likelihood-ratio test of conditional independence of immediate AZT use and AIDS symptom development has test statistic equal to 6.9, with $df = 1$, and the ML estimate of the conditional odds ratio equals $\exp(\hat{\beta}_1^X - \hat{\beta}_2^X) = 0.49$. The *CMH* statistic (3.2.1) equals 6.8 with $df = 1$, and the Mantel–Haenszel estimate (3.2.2) of a common odds ratio equals 0.49. Similarity of results among likelihood-ratio, Wald, and *CMH* (efficient score) tests usually happens when the sample size is large relative to the number of strata.

For model (5.4.3) with binary X, the odds ratio between X and Y is the same at each level of Z. In testing goodness of fit for this model, we are testing that this structure holds. That is, the goodness-of-fit tests also provide tests of the hypothesis of homogeneous odds ratios between X and Y at the K levels of Z. These large-sample tests have $df = K - 1$ and are alternatives to the Breslow–Day test presented in Section 3.2.4. For Table 5.5, for instance, the goodness-of-fit statistics equal $G^2 = 1.38$ and $X^2 = 1.39$ and the Breslow–Day statistic equals 1.39, all based on $df = 1$. They all indicate that homogeneity is plausible.

5.5 MULTIPLE LOGISTIC REGRESSION

The logistic regression model and other GLMs, like ordinary regression models for normal data, generalize to allow for several explanatory variables. The predictors can be quantitative, qualitative, or of both types.

Denote a set of k predictors for a binary response Y by $X_1, X_2, \ldots, X_k$. Model (5.1.1) for the logit of the probability π that $Y = 1$ generalizes to

$$\text{logit}(\pi) = \alpha + \beta_1 x_1 + \beta_2 x_2 + \cdots + \beta_k x_k. \tag{5.5.1}$$

The parameter β_i refers to the effect of X_i on the log odds that $Y = 1$, controlling the other Xs. For instance, $\exp(\beta_i)$ is the multiplicative effect on the odds of a 1-unit increase in X_i, at fixed levels of the other Xs.

5.5.1 Horseshoe Crab Example Using Color and Width Predictors

We continue our analysis of the horseshoe crab data of Table 4.2 (Section 4.3.2) by including both the female crab's width and color as predictors. Color has five categories: light, medium light, medium, medium dark, dark. Color is a surrogate for age, older crabs tending to be darker. The sample contained no light crabs, so our models use only the other four categories.

We first treat color in a qualitative manner by using three dummy variables to represent the four categories. The model is

$$\text{logit}(\pi) = \alpha + \beta_1 c_1 + \beta_2 c_2 + \beta_3 c_3 + \beta_4 x, \qquad (5.5.2)$$

where x denotes width and

$c_1 = 1$ for medium light color, and 0 otherwise,
$c_2 = 1$ for medium color, and 0 otherwise,
$c_3 = 1$ for medium dark color, and 0 otherwise.

The crab color is dark (category 4) when $c_1 = c_2 = c_3 = 0$. The ML estimates of the parameters are

$$\text{Intercept:} \quad \hat{\alpha} = -12.715, \ ASE = 2.762$$
$$c_1: \hat{\beta}_1 = \quad 1.330, \ ASE = 0.852$$
$$c_2: \hat{\beta}_2 = \quad 1.402, \ ASE = 0.548$$
$$c_3: \hat{\beta}_3 = \quad 1.106, \ ASE = 0.592$$
$$\text{width:} \ \hat{\beta}_4 = \quad 0.468, \ ASE = 0.106.$$

For instance, for dark crabs, $c_1 = c_2 = c_3 = 0$, and the prediction equation is $\text{logit}(\hat{\pi}) = -12.715 + 0.468x$; by contrast, for medium-light crabs, $c_1 = 1$, and $\text{logit}(\hat{\pi}) = (-12.715 + 1.330) + 0.468x = -11.385 + 0.468x$.

The model assumes a lack of interaction between color and width in their effects on the response. Width has the same effect (coefficient 0.468) for all colors, so the shapes of the curves relating width to $\pi = P(Y = 1)$ are identical. For each color, a 1-cm increase in width has a multiplicative effect of $\exp(0.468) = 1.60$ on the odds that $Y = 1$. Figure 5.4 displays the fitted model. Any one curve is simply any other curve shifted to the right or to the left. The parallelism of curves in the horizontal dimension implies that two curves never cross. At all width values, for instance, color 4 (dark) has a lower predicted probability of a satellite than the other colors.

The positive effect of width on the odds of having satellites is similar to the effect seen in Section 5.1.2 for the simpler model excluding color. One can calculate predicted probabilities of a satellite using prediction equations for the probabilities, extending the approach of Section 5.1.2. To illustrate, for a medium-light crab of average width (26.3 cm), the predicted probability is

$$\frac{\exp[-11.385 + 0.468(26.3)]}{1 + \exp[-11.385 + 0.468(26.3)]} = .715.$$

By comparison, a dark crab of average width has predicted probability $\exp[-12.715 + 0.468(26.3)]/\{1 + \exp[-12.715 + 0.468(26.3)]\} = .399$.

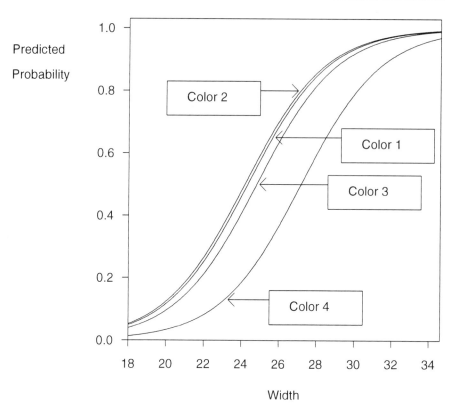

Figure 5.4   Logistic regression model using width and color predictors.

The exponentiated difference between two color parameter estimates is an odds ratio comparing those colors. For instance, the difference in color parameter estimates between medium-light crabs and dark crabs equals 1.330; at any given width, the estimated odds that a medium-light crab has a satellite are exp(1.330) = 3.8 times the estimated odds for a dark crab. Using the probabilities just calculated at width 26.3, the odds equal .715/.285 = 2.51 for a medium-light crab, and .399/.601 = 0.66 for a dark crab, for which 2.51/0.66 = 3.8. The color estimates indicate that, in this sample, dark crabs are less likely than crabs of other colors to have satellites.

5.5.2   Model Comparison

One can use the likelihood-ratio method to test hypotheses about parameters in multiple logistic regression models. For instance, to test whether color makes a significant contribution to model (5.5.2), we test $H_0 : \beta_1 = \beta_2 = \beta_3 = 0$. This hypothesis states that, controlling for width, the probability of a satellite is independent of color. We compare the maximized log-likelihood L_1 for the full model (5.5.2) to the maximized log-likelihood L_0 for the simpler model in which those parameters equal 0, using test statistic $-2(L_0 - L_1) = 7.0$. The chi-squared $df = 3$, the difference between the numbers of parameters in the two models. The P-value of .07 provides slight evidence of a color effect.

More generally, one can compare maximized log-likelihoods for any pair of models such that one is a special case of the other. One such comparison checks whether the model requires interaction terms. The test analyzes whether a better fitting model results from adding the interaction of color and width to model (5.5.2). This more complex model allows a separate width effect for each color. It has three additional terms, the cross-products of width with the color dummy variables. Fitting this model is equivalent to fitting the logistic regression model with width as the predictor separately for the crabs of each of the four colors. Each color then has a different-shaped curve relating width to the probability of a satellite, so a comparison of two colors varies according to the level of width. The likelihood-ratio statistic comparing the models with and without the interaction terms equals 4.4, based on $df = 3$. The evidence of interaction is not strong ($P = .22$).

The reduced model (5.5.2) has the advantage of simpler interpretations. In fact, this model fits adequately according to formal goodness-of-fit tests. For instance, the Hosmer–Lemeshow test with ten groups of predicted probabilities has a test statistic equal to 3.7, based on $df = 8$.

5.5.3 Quantitative Treatment of Ordinal Predictor

Color has a natural ordering of categories, from lightest to darkest. One can construct a simpler model yet by treating this ordinal predictor in a quantitative manner. Color may have a linear effect, for a set of monotone scores assigned to its categories.

To illustrate, we assign scores $c = \{1, 2, 3, 4\}$ to the color categories and fit the model

$$\text{logit}(\pi) = \alpha + \beta_1 c + \beta_2 x, \tag{5.5.3}$$

The prediction equation is

$$\text{logit}(\hat{\pi}) = -10.071 - 0.509c + 0.458x.$$

The color and width estimates have *ASE* values of 0.224 and 0.104, showing strong evidence of an effect for each. At a given width, for every one-category increase in color darkness, the estimated odds of a satellite multiply by $\exp(-0.509) = 0.60$. For instance, the estimated odds of a satellite for medium colored crabs are 60% of those for medium-light crabs.

A likelihood-ratio test compares the fit of this model to the more complex model (5.5.2) that has a separate parameter for each color. The test statistic equals $-2(L_0 - L_1) = 1.7$, based on $df = 2$. This statistic tests that the simpler model (5.5.3) holds, given that model (5.5.2) is adequate. It tests that the color parameters in (5.5.2), when plotted against the color scores, follow a linear trend. The simplification seems permissible ($P = .44$).

The estimates of the color parameters in the model (5.5.2) that treats color as qualitative are $(1.33, 1.40, 1.11, 0)$, the 0 value for the dark category reflecting the lack of a dummy variable for that category. Though these values do not depart significantly from a linear trend, the first three are quite similar compared to the last one. This suggests that another potential color scoring for model (5.5.3) is $\{1, 1, 1, 0\}$; that is, score $= 0$ for dark-colored crabs, and score $= 1$ otherwise. The likelihood-ratio statistic comparing model (5.5.3) with these binary scores to model (5.5.2) equals 0.5, based on $df = 2$, showing that this simpler model is also adequate ($P = .78$). This model has a width estimate of 0.478 ($ASE = 0.104$) and a color estimate of 1.300 ($ASE = 0.525$). At a given width, the estimated odds that a lighter-colored crab has a satellite are $\exp(1.300) = 3.7$ times the estimated odds for a dark crab.

In summary, the qualitative-color model, the ordinal model with color scores $\{1, 2, 3, 4\}$, and the model with binary color scores $\{1, 1, 1, 0\}$ all suggest that dark crabs are least likely to have satellites. It would require a much larger sample size to determine which of the two color scorings is more appropriate. It is advantageous to treat ordinal predictors in a quantitative manner when such models fit well. The advantages include that the model is simpler and easier to interpret, and that tests of the effect of the ordinal predictor are generally more powerful when it has a single parameter rather than several parameters.

5.5.4 Model Selection with Several Predictors

The horseshoe crab data set in Table 4.2 has four predictors: color (four categories), spine condition (three categories), weight, and width of the carapace shell. We next fit a logistic regression model using all these predictors.

Several model selection procedures exist, no one of which is "best." Cautions that apply to ordinary regression modeling of normal data hold for any generalized linear model. For instance, a model with several predictors has the potential for *multicollinearity*: strong correlations among predictors making it seem that no one variable is important when all the others are in the model. A variable may seem to have little effect simply because it "overlaps" considerably with other predictors in the model.

To illustrate, suppose we started by fitting a model containing main effects for the four predictors, treating color and spine condition as qualitative (factors). A likelihood-ratio test that the probability of a satellite is jointly independent of these four predictors simultaneously tests that all their parameters equal zero. The likelihood-ratio statistic based on comparing the main-effects model to the null model that has only an intercept term equals $-2(L_0 - L_1) = 40.6$ with $df = 7$. This statistic has P-value $< .0001$, extremely strong evidence that at least one predictor has an effect. Table 5.7 shows the parameter estimates and their ASE values. Even though the overall test is highly significant, the results in this table are not encouraging. The estimates for the quantitative predictors, weight and width, are only slightly larger than their ASE values. The estimates for the qualitative predictors compare each level to the final category as a baseline; that is, they set up dummy variables for the first three colors and for the first two spine conditions. For color, the largest difference

Table 5.7 Parameter Estimates for Main Effects Model
with Horseshoe Crab Data

Parameter	Estimate	ASE
Intercept	−9.273	3.838
Color(1)	1.609	0.936
Color(2)	1.506	0.567
Color(3)	1.120	0.593
Spine(1)	−0.400	0.503
Spine(2)	−0.496	0.629
Weight	0.826	0.704
Width	0.263	0.195

between estimates for two levels is between the first and fourth, which is less than two standard errors; for spine condition, the largest difference between estimates for two levels is between the second and third, which is less than a standard error.

The very small P-value for the overall test, yet the lack of significance shown in Table 5.7, is a warning signal of potential multicollinearity. Section 5.2 showed strong evidence of a width effect on the presence of satellites, yet, controlling for weight, color, and spine condition, little evidence exists of a partial width effect. Graphical exploration reveals, however, a strong linear component for the relationship between width and weight. The sample correlation between them equals 0.887. It does not make much sense to analyze an effect of width while controlling for weight, since weight naturally increases as width does.

For practical purposes, width and weight serve equally well as predictors, but it is redundant to use them both. In further analysis, we use width alone together with color and spine condition as predictors. Denote these predictors by W, C, and S. For simplicity, we symbolize various models by the highest order terms in the model, regarding C and S in the models as factors. For instance, $C + S + W$ denotes a model with main effects of the sort shown in Table 5.7, whereas $C + S * W$ denotes a model that also has an interaction between S and W.

5.5.5 Backward Elimination of Predictors

Table 5.8 summarizes results of fitting and comparing several logistic regression models. The *deviance* of a model is the G^2 test of goodness of fit based on comparing the model to the saturated model (Sections 4.5.3 and 5.3.2); the difference of deviances between two models is the likelihood-ratio statistic $-2(L_0 - L_1)$ for comparing them. To select a model, we use a *backward elimination procedure*, starting with a complex model and successively taking out terms. At each stage, we eliminate the term in the model that has the largest P-value when we test that its parameters equal zero. We test only the highest-order terms for each variable. It is inappropriate, for instance, to remove a main effect term if the model contains higher-order interactions involving that term.

Table 5.8   Results of Fitting Several Logistic Regression Models to Horseshoe Crab Data

Model	Predictors	Deviance	DF	Models Compared	Difference	$(Y, \hat{\pi})$ Correlation
(1)	$C * S * W$	170.44	152	—	—	0.526
(2)	$C * S + C * W + S * W$	173.68	155	(2)-(1)	3.2 ($df = 3$)	
(3a)	$C * S + S * W$	177.34	158	(3a)-(2)	3.7 ($df = 3$)	
(3b)	$C * W + S * W$	181.56	161	(3b)-(2)	7.9 ($df = 6$)	
(3c)	$C * S + C * W$	173.69	157	(3c)-(2)	0.0 ($df = 2$)	
(4a)	$S + C * W$	181.64	163	(4a)-(3c)	8.0 ($df = 6$)	
(4b)	$W + C * S$	177.61	160	(4b)-(3c)	3.9 ($df = 3$)	
(5)	$C + S + W$	186.61	166	(5)-(4b)	9.0 ($df = 6$)	
(6a)	$C + S$	208.83	167	(6a)-(5)	22.2 ($df = 1$)	
(6b)	$S + W$	194.42	169	(6b)-(5)	7.8 ($df = 3$)	
(6c)	$C + W$	187.46	168	(6c)-(5)	0.8 ($df = 2$)	0.452
(7a)	C	212.06	169	(7a)-(6c)	24.5 ($df = 1$)	0.285
(7b)	W	194.45	171	(7b)-(6c)	7.0 ($df = 3$)	0.402
(8)	C=dark $+ W$	187.96	170	(8)-(6c)	0.5 ($df = 2$)	0.447
(9)	None	225.76	172	(9)-(8)	37.8 ($df = 2$)	0.000

Note: C = color, S = spine condition, W = width

We begin with the most complex model, symbolized by $C * S * W$, listed as model (1) in Table 5.8. This model predicts the logit of the probability of a satellite using main effects for each term as well as the three two-factor interactions and the three-factor interaction. Removing the three-factor interaction term yields the simpler model $C * S + C * W + S * W$ containing two-factor interactions and main effects, model (2) in Table 5.8. To compare the fits, we test the hypothesis that the simpler model holds against the alternative that the more complex one holds. The likelihood-ratio statistic comparing the two models equals the difference in deviances, or 3.2 with $df = 3$. This does not suggest that the three-factor term is needed ($P = .36$), thank goodness, so we continue the simplification process.

The next stage considers the three models that remove a two-factor interaction. Of these models, $C * S + C * W$ gives essentially the same fit as the more complex model, so we drop the $S * W$ interaction from the model. Next, we consider dropping one of the other two-factor interactions. The model $S + C * W$ (i.e., dropping the $C * S$ interaction but maintaining the S main effect) has an increased deviance of 8.0 on $df = 6$ ($P = .24$); the model $W + C * S$, dropping the $C * W$ interaction, has an increased deviance of 3.9 on $df = 3$ ($P = .27$). Neither increase is important, suggesting that we can drop either one of them and proceed. In either case, dropping next the remaining interaction also seems permissible. For instance, dropping the $C * S$ interaction from model $W + C * S$, leaving the main-effects model $C + S + W$, increases the deviance by 9.0 on $df = 6$ ($P = .17$).

The working model now has the main effects alone. The next stage considers dropping one of them. Table 5.8 shows little consequence from removing S. Both remaining variables (C and W) then have nonnegligible effects. For instance, removing C increases the deviance (comparing models (7b) and (6c)) by 7.0 on $df = 3$ ($P = .07$). The analysis in Section 5.5.3 revealed a noticeable difference between dark crabs (category 4) and the others. The simpler model that has a single dummy variable for color, equaling 0 for dark crabs and 1 otherwise, fits essentially as well (the deviance difference between models (8) and (6c) equals 0.5, with $df = 2$). Further simplification results in large increases in deviance and is unjustified. As a final step, one can use the methods of Section 5.3 to check further the fit of this model.

Computerized variable selection procedures should be used with caution. When one considers a large number of terms for potential inclusion in a model, one or two of them that are not really important may look impressive simply due to chance. For instance, when all the true effects are weak, the largest sample effect may substantially overestimate its true effect. In addition, it often makes sense to include certain variables of special interest in a model and report their estimated effects even if they are not statistically significant at some level.

5.5.6 A Correlation Summary of Predictive Power*

It can be informative to compare various GLMs fitted to a data set in terms of their predictive power. The correlation R between the observed responses $\{Y_i\}$ and the model's fitted values $\{\hat{\mu}_i\}$ describes this. For least squares regression (i.e., a GLM with normal random component), R represents the *multiple correlation* between the response variable and the predictors; then R^2 describes the proportion of variation in Y that is explained by the predictors.

For logistic regression, R is the correlation between binary $Y = (0, 1)$ observations on the response and the predicted probabilities $\hat{\pi}$. For such models, R is a crude index of predictive power, and it does not have the nice properties it has for normal GLMs. For instance, R is not guaranteed to be nondecreasing as the model gets more complex. Also, like any correlation measure, its value can depend strongly on the range of observed values of explanatory variables. Nevertheless, R is useful for comparing fits of different models to the same data set.

Table 5.8 shows correlations between observed responses and estimated probabilities for a few of the models fitted to the horseshoe crab data. Width alone has $R = .402$, and adding color to the model increases R to .452. The simpler model that uses color merely to indicate whether a crab is dark does essentially as well, with $R = .447$. This model has 85% as high a correlation as the complex model containing color, spine condition, width, and all their two-way and three-way interactions ($R = .526$). Little is lost and much is gained by using the simpler model to describe the data.

To compare effects of quantitative predictors having different units in multiple logistic regression models, it can be helpful to report standardized coefficients. One can do this by fitting the model to standardized versions of the predictors, where each measurement on a predictor is replaced by its z-score, (measurement − mean)/(standard deviation). Then, each reported regression coefficient represents the effect of a one standard deviation change in a predictor, controlling for the other variables. See Problem 5.30.

5.6 SAMPLE SIZE AND POWER FOR LOGISTIC REGRESSION*

The major aim of many studies is to determine whether a particular variable X has an effect on a binary response. The study design should take into account the sample size N needed to provide a good chance of detecting an effect of a given size.

5.6.1 Sample Size for Comparing Two Proportions

When X also is binary, the study often refers to comparing two groups (i.e., the two levels of X) on the binary response. To test the hypothesis that the "success" probabilities π_1 and π_2 for the groups are identical, one might conduct a chi-squared test for the 2×2 table that cross-classifies group by response, rejecting the null hypothesis if the P-value is smaller than some fixed level, α. To determine sample size, one needs to specify the probability β of failing to detect a difference between π_1 and π_2 of some fixed size considered to be practically important. For this size of effect, β is the probability of failing to "reject H_0" at the α level. Then, α is the probability of Type I error and β is the probability of Type II error. The *power* of the test equals $1 - \beta$.

A study using equal sample sizes for each group requires approximately

$$N_1 = N_2 = \frac{(z_{\alpha/2} + z_\beta)^2 [\pi_1(1 - \pi_1) + \pi_2(1 - \pi_2)]}{(\pi_1 - \pi_2)^2}$$

in each group to achieve these error probabilities. This formula requires values for π_1 and π_2 and for the error probabilities α and β. To illustrate, for testing $H_0 : \pi_1 = \pi_2$ at the .05 level, suppose one wants a probability of Type II error equal to .10 if π_1 and π_2 are truly about .20 and .30. Then $\alpha = .05$, $\beta = .10$, the z-scores are $z_{.025} = 1.96$ and $z_{.10} = 1.28$, and the required sample sizes equal

$$N_1 = N_2 = \frac{(1.96 + 1.28)^2 [(.2)(.8) + (.3)(.7)]}{(.2 - .3)^2} = 389.$$

This formula also provides the sample sizes needed for a comparable confidence interval for $\pi_1 - \pi_2$. Then, α is the error probability for the interval and β equals the probability that the confidence interval indicates a plausible lack of effect, in the sense that it contains the value zero. For instance, based on the above calculation with $\alpha = .05$ and $\beta = .10$, one needs about 400 subjects in each group for a 95% confidence interval to have only a .10 chance of containing 0 when actually $\pi_1 = .20$ and $\pi_2 = .30$.

This sample-size formula is approximate and may underestimate slightly the actual required values. It is adequate for most practical work, though, since one normally has only rough conjectures for π_1 and π_2. Fleiss (1981) provided more precise formulas.

The null hypothesis for the test comparing two proportions in a 2×2 table corresponds to one for a parameter in a logistic regression model having form

$$\text{logit}(\pi) = \alpha^* + \beta^* x. \tag{5.6.1}$$

(We use the $*$ notation so as not to confuse the parameters with the probabilities of Type I and Type II error.) If we set $x = 1$ for Group 1 and $x = 0$ for Group 2, the logit of the success probability equals $\alpha^* + \beta^*$ for Group 1 and α^* for Group 2. Identical probabilities corresponds to identical odds and identical logits, or $\beta^* = 0$. Thus, this example relates to sample size determination for a simple logistic regression model.

5.6.2 Sample Size with a Quantitative Predictor

For models of form (5.6.1) in which x is quantitative rather than a qualitative indicator, the sample size needed to achieve a certain power for testing $H_0 : \beta^* = 0$ depends on the distribution of the x values. One needs to guess the probability of success $\bar{\pi}$ at the mean of x. The size of the effect is the odds ratio θ comparing the probability of success at that point to the probability of success one standard deviation above the mean of x. Let $\lambda = \log(\theta)$. F. Y. Hsieh (*Statist. Medic.*, 8: 795–802 (1989)) provided the approximate sample-size formula for a one-sided test,

$$N = \frac{[z_\alpha + z_\beta \exp(-\lambda^2/4)]^2(1 + 2\bar{\pi}\delta)}{\bar{\pi}\lambda^2},$$

where

$$\delta = \frac{1 + (1 + \lambda^2)\exp(5\lambda^2/4)}{1 + \exp(-\lambda^2/4)}.$$

We illustrate for modeling the dependence of the probability of severe heart disease on $x = $ cholesterol level for a middle-aged population. Consider the test of independence ($\beta^* = 0$) against the alternative $\beta^* > 0$ of increasing risk as cholesterol increases. Suppose previous studies have suggested that the probability of severe heart disease at an average level of cholesterol is about .08. Suppose we want the test to be sensitive to a 50% increase in this probability (i.e., to .12), for a standard deviation increase in cholesterol. The odds of severe heart disease at the mean cholesterol level equal $.08/.92 = 0.087$, and the odds one standard deviation above the mean equal $.12/.88 = 0.136$. The odds ratio equals $\theta = 0.136/0.087 = 1.57$, and $\lambda = \log(1.57) = 0.450$, $\lambda^2 = 0.202$. For a $\beta = .10$ chance of a Type II error in an $\alpha = .05$-level test, $z_\alpha = z_{.05} = 1.645$, $z_\beta = z_{.10} = 1.28$. Thus,

$$\delta = \frac{1 + (1.202)\exp(5 \times 0.202/4)}{1 + \exp(-0.202/4)} = \frac{2.548}{1.951} = 1.306,$$

and

$$N = \frac{[1.645 + 1.28\exp(-0.202/4)]^2(1 + 2(.08)(1.306))}{(.08)(0.202)} = 612.$$

The value N decreases as $\bar{\pi}$ gets closer to .5 and as $|\lambda|$ gets farther from the null value of 0. Its derivation assumes that the explanatory variable has approximately a normal distribution.

5.6.3 Sample Size in Multiple Logistic Regression

A multiple logistic regression model requires larger sample sizes to detect partial effects. Let R denote the multiple correlation between the predictor X of interest and the others in the model. One divides the above formula for N by $(1 - R^2)$. In that formula, $\bar{\pi}$ denotes the probability at the mean value of all the explanatory variables, and the odds ratio refers to the effect of the predictor of interest at the mean level of the others.

We illustrate by continuing the previous example. Consider a test for the effect of cholesterol on severe heart disease, while controlling for blood pressure level. If the correlation between cholesterol and blood pressure levels is .40, we need a sample size of roughly $612/[1 - (.40)^2] = 729$ for detecting the stated partial effect of cholesterol.

These formulas provide, at best, rough ballpark indications of sample size. In most applications, one has only a crude guess for indices such as $\bar{\pi}$ and R, and the explanatory variable may be far from normally distributed.

5.7 EXACT INFERENCE FOR LOGISTIC REGRESSION*

ML estimators of model parameters work best when the sample size is large compared to the number of parameters in the model. When the sample size is small, or when there are many parameters relative to the sample size, improved inference results using the method of *conditional maximum likelihood*.

5.7.1 Conditional Maximum Likelihood Inference

The conditional ML method bases inference for the primary parameters of interest on a *conditional likelihood* function that eliminates the other parameters. The technique uses a conditional probability distribution defined over data sets in which the values of certain "sufficient statistics" for the other parameters are fixed. This distribution is defined for potential samples that provide the same information about the other parameters that occurs in the observed sample. The distribution and the related conditional likelihood function depend only on the parameters of interest. The conditional ML method applies to any GLM that uses the *canonical link* (Section 4.1.3); for instance, the logit link with binomial data and the log link with Poisson data.

For binary data, conditional likelihood methods are especially useful when a logistic regression model contains a large number of "nuisance" parameters. For instance, models for matched-pairs data in matched case-control studies contain a separate parameter for each matched pair, so the number of parameters increases with the sample size. More accurate inference for the parameters of primary interest results from eliminating the nuisance parameters from the likelihood function. Section 9.2 discusses conditional ML methods for this type of model.

Conditional likelihood methods are also useful for small samples. One can perform *exact* inference for a parameter by using the conditional likelihood function that eliminates all the other parameters. Since that conditional likelihood does not involve unknown parameters, one can calculate probabilities such as P-values exactly rather than use crude approximations. This section introduces conditional ML methods for small-sample exact inference.

5.7.2 Exact Inference for Contingency Tables

Consider first the logistic regression model for a single explanatory variable,

$$\text{logit}[\pi(x)] = \alpha + \beta x.$$

When X can take only two values, the model applies to 2×2 tables for which the two rows are the levels of X and the two columns are the levels of Y. The usual sampling model treats the responses on Y at the separate x values as independent binomial variates. The row totals, which are the "numbers of trials" for those binomial variates, are naturally fixed.

For this model, the hypothesis of independence is $H_0 : \beta = 0$. The unknown parameter α refers to the relative number of response outcomes of the two types. One can eliminate α from the likelihood by conditioning also on the column marginal totals. These "sufficient statistics" for α represent the information that the data provide about α. Fixing both sets of marginal totals yields hypergeometric probabilities that do not depend on any unknown parameters. The resulting exact conditional test that $\beta = 0$ in 2×2 tables is simply Fisher's exact test (Section 2.6.1).

When X has I ordered levels, one can apply this logit model to the $I \times 2$ table by assigning scores to the rows. Again, conditioning on the column totals yields a conditional likelihood free of α, which one can use for exact inference about β. The exact test of $\beta = 0$ is an alternative to the large-sample trend test discussed in Section 2.5.5 (which is the "efficient score" test for the logit model) or the large-sample Wald or likelihood-ratio tests for β.

Next, suppose the model also contains a second explanatory factor, Z, having K levels. If Z is qualitative, a relevant model is

$$\text{logit}[\pi(x)] = \alpha + \beta x + \beta_k^Z.$$

When X is binary, this is the model presented in Section 5.4.4 for $2 \times 2 \times K$ contingency tables. The test of $H_0 : \beta = 0$ refers to the effect of X, controlling for Z. The exact test eliminates the other parameters by conditioning on the marginal totals in each of the partial tables. Section 3.3.1 discussed this exact test of conditional independence between X and Y, controlling for Z.

Computations for exact inference in logistic regression models are highly complex, but software is available. In fact, the manual for the software LogXact (Cytel Software) is a good source for discussion of this approach. When the sample size is small, the exact tests and related confidence intervals are more reliable than the ordinary large-sample ML inferences.

Exact methods are also useful when ordinary ML methods report an infinite parameter estimate. In such cases, exact methods can still yield P-values for tests and confidence intervals in which one endpoint is finite. An infinite estimate would occur in logistic regression model (5.1.1) for the effect of width on the presence of a satellite, for instance, if $Y = 0$ whenever width is 26.0 or less, and $Y = 1$ whenever width is at least 26.0. Generally, suppose there is no overlap in the sets of explanatory variable values having the two response outcomes, in the sense that a line (or plane) can pass through the data such that $Y = 1$ on one side of the line and $Y = 0$ on the other side. There is then *perfect discrimination*, as one can predict the sample outcomes perfectly by knowing the predictor values (except possibly at a boundary point). In such cases, an ML parameter estimate for the logistic regression model is infinite. Software may not always detect this, and may instead report a very large estimate with an extremely large standard error.

5.7.3 Diarrhea Example

Table 5.9 refers to a sample of 2493 patients having stays in a hospital. The response is whether they suffered from an acute form of diarrhea during their stay. The three predictors are age (1 for over 50 years old, 0 for under 50), length of stay in hospital (1 for more than one week, 0 for less than one week), and exposure to an antibiotic called Cephalexin (1 for yes, 0 for no). We discuss estimation of the effect of Cephalexin, controlling for age and length of stay, using a model containing only main effect terms.

The sample size is large, yet relatively few cases occurred of acute diarrhea. Moreover, all subjects having exposure to Cephalexin were also diarrhea cases. Such "boundary" situations in which none or all responses fall in one category cause infinite ML estimates of some model parameters. An ML estimate of ∞ for the Cephalexin effect means that the likelihood function continually increases as the parameter estimate for Cephalexin increases indefinitely. Software for logistic regression may or may not indicate this. Some software reports "no convergence," whereas other packages simply report a large value for the estimated effect of Cephalexin with an extremely large standard error (e.g., one procedure reports a ML estimate of 27.0 based on $ASE = 70{,}065.8$).

Table 5.9   Example for Exact Conditional Logistic Regression

Cephalexin	Age	Length of Stay	Cases of Diarrhea	Sample Size
0	0	0	0	385
0	0	1	5	233
0	1	0	3	789
0	1	1	47	1081
1	1	1	5	5

Source: Based on study by Dr. E. Jaffe and Dr. V. Chang, Cornell Medical Center, reported in manual for *LogXact* (Cambridge, MA: Cytel Software, 1993, p. 7-5).

To study the effect of Cephalexin, we conduct an exact analysis, conditioning on sufficient statistics for the other predictors. Though the estimate of the parameter for the effect of Cephalexin is infinite, a 95% confidence interval for the true value is $(2.95, \infty)$. Exponentiating yields a confidence interval of $(19, \infty)$ for the odds ratio between exposure to Cephalexin and the response. This suggests that subjects taking Cephalexin have odds of developing diarrhea at least 19 times the odds for subjects not taking it, controlling for age and length of stay in the hospital. Assuming that the main-effects model is valid, Cephalexin appears to have a strong effect. Similarly, we obtain $P < .0001$ for testing that the parameter for Cephalexin equals zero.

Though the confidence interval is wide, it provides information not available through ordinary ML methods. Results must be qualified somewhat, because there were no Cephalexin cases at the first three combinations of levels of age and length of stay. In fact, the first three rows of Table 5.9 make no contribution to the analysis (Problem 5.39). The data actually provide evidence about the effect of Cephalexin only for older subjects having a long stay.

PROBLEMS

5.1. For the 23 space shuttle flights that occurred before the Challenger mission disaster in 1986, Table 5.10 shows the temperature (°F) at the time of the flight and whether at least one primary O-ring suffered thermal distress.

Table 5.10

Ft	Temp	TD	Ft	Temp	TD	Ft	Temp	TD
1	66	0	9	57	1	17	70	0
2	70	1	10	63	1	18	81	0
3	69	0	11	70	1	19	76	0
4	68	0	12	78	0	20	79	0
5	67	0	13	67	0	21	75	1
6	72	0	14	53	1	22	76	0
7	73	0	15	67	0	23	58	1
8	70	0	16	75	0			

Note: Ft = flight no., Temp = temperature, TD = thermal distress (1 = yes, 0 = no).
Source: Data based on Table 1 in S. R. Dalal, E. B. Fowlkes, and B. Hoadley. *J. Amer. Statist. Assoc, 84*: 945–957 (1989). Reprinted with permission of the American Statistical Association.

a. Use logistic regression to model the effect of temperature on the probability of thermal distress. Interpret the model fit.

b. Calculate the predicted probability of thermal distress at 31°, the temperature at the time of the Challenger flight. At what temperature does the predicted probability equal .5? At that temperature, give a linear approximation for the change in the predicted probability per degree increase in temperature.

    c. Interpret the effect of temperature on the odds of thermal distress. Test the hypothesis that temperature has no effect, using (i) the Wald test, (ii) the likelihood-ratio test.

5.2. In the first nine decades of the twentieth century in baseball's National League, the percentage of times the starting pitcher pitched a complete game were: 72.7 (1900–1909), 63.4, 50.0, 44.3, 41.6, 32.8, 27.2, 22.5, 13.3 (1980–1989) (*Source:* George Will, *Newsweek*, April 10, 1989).

    a. Use a logistic regression fit to describe the time trend in these data. (For simplicity, suppose the number of games was the same in each decade.) Use the fitted model to predict the percentage of complete games for each of the next three decades.

    b. Repeat the analysis using a linear probability model. Are its future predictions realistic?

5.3. Refer to Table 4.1. Using scores $\{0, 2, 4, 5\}$ for levels of snoring, fit the logistic regression model. Interpret using fitted probabilities, linear approximations, and effects on the odds. Analyze goodness of fit.

5.4. Refer to Table 4.6. Use logistic regression to analyze these data.

5.5. Hastie and Tibshirani (1990, p. 282) described a study to determine risk factors for kyphosis, severe forward flexion of the spine following corrective spinal surgery. The age in months at the time of the operation for the 18 subjects for whom kyphosis was present were 12, 15, 42, 52, 59, 73, 82, 91, 96, 105, 114, 120, 121, 128, 130, 139, 139, 157 and for 22 of the subjects for whom kyphosis was absent were 1, 1, 2, 8, 11, 18, 22, 31, 37, 61, 72, 81, 97, 112, 118, 127, 131, 140, 151, 159, 177, 206.

    a. Fit a logistic regression model using age as a predictor of whether kyphosis is present. Test whether age has a significant effect.

    b. Plot the data. Note the difference in dispersion on age at the two levels of kyphosis.

    c. Fit the model $\text{logit}[\pi(x)] = \alpha + \beta_1 x + \beta_2 x^2$. Test the significance of the squared age term, plot the fit, and interpret.

5.6. Refer to Table 2.5. To investigate the association, one can use a logistic regression model, but interchange roles of response and explanatory variables.

    a. Fit the model, treating party identification as nominal. Note that it is saturated. (The likelihood-ratio test for an absence of a party effect in this model is identical to the G^2 test of independence in Section 2.4.)

    b. Fit a model that treats party identification as ordinal with equally spaced scores, and use the model to test independence. Interpret. Show that the model implies that the odds ratio for the first two columns is identical to that for the last two columns. Report this estimate, and interpret.

5.7. Table 5.11 contains results of a case-control study on the relationship between smoking and myocardial infarction (MI). The sample consisted of young and middle-aged women admitted to 30 coronary units in northern Italy and controls admitted to the same hospitals with other acute disorders. The table classifies

Table 5.11

Cigarettes/Day	Cases	Controls
None	90	346
1–14	57	91
15–24	65	48
≥ 25	40	18

Source: A. Gramenzi et al., *J. Epidemiol. and Commun. Health, 43*: 214–217 (1989).
Reprinted with permission of BMJ Publishing Group.

the cases and the controls by their smoking histories, measured in terms of average number of cigarettes per day. Using scores $\{0, 7.5, 19.5, 30\}$ for smoking level, use logistic regression to describe and make inferences about the effect. Obtain Pearson or adjusted residuals for the model fit. Do they show any lack of fit?

5.8. Refer to Table 4.2. Using weight as the predictor, fit the logistic regression model for the probability of a satellite.

a. Report and plot the ML fit, and find predicted probabilities at the values 1.20, 2.44, and 5.20 kg of weight, which are the sample minimum, mean, and maximum.

b. Find the weight at which the predicted probability equals .5. At that value, give a linear approximation for the estimated effect of a 1-kg increase in weight. This represents a relatively large increase, so interpret in terms of (i) a 0.1-kg increase, (ii) a standard deviation increase in weight.

c. Provide an odds interpretation for the association. Construct a 95% confidence interval to describe the effect of weight on the odds of a satellite.

d. Using a Wald test, find the P-value for the hypothesis that weight has no effect. Perform also a likelihood-ratio test for the effect. Interpret.

5.9. Refer to the previous problem. By categorizing weight, test whether the model is adequate for predicting the presence of satellites. Obtain Pearson or adjusted residuals, and interpret. Obtain influence diagnostics, and interpret.

5.10. For Table 4.2, fit a logistic regression model for the probability of a satellite, using color alone as the predictor.

a. Treat color as qualitative (nominal). Conduct a likelihood-ratio test of the hypothesis that color has no effect. Compare the result to the test for color conducted in Section 5.5.2 for the model that also contains width. Explain how the relation between the tests is analogous to that for a continuous response between a test for a factor in one-way ANOVA and in analysis of covariance.

b. Treating color in a quantitative manner, obtain a prediction equation. Interpret the fit, and test the hypothesis that color has no effect.

c. Test goodness of fit of the model in **(b)**. Interpret, and use Pearson or adjusted residuals to describe any lack of fit.

     d. The data are binomial responses at four color levels. Explain why the model in (**a**) is saturated. Express its parameter estimates in terms of the sample logits for each color.

5.11. A recent article (D. J. Moritz and W. A. Satariano, *J. Clin. Epidemiol., 46*: 443–454 (1993)) used multiple logistic regression to predict whether the stage of breast cancer at diagnosis was advanced or was local for a sample of 444 middle-aged and elderly women. A table referring to a particular set of demographic factors reported the estimated odds ratio for the effect of living arrangement (three categories) as 2.02 for spouse versus alone and 1.71 for others versus alone; it reported the effect of income (three categories) as 0.72 for $10,000–24,999 versus < $10,000 and 0.41 for $25,000+ versus < $10,000. Interpret these results, and estimate the odds ratios for the third pair of categories for each factor.

5.12. Fit model (5.4.1) to Table 5.5.

     a. For black veterans without immediate AZT use, obtain the predicted probabilities and fitted values for (yes, no) AIDS symptoms.

     b. Construct a 95% confidence interval for the conditional odds ratio between AZT use and the development of symptoms.

     c. Describe and test for the effect of race in this model.

5.13. Table 5.12 appeared in a national study of 15 and 16 year-old adolescents. The event of interest is ever having sexual intercourse. Analyze these data, including description and inference about the effects of gender and race, goodness-of-fit and residual analyses, and summary interpretations.

Table 5.12

		Intercourse	
Race	Gender	Yes	No
White	Male	43	134
	Female	26	149
Black	Male	29	23
	Female	22	36

Source: S. F. Morgan and J. D. Teachman, *J. Marriage & Fam., 50*: 929–936 (1988). Reprinted with permission of The National Council on Family Relations.

5.14. The U. S. National Collegiate Athletic Association (NCAA) conducted a study of graduation rates for student athletes who were freshmen during the 1984–1985 academic year. Table 5.13 shows the data. Analyze and interpret.

5.15. According to the *Independent* newspaper (London, March 8, 1994), the Metropolitan Police in London reported 30,475 people as missing in the year ending March 1993. For those of age 13 or less, 33 of 3271 missing males and 38 of 2486 missing females were still missing a year later. For ages 14–18, the values were 63 of 7256 males and 108 of 8877 females; for ages 19 and

Table 5.13

Group	Sample Size	Graduates
White females	796	498
White males	1625	878
Black females	143	54
Black males	660	197

Source: J. J. McArdle and F. Hamagami, *J. Am. Statist. Assoc.*, *89*: 1107–1123 (1994).
Reprinted with permission of the American Statistical Association.

above, the values were 157 of 5065 males and 159 of 3520 females. Analyze and interpret these data. (Thanks to Dr. P. M. E. Altham for showing me these data.)

5.16. Refer to Table 3.1.

  a. Fit model (5.4.1). Interpret the parameter estimates using conditional odds ratios.

  b. Test goodness of fit. Interpret.

  c. Test the effect of defendant's race on the death penalty, controlling for victims' race. Interpret.

5.17. Refer to Table 3.3.

  a. Fit model (5.4.3). Conduct Pearson and likelihood-ratio tests of goodness of fit, and interpret. Since the tests provide tests of equality of odds ratios, compare results to those given by the Breslow–Day test of Section 3.2.4.

  b. Check Pearson or adjusted residuals to analyze further the quality of fit. Interpret.

  c. For this model, conduct the likelihood-ratio test of conditional independence of smoking and lung cancer, given site. Estimate the conditional odds ratio between smoking and lung cancer. Interpret.

  d. Compare results of (**c**) to those obtained in Section 3.2 using *CMH* methods.

5.18. Refer to Table 3.5. Analyze these data using logistic regression, with lung cancer as the response. Can one fit a linear probability model or a probit model to these data? Explain.

5.19. Refer to Table 3.6. Using logistic regression, conduct analogous analyses to those in Problem 3.10. Compare results.

5.20. Refer to Problem 3.1. Analyze these data using logistic regression, treating death penalty as the response. Interpret results.

5.21. In a study designed to evaluate whether an educational program makes sexually active adolescents more likely to obtain condoms, adolescents were randomly assigned to two experimental groups. The educational program, involving a lecture and videotape about transmission of the HIV virus, was provided to one group but not the other. In logistic regression models, factors observed to influence a teenager to obtain condoms were gender, socioeconomic status, lifetime number of partners, and the experimental group. Study results are summarized in Table 5.14.

Table 5.14

Variables	Odds Ratio	95% Confidence Interval
Group (Education vs. None)	4.04	(1.17, 13.9)
Gender (Males vs. Females)	1.38	(1.23, 12.88)
SES (High vs. Low)	5.82	(1.87, 18.28)
Lifetime No. of Partners	3.22	(1.08, 11.31)

Source: V. I. Rickert et al., *Clin. Pediatrics 31*: 205–210 (1992).

   a. Interpret the odds ratio and the related confidence interval for the effect of group.
   b. Find the parameter estimates for the fitted model, using $(1, 0)$ dummy variables for the first three predictors. Based on the corresponding confidence interval for the log odds ratio, determine the standard error for the group effect.
   c. Explain why either the estimate of 1.38 for the odds ratio for gender or the corresponding confidence interval is incorrect. Show that if the reported interval is correct, then 1.38 is actually the *log* odds ratio, and the estimated odds ratio equals 3.98.

5.22. Table 5.15 refers to results of a case-control study about effects of cigarette smoking and coffee drinking on myocardial infarction (MI) for a sample of men under 55 years of age.

Table 5.15

Cups Coffee per Day	Cigarettes per Day							
	0		1–24		25–34		≥ 35	
	Cases	Controls	Cases	Controls	Cases	Controls	Cases	Controls
0	66	123	30	52	15	12	36	13
1–2	141	179	59	45	53	22	69	25
3–4	113	106	63	65	55	16	119	30
≥ 5	129	80	102	58	118	44	373	85

Source: L. Rosenberg et al., *Am. J. Epidemiol., 128*: 570–578 (1988).

   a. Fit a logit model, treating coffee drinking and cigarette smoking as qualitative factors. Interpret effects, and conduct likelihood-ratio tests of those effects.
   b. Fit another model, treating the ordinal predictors as quantitative by assigning category scores. Interpret, and conduct tests of the predictor effects.
   c. Use goodness-of-fit statistics to check the model fits in (**a**) and (**b**). Conduct residual analyses to describe lack of fit. Test whether there is a significant difference in the fits of the models that treat the predictors as nominal and ordinal. Interpret.

5.23. A sample of subjects were asked their opinion about current laws legalizing abortion (support, oppose). For the explanatory variables gender (female, male), religious affiliation (Protestant, Catholic, Jewish), and political party affiliation (Democrat, Republican, Independent), the model for the probability π of supporting legalized abortion,

$$\text{logit}(\pi) = \alpha + \beta_h^G + \beta_i^R + \beta_j^P,$$

has reported parameter estimates $\hat{\alpha} = 0.62$, $\hat{\beta}_1^G = 0.08$, $\hat{\beta}_2^G = -0.08$, $\hat{\beta}_1^R = -0.16$, $\hat{\beta}_2^R = -0.25$, $\hat{\beta}_3^R = 0.41$, $\hat{\beta}_1^P = 0.87$, $\hat{\beta}_2^P = -1.27$, $\hat{\beta}_3^P = 0.40$.
   a. Interpret how the odds of supporting legalized abortion depend on gender.
   b. Find the estimated probability of supporting legalized abortion for (i) Male Catholic Republicans, (ii) Female Jewish Democrats.
   c. Redefining parameters using constraints $\beta_1^G = \beta_1^R = \beta_1^P = 0$, report estimates of the other parameters, and explain how to interpret them.
5.24. Table 5.16 shows estimated effects for a fitted logistic regression model with squamous cell esophageal cancer ($Y = 1$, yes; $Y = 0$, no) as the response variable. Smoking status (S) equals 1 for at least one pack per day and 0 otherwise, alcohol consumption (A) equals the average number of alcoholic drinks consumed per day, and race (R) equals 1 for blacks and 0 for whites.

Table 5.16

Variable	Effect	P-value
Intercept	−7.00	<.01
Alcohol use	0.10	.03
Smoking	1.20	<.01
Race	0.30	.02
Race × Smoking	0.20	.04

   a. To describe the race-by-smoking interaction, construct the prediction equation when $R = 1$ and again when $R = 0$. Find the fitted Y-S conditional odds ratio for each case. Similarly, construct the prediction equation when $S = 1$ and again when $S = 0$. Find the fitted Y-R conditional odds ratio for each case. Note that, for each association, the coefficient of the cross-product term is the difference between the log odds ratios at the two fixed levels for the other variable.
   b. In Table 5.16, explain what the coefficients of R and S represent, for the coding as given above. What hypotheses do the P-values refer to for these variables?
   c. Suppose the model also contained an A-R interaction term, with coefficient 0.04. In the prediction equation, show that this represents the difference between the effect of A for blacks and for whites.
5.25. Refer to model (5.5.2). Explain how this model is analogous to the analysis of covariance model with no interaction for a continuous response, whereby the

slope effect of a quantitative predictor is the same at all levels of a qualitative factor. Fit the model, using weight instead of width as x. Interpret the parameter estimates. Controlling for weight, test the hypothesis that the probability of a satellite is independent of color. Interpret.

5.26. For the horseshoe crab data with color and width as predictors (Section 5.5.1), fit the logistic regression model permitting interaction.

   a. Report the prediction equations relating width to the probability of a satellite, for each of the four colors. Plot them, and interpret.

   b. Test for a difference between the slope parameters for medium-dark and dark crabs.

   c. Test whether the interaction model gives a better fit than the simpler model lacking the interaction terms.

5.27. Using models that treat color in a quantitative manner, repeat the analyses in (i) Problem 5.25, (ii) Problem 5.26c.

5.28. Refer to model (5.5.2). At width $= 20$ cm, find the predicted probabilities of a satellite for the colors medium dark and dark. Find the odds for each, and show that the odds ratio equals $\exp(1.106) = 3.02$. When each probability is close to zero, the odds ratio is similar to the ratio of probabilities, providing another interpretation for logistic regression parameters. For widths at which it is small, the predicted probability for medium-dark crabs is about three times that for dark crabs.

5.29. For the horseshoe crab data, fit a model using weight and width as predictors.

   a. Report the prediction equation. Interpret effects.

   b. Conduct a likelihood-ratio test of the hypothesis $H_0 : \beta_1 = \beta_2 = 0$ that neither variable affects the response. Interpret.

   c. Conduct separate tests for the partial effects of each variable. Why does neither test show evidence of an effect when the test in (**b**) does?

5.30. Refer to the prediction equation $\text{logit}(\hat{\pi}) = -10.071 - 0.509c + 0.458x$, using width and quantitative color for the horseshoe crab data. Color has a mean of 2.44 and standard deviation of 0.80, and width has a mean of 26.30 and standard deviation of 2.11. For standardized versions of the predictors, explain why the estimated coefficients equal $(0.80)(-.509) = -0.41$ and $(2.11)(.458) = 0.97$. Interpret these by comparing the partial effects on the odds of a one standard deviation increase in each predictor.

5.31. For the horseshoe crab data, use backward elimination to select a good model, when weight, spine condition, and color are the predictors for the probability of a satellite.

5.32. An alternative goodness-of-fit test when at least one predictor is continuous partitions values for the explanatory variables into a set of regions and adds a dummy variable to the model for each region. The test statistic compares the fit of this model to the simpler one, testing that the extra parameters are not needed. Doing this for model (5.5.2) by partitioning according to the eight width regions in Table 5.2, show that the likelihood-ratio statistic for testing that the extra parameters are unneeded equals 7.5, based on $df = 7$.

5.33. Refer to Problem 6.12 and Table 6.16 in Chapter 6. Treating opinion about premarital sex as the response variable, use backward elimination to select a model. Interpret that model.

5.34. Refer to Table 7.1, treating marijuana use as the response variable. Analyze these data using logit models. Interpret estimated effects.

5.35. About how large a sample is needed to test the hypothesis of equal proportions so that P(Type II error) $= .05$ when $\pi_1 = .40$ and $\pi_2 = .60$, if the hypothesis is rejected when the P-value is less than .01?

5.36. We expect two proportions to be about .20 and .30, and we want an 80% chance of detecting a difference between them using a 90% confidence interval. Assuming equal sample sizes, how large should they be? Compare results to the sample sizes required for a 90% interval with power 90%, a 95% interval with power 80%, and a 95% interval with power 90%.

5.37. The horseshoe crab width values in Table 4.2 have a mean of 26.3 and standard deviation of 2.1. If the true relationship were similar to the fitted equation reported in Section 5.1.2, how large a sample yields P(Type II error) $= .10$ in an $\alpha = .05$-level test of independence against the alternative of a positive effect of width on the probability of a satellite?

5.38. Refer to Table 2.7.

   a. Fit the logistic regression model using scores $\{0, .5, 1.5, 4, 7\}$ for alcohol consumption. Check goodness of fit.

   b. Test independence, using the likelihood-ratio test and the Wald test for the model in **(a)**. (*Note:* The trend test of Section 2.5 is the "efficient score" test for this model.)

   c. Conduct the tests of independence using scores $\{1, 2, 3, 4, 5\}$. Compare results to **(b)**, noting that results for highly unbalanced data can be sensitive to the choice of scores.

   d. Check the sensitivity of the choice of model by repeating the analyses of **(a)**–**(c)** using a different link function. Compare results to those obtained using the logit link.

   e. The table has some small counts, and exact methods have greater validity than large-sample ones. Show that the exact test of independence using the scores in **(a)** has one-sided P-value of .016.

5.39. Refer to Table 5.9. Apply exact logistic regression to the model discussed in Section 5.7.3 for Cephalexin Use (*C*), Age (*A*), Length of Stay (*L*), and Diarrhea Outcome (*D*).

   a. Obtain an exact P-value for testing the hypothesis of no *C* effect against the alternative of a positive effect, and construct a 95% "exact" confidence interval for the conditional *C–D* odds ratio.

   b. Construct the four partial tables relating *C* to *D*, for the combinations of levels of (A, L). Note that three of the tables have no data for the row with

$C = 1$. For the sole partial table having observations at both levels of C, find a 95% "exact" confidence interval for the odds ratio, and find an exact one-sided P-value. Compare to results using the entire data set. Comment about the contribution to inference of tables having only a single positive row total or a single positive column total.

c. Obtain the ordinary ML fit of the logistic regression model. To investigate the sensitivity of the estimated C effect, find the change in the estimate and standard error after adding one observation to the data set, a case with no diarrhea when $(C, A, L) = (1, 1, 1)$.

5.40. Refer to Table 7.10 in Chapter 7. Applying software for exact logistic regression, obtain exact P-values for testing the effects of each predictor in the multiple logistic regression model. Interpret results.

5.41. Refer to the logistic regression model (5.1.1). Show that e^α equals the odds of success when $X = 0$. Construct the odds of success when $X = 1, X = 2$, and $X = 3$. Use this to provide an interpretation of β. Generalize these results to the multiple logistic regression model (5.5.1).

5.42. The slope of the line drawn tangent to the probit regression curve at a particular x value equals $(.40)\beta \exp[-(\alpha + \beta x)^2/2]$. Show this is highest when $x = -\alpha/\beta$, where it equals 0.40β. At this point, $\pi(x) = \frac{1}{2}$. For the fit of the probit model to the crab data using width as a predictor, where does the predicted probability of a satellite equal $\frac{1}{2}$? Find the rate of change in the probability per 1-cm increase in width at that point. Compare results to those obtained with logistic regression.

5.43. The logistic regression curve (5.1.1) has the shape of the cdf of a logistic distribution with mean $\mu = -\alpha/\beta$ and standard deviation $\sigma = 1.814/(|\beta|)$. Show that for the ungrouped horseshoe crab data with width as a predictor, the curve has the shape of a logistic cdf with mean 24.8 and standard deviation 3.6. Since about 95% of a bell-shaped distribution occurs within two standard deviations of the mean, argue that the probability of a satellite increases from near 0 to near 1 as width increases from about 17 to 32 cm.

5.44. Refer to the models discussed in Section 5.3.2 for the crab data. Show that the value of L_S, and hence $G^2(M_0)$ and $G^2(M_1)$, depends on whether the data are entered as individual binary observations or summary binomial counts. However, L_0 and L_1, and hence $G^2(M_0 \mid M_1)$, do not depend on this choice.

CHAPTER 6

Loglinear Models for Contingency Tables

Section 4.3 introduced loglinear models as generalized linear models (GLMs) for Poisson-distributed data. Their most common use is the modeling of cell counts in contingency tables. The models specify how the size of a cell count depends on the levels of the categorical variables for that cell. The nature of this specification relates to the association and interaction structure among the variables. Loglinear models describe association and interaction patterns among a set of categorical variables.

Section 6.1 introduces loglinear models for two-way contingency tables, and Section 6.2 extends them to three-way tables. Section 6.3 discusses statistical inference for model parameters and the checking of model adequacy. Section 6.4 extends results to higher-way tables.

When one variable is a binary response and the others are explanatory variables, logit models for that response variable are equivalent to certain loglinear models. Section 6.5 presents the connection between loglinear and logit models and discusses model selection. We shall see that loglinear models are mainly of use when at least two variables in a contingency table are response variables; when there is a single response, it is simpler and more natural to use logit models.

6.1 LOGLINEAR MODELS FOR TWO-WAY TABLES

Consider an $I \times J$ contingency table that cross-classifies n subjects on two categorical responses. When those variables are statistically independent, the joint probabilities $\{\pi_{ij}\}$ for the cells in that table are determined by the row and column marginal totals,

$$\pi_{ij} = \pi_{i+} \pi_{+j}, \qquad i = 1, \ldots, I, \quad j = 1, \ldots, J.$$

The related expression for the expected frequencies $\{\mu_{ij} = n\pi_{ij}\}$ is $\mu_{ij} = n\pi_{i+}\pi_{+j}$ for all i and j. These formulas refer to cell probabilities $\{\pi_{ij}\}$, which are parameters for binomial and multinomial distributions. Loglinear model formulas use $\{\mu_{ij}\}$ rather than $\{\pi_{ij}\}$, so they also apply to Poisson sampling for cell counts with expectations $\{\mu_{ij}\}$.

6.1.1   Independence Model

Denote the row variable by X and the column variable by Y. The condition $\mu_{ij} = n\pi_{i+}\pi_{+j}$ of independence between X and Y is multiplicative. Taking the log of both sides of the equation yields an additive relation. Namely, $\log \mu_{ij}$ depends additively on a term based on the sample size, a term based on the probability in row i, and a term based on the probability in column j. Thus, independence has form

$$\log \mu_{ij} = \lambda + \lambda_i^X + \lambda_j^Y, \tag{6.1.1}$$

whereby the log expected frequency is an additive function of a row effect λ_i^X and a column effect λ_j^Y (The X and Y superscripts are simply labels, not "power" exponents). This model is called the *loglinear model of independence* for two-way contingency tables.

The parameter λ_i^X represents the effect of classification in row i for variable X. The larger the value of λ_i^X, the larger each expected frequency is in row i of the table. When $\lambda_h^X = \lambda_i^X$, each expected frequency in row h equals the corresponding expected frequency in row i. Similarly, the parameter λ_j^Y represents the effect of classification in column j for variable Y.

The null hypothesis of independence between two categorical variables is simply the hypothesis that this loglinear model holds. The fitted values that satisfy the model are $\{\hat{\mu}_{ij} = n_{i+}n_{+j}/n\}$, the estimated expected frequencies for chi-squared tests of independence. Chi-squared tests using X^2 and G^2 ((2.4.3) in Section 2.4) are also goodness-of-fit tests of this loglinear model.

6.1.2   Belief in Afterlife Example

Table 6.1 shows a 2×2 table of cell counts and the corresponding fitted values $\{\hat{\mu}_{ij}\}$ for the loglinear model of independence (6.1.1). The cell counts are taken from Table 2.1, a cross classification of X = gender and Y = belief in the afterlife. The fitted values are also the estimated expected frequencies for testing independence.

Since the fitted values satisfy independence, they have an odds ratio of 1. They are close to the observed counts, and the goodness-of-fit statistics for testing that the independence loglinear model holds equal $X^2 = 0.2$ and $G^2 = 0.2$, with $df = 1$. These are simply the test statistics (2.4.3) for testing the hypothesis of independence in two-way tables. The independence model is plausible for these data.

6.1.3   Interpretation of Parameters

Loglinear models for contingency tables are examples of generalized linear models. For $I \times J$ tables, this GLM treats the $N = IJ$ cell counts as N independent observations of a Poisson random component. For loglinear GLMs, the data are the N cell counts

Table 6.1   Loglinear Parameter Estimates for the Independence Model Fitted to Table 2.1, Relating Belief in the Afterlife (Columns) to Gender (Rows)

Observed Frequency		Fitted Value		Log Fitted Value	
435	147	432.10	149.90	6.069	5.010
375	134	377.90	131.10	5.935	4.876

Parameter	Set 1	Set 2	Set 3
λ	4.876	6.069	5.472
λ_1^X	0.134	0	0.067
λ_2^X	0	−0.134	−0.067
λ_1^Y	1.059	0	0.529
λ_2^Y	0	−1.059	−0.529

rather than the individual classifications of the n subjects. The expectations $\{\mu_{ij}\}$ of the cell counts are linked to the explanatory terms using the log link.

As formula (6.1.1) illustrates, loglinear models for contingency tables do not distinguish between response and explanatory classification variables. They treat all variables jointly as responses, modeling the cell count for all combinations of their levels. Most studies, however, identify some classifications as response variables and others as explanatory factors. This can influence the interpretation of parameters and, as Section 6.5 shows, the choice of model.

Parameter interpretation is simplest for binary responses. For instance, consider the independence model (6.1.1) for $I \times 2$ tables, with the two columns being levels of a response Y. In row i, the logit for the probability π that $Y = 1$ equals

$$\log \left(\frac{\pi}{1 - \pi} \right) = \log \left(\frac{\mu_{i1}}{\mu_{i2}} \right) = \log \mu_{i1} - \log \mu_{i2}$$

$$= (\lambda + \lambda_i^X + \lambda_1^Y) - (\lambda + \lambda_i^X + \lambda_2^Y)$$

$$= \lambda_1^Y - \lambda_2^Y.$$

The final term is a constant that does not depend on i; that is, the logit for Y does not depend on the level of X. The loglinear model of independence corresponds to the simple model having form logit$(\pi) = \alpha$, whereby the logit for the probability π of response in column 1 takes the same value in every row. In each row, the odds of response in column 1 equal $\exp(\alpha) = \exp(\lambda_1^Y - \lambda_2^Y)$. The chance of classification in a particular column is the same in all rows.

In loglinear model (6.1.1), differences between two parameters for a given variable relate to the log odds of making one response, relative to another, on that variable. Of course, when there is a single response variable that is binary, one can apply logit models directly, and loglinear models are unneeded. Loglinear models are primarily useful for modeling relationships among two or more categorical response variables.

6.1.4 Parameter Constraints

We now discuss parameter identification and interpretation for Table 6.1. That table also exhibits the logs of the fitted values for the independence model. The $\{\log \hat{\mu}_{ij}\}$ satisfy formula (6.1.1) with estimated parameter values $\hat{\lambda} = 4.876$, $\hat{\lambda}_1^X = 0.134$, $\hat{\lambda}_2^X = 0$, $\hat{\lambda}_1^Y = 1.059$, $\hat{\lambda}_2^Y = 0$. For instance, $\hat{\mu}_{12} = 149.90$, and from (6.1.1), $\log \hat{\mu}_{12} = \hat{\lambda} + \hat{\lambda}_1^X + \hat{\lambda}_2^Y = 4.876 + 0.134 + 0 = 5.010 = \log(149.90)$. The model fit also satisfies other possible values for the estimated parameters, such as the lower panel of Table 6.1 exhibits. There is no unique way of specifying the parameters or their estimates, and Table 6.1 shows three possible sets of estimates.

For the independence model, one $\{\lambda_i^X\}$ parameter is redundant, and one $\{\lambda_j^Y\}$ parameter is redundant. This is analogous to ANOVA and multiple regression models with factors, which require one fewer dummy variable than the number of factor levels. Section 5.4.3 showed ways of eliminating the redundancy. One can set the parameter for the last level of each factor equal to 0, as in the first set of estimates in Table 6.1. Or, one can set the parameter for the first level of each factor equal to 0, as in the second set of estimates. Another approach lets the parameters for each factor sum to 0; for instance, for 2×2 tables,

$$\lambda_1^X + \lambda_2^X = 0 \quad \text{and} \quad \lambda_1^Y + \lambda_2^Y = 0,$$

as in the third set. The choice of constraints is arbitrary. Some software for loglinear models, such as SAS-GENMOD and GLIM, set a parameter equal to 0; others, such as SAS-CATMOD, have parameters sum to zero.

What *is* unique is the *difference* between two main effect parameters of a particular type. For instance, for each set of estimates in Table 6.1, $\hat{\lambda}_1^Y - \hat{\lambda}_2^Y = 1.059$. For each gender, the estimated log odds of response in column 1 instead of column 2 equals 1.059; that is, the estimated odds of belief in the afterlife equal $\exp(1.059) = 2.9$. The estimated probability of belief in the afterlife equals $\exp(1.059)/[1 + \exp(1.059)] = 0.74$.

6.1.5 Saturated Model

Variables that are statistically dependent rather than independent satisfy a more complex loglinear model,

$$\log \mu_{ij} = \lambda + \lambda_i^X + \lambda_j^Y + \lambda_{ij}^{XY}. \tag{6.1.2}$$

The $\{\lambda_{ij}^{XY}\}$ parameters are association terms that reflect deviations from independence of X and Y. The parameters represent interactions between X and Y, whereby the effect of one variable on the expected cell count depends on the level of the other variable. This model describes perfectly any set of expected frequencies. Called the *saturated loglinear model*, it is the most general model for two-way contingency tables. The independence model (6.1.1) is the special case in which all $\lambda_{ij}^{XY} = 0$.

Direct relationships exist between log odds ratios and the $\{\lambda_{ij}^{XY}\}$ association parameters. For instance, the saturated model for 2×2 tables has log odds ratio

$$
\begin{aligned}
\log \theta = \log \left(\frac{\mu_{11}\mu_{22}}{\mu_{12}\mu_{21}} \right) &= \log \mu_{11} + \log \mu_{22} - \log \mu_{12} - \log \mu_{21} \\
&= (\lambda + \lambda_1^X + \lambda_1^Y + \lambda_{11}^{XY}) + (\lambda + \lambda_2^X + \lambda_2^Y + \lambda_{22}^{XY}) \\
&\quad - (\lambda + \lambda_1^X + \lambda_2^Y + \lambda_{12}^{XY}) - (\lambda + \lambda_2^X + \lambda_1^Y + \lambda_{21}^{XY}) \\
&= \lambda_{11}^{XY} + \lambda_{22}^{XY} - \lambda_{12}^{XY} - \lambda_{21}^{XY}.
\end{aligned} \tag{6.1.3}
$$

Thus, $\{\lambda_{ij}^{XY}\}$ determine the log odds ratio. When these parameters equal zero, the log odds ratio is zero. The odds ratio then equals 1, and X and Y are independent.

The observed counts in Table 6.1 have an odds ratio of $(435 \times 134)/(147 \times 375) = 1.057$, and a log odds ratio of $\log(1.057) = 0.056$. Table 6.2 shows possible association parameter estimates that have a log odds ratio of

$$
\hat{\lambda}_{11}^{XY} + \hat{\lambda}_{22}^{XY} - \hat{\lambda}_{12}^{XY} - \hat{\lambda}_{21}^{XY} = 0.056. \tag{6.1.4}
$$

One can specify association parameters so that the ones in the last row and in the last column are zero, or so that the ones in the first row and in the first column are zero. Or, one can let the sum of the parameters equal 0 within each row and within each column. The combination that is identical for each potential definition is (6.1.4).

In $I \times J$ tables, only $(I-1)(J-1)$ association parameters are nonredundant. These "interaction" parameters in the saturated model are coefficients of cross-products of $(I-1)$ dummy variables for X with $(J-1)$ dummy variables for Y. Tests of independence analyze whether these $(I-1)(J-1)$ parameters equal zero, so they have residual $df = (I-1)(J-1)$. When $I = J = 2$, as in Table 6.1, a single nonredundant parameter determines the odds ratio.

The saturated model (6.1.2) has a single constant parameter (λ), $(I-1)$ nonredundant parameters of form λ_i^X, $(J-1)$ nonredundant parameters of form λ_j^Y, and $(I-1)(J-1)$ nonredundant parameters of form λ_{ij}^{XY}. The total number of nonredundant parameters equals $1 + (I-1) + (J-1) + (I-1)(J-1) = IJ$. The model has as many parameters as it has Poisson observations, which is why it gives a perfect fit. It has the maximum possible number of parameters, and in this sense the model is said to be *saturated*.

Table 6.2 Equivalent Association Parameter Estimates for Saturated Loglinear Model

Association Parameter	Set 1	Set 2	Set 3
λ_{11}^{XY}	0.056	0.014	0.0
λ_{12}^{XY}	0	−0.014	0.0
λ_{21}^{XY}	0	−0.014	0.0
λ_{22}^{XY}	0	0.014	0.056

In practice, we try to analyze data using unsaturated models, since their fit smooths the sample data and yields simpler interpretations. For three-way and higher-dimensional tables, unsaturated models can include association terms. Then, loglinear models are more commonly used to describe associations (through two-factor terms) than to describe odds (through single-factor terms). After fitting loglinear models, we convert estimates of two-factor parameters to estimates of conditional odds ratios between pairs of variables.

Model (6.1.2) and others used in this text are called *hierarchical models*. This means that the model includes all lower-order terms composed from variables contained in a higher-order term in the model. When the model contains λ_{ij}^{XY}, for instance, it also contains λ_i^X and λ_j^Y. Similarly, when an ordinary regression or logistic regression model contains an interaction term (form $\beta x_1 x_2$) for two predictors or a quadratic term (form βx_1^2), it also contains the main effect terms (form $\beta_1 x_1$ and $\beta_2 x_2$). A reason for including lower-order terms is that, otherwise, the statistical significance and practical interpretation of a higher-order term depends on how the variables are coded. This is undesirable, and with hierarchical models one gets the same results no matter how variables are coded.

When a model has two-factor terms, one should be cautious in interpreting the lower-order single-factor terms. By analogy with two-way ANOVA, when there is two-factor interaction, it can be misleading to report main effects. The estimates of the main effect terms depend on the coding scheme used for the higher-order effects, and the interpretation also depends on that scheme. Normally, we restrict our attention to the highest-order terms for a variable, as the next section illustrates.

6.2 LOGLINEAR MODELS FOR THREE-WAY TABLES

We next introduce loglinear models for three-way contingency tables. Different models have different independence and association patterns. A good-fitting model helps us describe those patterns and estimate the strength of pairwise odds ratios.

6.2.1 Loglinear Models and Independence Structure

Denote the cell expected frequencies in the contingency table by $\{\mu_{ijk}\}$. Single-factor terms in loglinear models for $\{\mu_{ijk}\}$ represent marginal distributions. For instance, including λ_i^X in a model forces the fitted values $\{\hat{\mu}_{ijk}\}$ to have the same totals at the various levels of X as do the observed data. The information of primary use in loglinear models relates to higher-order parameters. Partial associations between variables correspond to two-factor terms.

For instance, consider loglinear model

$$\log \mu_{ijk} = \lambda + \lambda_i^X + \lambda_j^Y + \lambda_k^Z + \lambda_{ik}^{XZ} + \lambda_{jk}^{YZ}. \tag{6.2.1}$$

Since it contains an X-Z two-factor term, it permits association between X and Z, controlling for Y. This model also permits a Y-Z association, controlling for X. It does not contain a two-factor term for an *X-Y* association. This loglinear model specifies conditional independence between X and Y, controlling for Z.

We symbolize this loglinear model of X-Y conditional independence by (XZ, YZ). The symbol lists the highest-order terms in the model for each variable. This model is an important one. It holds, for instance, if an association between X and Y disappears when we control for Z. For $2 \times 2 \times K$ tables, this model corresponds to the hypothesis tested using the Cochran–Mantel–Haenszel statistic in Section 3.2.

Simpler models than (6.2.1) that delete other two-factor association terms are too simple to fit most data sets well. For instance, the model that contains only single-factor terms, denoted by (X, Y, Z), is called the *mutual independence model*. It treats each pair of variables as independent. When variables are chosen wisely for a study, this model is rarely appropriate.

A model that permits all three pairs of variables to be conditionally dependent is

$$\log \mu_{ijk} = \lambda + \lambda_i^X + \lambda_j^Y + \lambda_k^Z + \lambda_{ij}^{XY} + \lambda_{ik}^{XZ} + \lambda_{jk}^{YZ}. \qquad (6.2.2)$$

For this model, the next subsection shows that conditional odds ratios between any two variables are identical at each level of the third variable. This is the property of homogeneous association (Section 3.1.5). We refer to this loglinear model as the *homogeneous association model* and symbolize it by (XY, XZ, YZ).

6.2.2 Interpreting Model Parameters

The most general loglinear model for three-way tables is

$$\log \mu_{ijk} = \lambda + \lambda_i^X + \lambda_j^Y + \lambda_k^Z + \lambda_{ij}^{XY} + \lambda_{ik}^{XZ} + \lambda_{jk}^{YZ} + \lambda_{ijk}^{XYZ}. \qquad (6.2.3)$$

The three-factor term pertains to three-factor interaction. Denoted by (XYZ), this model permits the odds ratio between any two variables to vary across levels of the third variable. This is the saturated model for a three-way table, and it provides a perfect fit. The homogeneous association model (6.2.2) that deletes the three-factor term is often called the *no three-factor interaction model*. The right-hand side of (6.2.3) has the same structure as models for a three-way factorial ANOVA. The single-factor terms are main effects, and the two-factor interactions represent partial associations.

Interpretations of loglinear model parameters refer to their highest-order terms. For instance, interpretations for the homogeneous association model (6.2.2) use the two-factor terms to describe associations. In their presence, the main effect estimates are irrelevant; they can take dramatically different values depending on the parameter constraints for the higher-order terms. The two-factor parameters relate directly to conditional odds ratios, such as those introduced in Section 3.1.3 for three-way tables.

To illustrate, for $2 \times 2 \times K$ tables, consider model (XY, XZ, YZ) or any simpler model containing the *X-Y* term. The X-Y conditional odds ratio $\theta_{XY(k)}$ describes association between X and Y in the kth partial table. Using an argument similar to

Table 6.3   Alcohol (A), Cigarette (C), and Marijuana (M) Use for High School Seniors

Alcohol	Cigarette	Marijuana Use	
Use	Use	Yes	No
Yes	Yes	911	538
	No	44	456
No	Yes	3	43
	No	2	279

Source: I am grateful to Prof. Harry Khamis, Wright State University, for supplying these data.

that in Section 6.1.5, one can show that

$$\log \theta_{XY(k)} = \log \left(\frac{\mu_{11k}\mu_{22k}}{\mu_{12k}\mu_{21k}} \right) = \lambda_{11}^{XY} + \lambda_{22}^{XY} - \lambda_{12}^{XY} - \lambda_{21}^{XY}. \tag{6.2.4}$$

The right-hand side of this expression does not depend on k, so the odds ratio is the same at every level of Z. Similarly, model (XY, XZ, YZ) also has equal X-Z odds ratios at different levels of Y, and equal Y-Z odds ratios at different levels of X. Any model not having the three-factor interaction term has homogeneous association structure.

6.2.3   Alcohol, Cigarette, and Marijuana Use Example

Table 6.3 refers to a survey conducted in 1992 by the Wright State University School of Medicine and the United Health Services in Dayton, Ohio (Thanks to Professor Harry Khamis for providing these data). Among other things, the survey asked students in their final year of high school in a nonurban area near Dayton, Ohio whether they had ever used alcohol, cigarettes, or marijuana. Denote the variables in this $2 \times 2 \times 2$ table by A for alcohol use, C for cigarette use, and M for marijuana use.

Widely available software can fit loglinear models. Our discussion emphasizes interpretive aspects rather than calculations. Table 6.4 shows fitted values for several

Table 6.4   Fitted Values for Loglinear Models Applied to Table 6.3

Alcohol	Cigarette	Marijuana	Loglinear Model				
Use	Use	Use	(A, C, M)	(AC, M)	(AM, CM)	(AC, AM, CM)	(ACM)
Yes	Yes	Yes	540.0	611.2	909.24	910.4	911
		No	740.2	837.8	438.84	538.6	538
	No	Yes	282.1	210.9	45.76	44.6	44
		No	386.7	289.1	555.16	455.4	456
No	Yes	Yes	90.6	19.4	4.76	3.6	3
		No	124.2	26.6	142.16	42.4	43
	No	Yes	47.3	118.5	0.24	1.4	2
		No	64.9	162.5	179.84	279.6	279

Table 6.5 Estimated Odds Ratios for Loglinear Models in Table 6.4

Model	Conditional Association			Marginal Association		
	A-C	A-M	C-M	A-C	A-M	C-M
(A, C, M)	1.0	1.0	1.0	1.0	1.0	1.0
(AC, M)	17.7	1.0	1.0	17.7	1.0	1.0
(AM, CM)	1.0	61.9	25.1	2.7	61.9	25.1
(AC, AM, CM)	7.8	19.8	17.3	17.7	61.9	25.1
(ACM) Level 1	13.8	24.3	17.5	17.7	61.9	25.1
(ACM) Level 2	7.7	13.5	9.7			

loglinear models. The fit for model (AC, AM, CM) is close to the observed data, which are the fitted values for model (ACM). The other models seem to fit poorly.

Table 6.5 illustrates association patterns for these models by presenting estimated odds ratios for their marginal and conditional associations. For example, the entry 1.0 for the A-C conditional association for model (AM, CM) is the common value of the A-C fitted odds ratios at the two levels of M,

$$1.0 = \frac{909.24 \times 0.24}{45.76 \times 4.76} = \frac{438.84 \times 179.84}{555.16 \times 142.16}.$$

This model implies conditional independence between alcohol use and cigarette use, controlling for marijuana use, and yields odds ratios of 1.0 for the A-C conditional association. The entry 2.7 for the A-C marginal association for this model is the fitted odds ratio for the marginal A-C fitted table,

$$2.7 = \frac{(909.24 + 438.84)(0.24 + 179.84)}{(45.76 + 555.16)(4.76 + 142.16)}.$$

The odds ratios for the observed data are those reported for the saturated model (ACM), which provides a perfect fit.

Model (AC, AM, CM) permits all pairwise associations but maintains homogeneous odds ratios between two variables at each level of the third variable. The A-C fitted conditional odds ratios for this model equal 7.8; for each level of M, students who have smoked cigarettes have estimated odds of having drunk alcohol that are 7.8 times the estimated odds for students who have not smoked cigarettes. One can calculate this odds ratio using the model's fitted values at either level of M, or (from (6.2.4)) using the parameter estimates to construct $\exp(\hat{\lambda}_{11}^{AC} + \hat{\lambda}_{22}^{AC} - \hat{\lambda}_{12}^{AC} - \hat{\lambda}_{21}^{AC})$. The A-C marginal odds ratio of 17.7 ignores the third factor (M), whereas the conditional odds ratio of 7.8 controls for it.

Table 6.5 shows that estimated conditional odds ratios equal 1.0 for each pairwise term not appearing in a model, such as the A-C association in model (AM, CM). For that model, the estimated marginal A-C odds ratio differs from 1.0; Section 3.1.4 noted that conditional independence does not imply marginal independence. Some models have conditional associations that are necessarily the same as the corresponding marginal associations. Section 7.1.2 presents a condition guaranteeing this. This equality does not normally happen for loglinear models containing all pairwise associations.

Table 6.5 shows that estimates of conditional and marginal odds ratios are highly dependent on the model. This highlights the importance of good model selection. An estimate from this table is informative only to the extent that its model fits well. The next section shows how to check model goodness of fit.

6.3 INFERENCE FOR LOGLINEAR MODELS

A good-fitting loglinear model provides a basis for describing and making inferences about the true association structure among a set of categorical response variables. We now present ways of checking the goodness of fit of loglinear models and conducting inference about model parameters.

6.3.1 Fitting Loglinear Models

Some loglinear models have explicit formulas for fitted values $\{\hat{\mu}_{ijk}\}$ in terms of the sample cell counts $\{n_{ijk}\}$. The estimates are then said to be *direct*. For example, the model (XZ, YZ) of X-Y conditional independence has $\hat{\mu}_{ijk} = n_{i+k}n_{+jk}/n_{++k}$. The fitted value for a cell in partial table k equals its row total n_{i+k} times its column total n_{+jk}, divided by the sample size n_{++k} for that partial table. This equals the fitted value for the model of independence applied to that partial table alone.

Many loglinear models do not have direct estimates. ML estimation then requires iterative methods, such as the Newton–Raphson algorithm (Section 4.5.1). In practice, it is unimportant to know which models have direct estimates. Iterative methods for models not having direct estimates also work for models having them. Most software packages use iterative routines for all cases.

6.3.2 Chi-Squared Goodness-of-Fit Tests

Consider the null hypothesis that the expected frequencies for a three-way table satisfy a given loglinear model. As usual, large-sample chi-squared statistics assess model goodness of fit by comparing the cell fitted values to the observed counts. The likelihood-ratio and Pearson statistics are

$$G^2 = 2 \sum n_{ijk} \log \left(\frac{n_{ijk}}{\hat{\mu}_{ijk}} \right), \qquad X^2 = \sum \frac{(n_{ijk} - \hat{\mu}_{ijk})^2}{\hat{\mu}_{ijk}}.$$

The degrees of freedom equal the number of cell counts minus the number of nonredundant parameters in the model. The df value decreases as the model gets more complex. The saturated model has $df = 0$.

For the Dayton drug survey (Table 6.3), Table 6.6 presents results of tests of fit for several loglinear models. For a given df, larger G^2 or X^2 values have smaller P-values

Table 6.6   Goodness-of-Fit Tests for Loglinear Models Relating Alcohol (A), Cigarette (C), and Marijuana (M) Use

Model	G^2	X^2	df	P-value[a]
(A, C, M)	1286.0	1411.4	4	$< .001$
(A, CM)	534.2	505.6	3	$< .001$
(C, AM)	939.6	824.2	3	$< .001$
(M, AC)	843.8	704.9	3	$< .001$
(AC, AM)	497.4	443.8	2	$< .001$
(AC, CM)	92.0	80.8	2	$< .001$
(AM, CM)	187.8	177.6	2	$< .001$
(AC, AM, CM)	0.4	0.4	1	.54
(ACM)	0.0	0.0	0	—

[a]P-value for G^2 statistic.

and indicate poorer fits. The models that lack any of the two-way association terms fit poorly, having P-values below .001. The model (AC, AM, CM) that permits all pairwise associations but assumes no three-factor interaction fits well ($P = .54$).

6.3.3   Loglinear Residuals

One can inspect quality of fit more closely by studying cell residuals. They can show why a model fits poorly, perhaps suggesting an alternative model, or highlight cells that display lack of fit in a generally good-fitting model. When a table has many cells, some residuals may be large purely by chance, of course.

Section 2.4.5 introduced adjusted residuals for the independence model, and Section 4.4.3 discussed them for arbitrary Poisson GLMs. They divide differences between observed and fitted counts by their standard errors. When the model holds, adjusted residuals have approximately a standard normal distribution. Thus, absolute values of adjusted residuals larger than about 2 when there are few cells or 3 when there are many cells indicate lack of fit.

Table 6.7 shows adjusted residuals for the fit of the model (AM, CM) of A-C conditional independence to Table 6.3. Since this model corresponds to independence between A and C at each level of M, one can calculate these by applying formula (2.4.4) for two-way tables separately for the partial table at each level of M. This model has $df = 2$ for testing fit, and the two nonredundant residuals refer to checking A-C independence at each level of M. The large residuals reflect the overall poor fit (In fact, X^2 is related to the two nonredundant residuals by $X^2 = (3.70)^2 + (12.80)^2 = 177.6$). Extremely large residuals occur for students who have not smoked marijuana. For them, the positive residuals occur when A and C are both "yes" or both "no." More of these students have used both or neither of alcohol and cigarettes than one would expect if their usage were independent. The same pattern persists for students who have smoked marijuana, but the differences between observed and fitted counts are then not as striking.

Table 6.7   Adjusted Residuals for Two Loglinear Models

Drug Use			Observed Count	Model (AM, CM)		Model (AC, AM, CM)	
A	C	M		Fitted Count	Adjusted Residual	Fitted Count	Adjusted Residual
Yes	Yes	Yes	911	909.2	3.70	910.4	0.63
		No	538	438.8	12.80	538.6	−0.63
	No	Yes	44	45.8	−3.70	44.6	−0.63
		No	456	555.2	−12.80	455.4	0.63
No	Yes	Yes	3	4.8	−3.70	3.6	−0.63
		No	43	142.2	−12.80	42.4	0.63
	No	Yes	2	0.2	3.70	1.4	0.63
		No	279	179.8	12.80	279.6	−0.63

Table 6.7 also shows adjusted residuals for model (AC, AM, CM). Since $df = 1$ for this model, only one residual value is nonredundant. Both G^2 and X^2 are small, so it is not surprising that these residuals indicate a good fit (In fact, when $df = 1$, X^2 is simply the square of each adjusted residual). Pearson residuals (Section 4.4.3) provide similar conclusions. For Poisson loglinear models, they have form $e_{ijk} = (n_{ijk} - \hat{\mu}_{ijk})/\sqrt{\hat{\mu}_{ijk}}$.

6.3.4   Tests about Partial Associations

One can conduct tests about partial associations by comparing different loglinear models. For instance, for model (AC, AM, CM), the null hypothesis of no partial association between alcohol use and cigarette smoking states that the λ^{AC} term equals zero. The test analyzes whether the simpler model (AM, CM) of A-C conditional independence holds, against the alternative that model (AC, AM, CM) holds.

The likelihood-ratio test of the hypothesis that a loglinear model term equals zero is simple. The likelihood-ratio statistic $-2(L_0 - L_1)$ is identical to the difference between the goodness-of-fit G^2 statistics (i.e., the deviances) for the model without that term and the model containing the term. The df equal the difference between the corresponding df values. Section 5.3.2 presented this procedure for logistic regression modeling.

The test statistic for testing that $\lambda^{AC} = 0$ in model (AC, AM, CM) is the difference

$$G^2[(AM, CM) \mid (AC, AM, CM)] = G^2(AM, CM) - G^2(AC, AM, CM)$$

between $G^2 = 187.8$ ($df = 2$) for model (AM, CM) and $G^2 = 0.4$ ($df = 1$) for model (AC, AM, CM). The difference of 187.4 is based on $df = 2 - 1 = 1$ ($P < .001$). The small P-value provides strong evidence against the null hypothesis and in favor of an A-C partial association. The statistics comparing models (AC, CM) and (AC, AM) with model (AC, AM, CM) also provide strong evidence of A-M and C-M partial associations. Further analyses of Table 6.3 should use model (AC, AM, CM) rather than any of the simpler models.

The sample size can strongly influence results of any inferential procedure. One is more likely to detect an effect of given size as the sample size increases. This suggests a cautionary remark. For small sample sizes, reality may be much more complex than indicated by the simplest model that passes a goodness-of-fit test. By contrast, for large sample sizes, Section 6.4.4 shows that statistically significant effects can be weak and unimportant. This remark reflects limitations of hypothesis testing. A more relevant concern is whether the difference between true parameters and null hypothesis values is large enough to be substantively important. Estimation, particularly confidence intervals, is more useful than significance tests for assessing the importance of effects.

6.3.5 Confidence Intervals for Odds Ratios

ML estimators of loglinear model parameters have large-sample normal distributions. Software for loglinear models reports the estimates and their standard errors. For models in which the highest-order terms are two-factor associations, the estimates refer to conditional log odds ratios. One can use the estimates and their standard errors to construct confidence intervals for true log odds ratios and then exponentiate them to form intervals for odds ratios.

To illustrate, we estimate the conditional odds ratio between alcohol use and cigarette use for Table 6.3, assuming that model (AC, AM, CM) holds. Software that sets redundant association parameters in the last row and the last column equal to zero (such as PROC GENMOD in SAS) reports $\hat{\lambda}_{11}^{AC} = 2.054$, with $ASE = 0.174$. For that approach, the lone nonzero term equals the estimated conditional log odds ratio. Software for which parameters sum to zero across levels of each index (such as CATMOD in SAS) reports $\hat{\lambda}_{11}^{AC} = \hat{\lambda}_{22}^{AC} = 0.514$ and $\hat{\lambda}_{12}^{AC} = \hat{\lambda}_{21}^{AC} = -0.514$, with $ASE = 0.044$ for each of the four terms. For that approach, from (6.2.4), the estimated conditional log odds ratio equals

$$\hat{\lambda}_{11}^{AC} + \hat{\lambda}_{22}^{AC} - \hat{\lambda}_{12}^{AC} - \hat{\lambda}_{21}^{AC} = 2.054 = 4\hat{\lambda}_{11}^{AC}.$$

Similarly, the ASE of that conditional log odds ratio estimate equals four times the ASE reported for $\hat{\lambda}_{11}^{AC}$, or 0.174.

Thus, regardless of the parameter constraints, the estimated A-C conditional log odds ratio equals 2.054, with $ASE = 0.174$. A 95% confidence interval for the true conditional log odds ratio is $2.054 \pm 1.96(0.174)$ or $(1.71, 2.39)$, yielding $(e^{1.71}, e^{2.39}) = (5.5, 11.0)$ for the odds ratio. There is a strong positive association between cigarette use and alcohol use, both for users and nonusers of marijuana.

For model (AC, AM, CM), the 95% confidence intervals are $(8.0, 49.2)$ for the A-M conditional odds ratio and $(12.5, 23.8)$ for the C-M conditional odds ratio. The intervals are wide, but these associations also are strong. In summary, estimation using this loglinear model reveals strong conditional associations for each pair of drugs; there is a strong tendency for users of one drug to be users of a second drug, and this is true both for users and for nonusers of the third drug. Inspection in Table 6.5 of the estimated marginal associations for this model shows that they are even stronger; controlling for outcome on one drug moderates the association somewhat between the other two drugs.

The analyses in this section pertain to association structure. A different type of analysis pertains to comparing marginal distributions, for instance to determine if there is more usage of one drug than the others. Section 9.1 discusses this type of analysis.

6.4   LOGLINEAR MODELS FOR HIGHER DIMENSIONS

Loglinear models for three-way tables are more complex than for two-way tables, because of the variety of potential association patterns. Basic concepts for loglinear models with three-way tables extend readily, however, to multi-way tables.

6.4.1   Four-Way Tables

We illustrate models for higher dimensions using a four-way table with variables W, X, Y, and Z. Interpretations are simplest when the model has no three-factor interaction terms. Such models are special cases of the model denoted by (WX, WY, WZ, XY, XZ, YZ). This model has homogenous association structure. Each pair of variables is conditionally dependent, with the same odds ratios at each combination of levels of the other two variables. An absence of a two-factor term implies conditional independence for those variables. Model (WX, WY, WZ, XZ, YZ), for instance, does not contain an X-Y term, so it treats X and Y as conditionally independent at each combination of levels of W and Z.

A variety of models exhibit three-factor interaction. A model could contain any of four possible terms: WXY, WXZ, WYZ, or XYZ. The XYZ term permits the association between any pair of those three variables to vary across levels of the third variable, at each fixed level of W. The saturated model contains all these terms plus a four-factor interaction term.

6.4.2   Automobile Accident Example

Table 6.8 refers to observations of 68,694 passengers in autos and light trucks involved in accidents in the state of Maine in 1991. The table classifies passengers by gender (G), location of accident (L), seat-belt use (S), and injury (I). Table 6.8 reports the sample proportion of passengers who were injured. For each G-L combination, the proportion of injuries was about halved for passengers wearing seat belts.

Table 6.9 displays tests of fit for several loglinear models. To investigate the complexity of model needed, we consider model (G, I, L, S) containing only single-factor terms, model (GI, GL, GS, IL, IS, LS) containing those and all the two-factor terms, and model (GIL, GIS, GLS, ILS) containing those and all the three-factor terms. Model (G, I, L, S), which implies mutual independence of the four variables, fits very poorly $(G^2 = 2792.8, df = 11)$. Model (GI, GL, GS, IL, IS, LS) fits much better $(G^2 = 23.4, df = 5)$, but still has lack of fit $(P < .001)$. Model (GIL, GIS, GLS, ILS)

Table 6.8 Injury (*I*) by Gender (*G*), Location (*L*), and Seat Belt Use (*S*), with Fit of Models (*GI*, *GL*, *GS*, *IL*, *IS*, *LS*) and (*GLS*, *GI*, *IL*, *IS*)

Gender	Location	Seat Belt	Injury		(*GI*, *GL*, *GS*, *IL*, *IS*, *LS*)		(*GLS*, *GI*, *IL*, *IS*)		Sample Prop. Yes
			No	Yes	No	Yes	No	Yes	
Female	Urban	No	7287	996	7166.4	993.0	7273.2	1009.8	.12
		Yes	11587	759	11748.3	721.3	11632.6	713.4	.06
	Rural	No	3246	973	3353.8	988.8	3254.7	964.3	.23
		Yes	6134	757	5985.5	781.9	6093.5	797.5	.11
Male	Urban	No	10381	812	10471.5	845.1	10358.9	834.1	.07
		Yes	10969	380	10837.8	387.6	10959.2	389.8	.03
	Rural	No	6123	1084	6045.3	1038.1	6150.2	1056.8	.15
		Yes	6693	513	6811.4	518.2	6697.6	508.4	.07

Source: I am grateful to Dr. Cristanna Cook, Medical Care Development, Augusta, Maine, for supplying these data.

Table 6.9 Goodness-of-Fit Tests for Loglinear Models Relating Injury (I), Gender (G), Location (L), and Seat-Belt Use (S)

Model	G^2	df	P-value
(G, I, L, S)	2792.8	11	$< .0001$
(GI, GL, GS, IL, IS, LS)	23.4	5	$< .001$
(GIL, GIS, GLS, ILS)	1.3	1	.25
(GIL, GS, IS, LS)	18.6	4	.001
(GIS, GL, IL, LS)	22.8	4	$< .001$
(GLS, GI, IL, IS)	7.5	4	.11
(ILS, GI, GL, GS)	20.6	4	$< .001$

seems to fit well ($G^2 = 1.3$, $df = 1$), but is quite complex and difficult to interpret. This suggests studying models that are more complex than (GI, GL, GS, IL, IS, LS) but simpler than (GIL, GIS, GLS, ILS). We do this in the next subsection, but first we analyze model (GI, GL, GS, IL, IS, LS).

Table 6.8 displays the fitted values for model (GI, GL, GS, IL, IS, LS). Since the model contains no three-factor interaction terms, it assumes homogeneous conditional odds ratios for each pair of variables. Table 6.10 reports the model-based estimates of these odds ratios. One can obtain them directly using the fitted values for partial tables relating two variables at any combination of levels of the other two; to illustrate, for the I-S odds ratio, females in urban accidents have a value of $(7166.4)(721.3)/(993.0)(11748.3) = 0.44$. The log odds ratios also follow directly from loglinear parameter estimates; for instance, $\log(0.44) = -0.814 = \hat{\lambda}^{IS}_{11} + \hat{\lambda}^{IS}_{22} - \hat{\lambda}^{IS}_{12} - \hat{\lambda}^{IS}_{21}$.

Since the sample size is large, the estimates of odds ratios are quite precise. For instance, the ASE of the estimated I-S conditional log odds ratio is 0.028. A 95% confidence interval for the true log odds ratio is $-0.814 \pm 1.96(0.028)$, or $(-0.868, -0.760)$. This translates to a confidence interval of $(0.42, 0.47)$ for the odds ratio. This model predicts that the odds of injury for passengers wearing seat belts were less than half the odds for passengers not wearing them, for each gender–location combination. The fitted odds ratios in Table 6.10 also suggest that, other factors being fixed, injury was more likely in rural than urban accidents and more likely for females than males; also, the estimated odds that males used seat belts are only 0.63 times the estimated odds for females.

6.4.3 Three-Factor Interaction

Interpretations are more complicated when a model contains three-factor interaction terms. Table 6.9 shows results of adding a single three-factor interaction term to model (GI, GL, GS, IL, IS, LS). Of the four possible models, (GLS, GI, IL, IS) appears to fit best. Table 6.8 also displays its fit. Given the large sample size, its G^2 value suggests that it fits quite well.

For model (GLS, GI, IL, IS), each pair of variables is conditionally dependent, and at each level of I the association between G and L or between G and S or between L and

S varies across the levels of the remaining variable. For this model, it is inappropriate to try to interpret the *G-L*, *G-S*, and *L-S* two-factor terms on their own. For instance, one would not convert $\{\hat{\lambda}^{GS}\}$ to a fitted *G-S* odds ratio, because the presence of the *GLS* three-factor interaction term implies that the *G-S* odds ratio varies across the levels of *L*. Since *I* does not occur in a three-factor interaction, the conditional odds ratio between *I* and each variable is the same at each combination of levels of the other two variables. The top portion of Table 6.10 reports the fitted odds ratios for the *G-I*, *I-L*, and *I-S* associations.

When a model has a three-factor interaction term but no term of higher-order than that, one can study the interaction by calculating fitted odds ratios between two variables at each level of the third. One can do this at any levels of remaining variables not involved in the interaction. The bottom portion of Table 6.10 illustrates this for model (*GLS, GI, IL, IS*). For instance, the fitted *G-S* odds ratio of 0.66 for (*L* = urban) refers to four fitted values for urban accidents, both the four with (injury = no) and the four with (injury = yes); that is,

$$0.66 = (7273.2)(10,959.2)/(11632.6)(10,358.9)$$

$$= (1009.8)(389.8)/(713.4)(834.1).$$

This model permits the association between any two of *G, L*, and *S* to vary across levels of the third. Model (*GI, GL, GS, IL, IS, LS*) forces such odds ratios to be identical, as Table 6.10 shows.

6.4.4 Large Samples and Statistical versus Practical Significance

Model (*GLS, GI, IL, IS*) seems to fit much better than (*GI, GL, GS, IL, IS, LS*), the difference in G^2 values of $23.4 - 7.5 = 15.9$ being based on $df = 5 - 4 = 1$ ($P = .0001$). The fitted odds ratios in Table 6.10, however, show that the degree of three-factor interaction is weak. The fitted odds ratio between any two of *G, L*, and *S* is similar at both levels of the third variable. The significantly better fit of model

Table 6.10   Estimated Conditional Odds Ratios for Two Loglinear Models

Odds Ratio	Loglinear Model	
	(*GI, GL, GS, IL, IS, LS*)	(*GLS, GI, IL, IS*)
G-I	0.58	0.58
I-L	2.13	2.13
I-S	0.44	0.44
G-L (*S* = no)	1.23	1.33
G-L (*S* = yes)	1.23	1.17
G-S (*L* = urban)	0.63	0.66
G-S (*L* = rural)	0.63	0.58
L-S (*G* = female)	1.09	1.17
L-S (*G* = male)	1.09	1.03

(GLS, GI, IL, IS) mainly reflects the enormous sample size. Though the three-factor interaction is weak, it is significant because the large sample provides small standard errors.

A statistically significant effect need not be important in a practical sense. With huge samples, it is crucial to focus on estimation rather than hypothesis testing. For instance, a comparison of fitted odds ratios for the two models in Table 6.10 suggests that the simpler model (GI, GL, GS, IL, IS, LS) is adequate for practical purposes. Simpler models are easier to summarize. One should not use goodness-of-fit tests alone to select a final model.

For a table of arbitrary dimension with cell counts $\{n_i = np_i\}$ and fitted values $\{\hat{\mu}_i = n\hat{\pi}_i\}$, one can summarize the closeness of a model fit to the sample data by the *dissimilarity index*,

$$D = \sum |n_i - \hat{\mu}_i|/2n = \sum |p_i - \hat{\pi}_i|/2.$$

This index takes values between 0 and 1, with smaller values representing a better fit. It represents the proportion of sample cases that must move to different cells in order for the model to achieve a perfect fit.

The dissimilarity index D estimates a corresponding index Δ that describes model lack-of-fit in the population sampled. The value $\Delta = 0$ occurs when the model holds perfectly. In that case D overestimates Δ, substantially so for small samples, because of sampling variation. In practice, models rarely (if ever) hold perfectly, so $\Delta > 0$. When the model does not hold, for sufficiently large n, the goodness-of-fit statistics G^2 and X^2 will be large, showing lack of fit. The estimator D then reveals whether the lack of fit suggested by those statistics is important in a practical sense. A value of D less than about .03 suggests that the sample data follow the model pattern quite closely, even though the model is not "perfect."

For Table 6.8, model (GI, GL, GS, IL, IS, LS) has $D = .008$, and model (GLS, GI, IL, IS) has $D = .003$. These values are very small. For either model, moving less than 1% of the data yields a perfect fit. The relatively large value of G^2 for model (GI, GL, GS, IL, IS, LS) indicated that the model does not truly hold. Nevertheless, the small value for D suggests that, in practical terms, the model provides a decent fit. This is also suggested by the correlation between the observed cell counts and the model's fitted values, which equals .9998.

6.5 THE LOGLINEAR–LOGIT CONNECTION

Loglinear models for contingency tables do not distinguish between response and explanatory variables. In essence, they treat all variables as response variables. Logit models, on the other hand, describe how a binary response depends on a set of explanatory variables. Though the model types seem distinct, strong connections exist between them. For a loglinear model, one can construct logits for one response to help interpret the model. Moreover, logit models with categorical explanatory variables have equivalent loglinear models.

6.5.1 Using Logit Models to Interpret Loglinear Models

To understand implications of a loglinear model formula, it can help to form a logit that treats one variable as a response and the others as explanatory. We illustrate with the loglinear model of homogeneous association in three-way tables,

$$\log \mu_{ijk} = \lambda + \lambda_i^X + \lambda_j^Y + \lambda_k^Z + \lambda_{ij}^{XY} + \lambda_{ik}^{XZ} + \lambda_{jk}^{YZ}. \tag{6.5.1}$$

Suppose Y is binary, and we treat it as a response and X and Z as explanatory. Let π denote the probability that $Y = 1$, which depends on the levels of X and Z. The logit for Y is

$$\text{logit}(\pi) = \log\left(\frac{\pi}{1 - \pi}\right) = \log\left(\frac{P(Y = 1 \mid X = i, Z = k)}{P(Y = 2 \mid X = i, Z = k)}\right)$$

$$= \log\left(\frac{\mu_{i1k}}{\mu_{i2k}}\right) = \log(\mu_{i1k}) - \log(\mu_{i2k})$$

$$= \left(\lambda + \lambda_i^X + \lambda_1^Y + \lambda_k^Z + \lambda_{i1}^{XY} + \lambda_{ik}^{XZ} + \lambda_{1k}^{YZ}\right)$$

$$\quad - \left(\lambda + \lambda_i^X + \lambda_2^Y + \lambda_k^Z + \lambda_{i2}^{XY} + \lambda_{ik}^{XZ} + \lambda_{2k}^{YZ}\right)$$

$$= \left(\lambda_1^Y - \lambda_2^Y\right) + \left(\lambda_{i1}^{XY} - \lambda_{i2}^{XY}\right) + \left(\lambda_{1k}^{YZ} - \lambda_{2k}^{YZ}\right).$$

The first parenthetical term is a constant; that is, it does not depend on i or k. The second parenthetical term depends on the level i of X. The third parenthetical term depends on the level k of Z. The logit has the additive form

$$\text{logit}(\pi) = \alpha + \beta_i^X + \beta_k^Z. \tag{6.5.2}$$

Section 5.4.3 discussed this model, in which the logit depends on the level of X and the level of Z in an additive manner. The effect of X on the logit is the same at each level of Z, and the effect of Z is the same at each level of X. Additivity on the logit scale is the standard definition of "no interaction" for categorical variables. When Y is binary, the loglinear model of homogeneous association (no three-factor interaction) is equivalent to this logit model.

When X is also binary, logit model (6.5.2) and loglinear model (XY, XZ, YZ) are characterized by equal odds ratios between X and Y at each of the K levels of Z. This is the condition tested by the Breslow–Day test (Section 3.2.4). The G^2 or X^2 goodness-of-fit statistics for these models provide alternative ways of testing for a common odds ratio. When the sample size is large relative to K, they also have approximate chi-squared distributions with $df = K - 1$.

6.5.2 Accident Data Revisited

For the data on Maine auto accidents (Table 6.8), Section 6.4.3 showed that loglinear model (GLS, GI, LI, IS) fits well. That model has form

$$\log \mu_{gils} = \lambda + \lambda_g^G + \lambda_i^I + \lambda_l^L + \lambda_s^S + \lambda_{gi}^{GI} + \lambda_{gl}^{GL} + \lambda_{gs}^{GS} + \lambda_{il}^{IL} + \lambda_{is}^{IS} + \lambda_{ls}^{LS} + \lambda_{gls}^{GLS}. \quad (6.5.3)$$

For these data, one could treat injury (I) as a response variable and gender (G), location (L), and seat-belt use (S) as explanatory variables. Let π denote the probability of injury. Forming logit(π) at each combination of levels of G, L, and S, one can show that loglinear model (6.5.3) is equivalent to logit model

$$\text{logit}(\pi) = \alpha + \beta_g^G + \beta_l^L + \beta_s^S. \quad (6.5.4)$$

Here, G, L, and S all affect I, but without interacting. The parameters in the two models are related by $\beta_g^G = \lambda_{g1}^{GI} - \lambda_{g2}^{GI}$, $\beta_l^L = \lambda_{1l}^{IL} - \lambda_{2l}^{IL}$, and $\beta_s^S = \lambda_{1s}^{IS} - \lambda_{2s}^{IS}$. In the logit calculation, all terms in the loglinear model not having the injury index i in the subscript cancel.

Odds ratios relate to two-factor loglinear parameters and main-effect logit parameters. For instance, in logit model (6.5.4), the log odds ratio for the effect of S on I equals $\beta_1^S - \beta_2^S$. This equals $\lambda_{11}^{IS} + \lambda_{22}^{IS} - \lambda_{12}^{IS} - \lambda_{21}^{IS}$ in the loglinear model. These values are the same no matter how one's software sets up constraints for the nonredundant parameters. For Table 6.8, for instance, $\hat{\beta}_1^S - \hat{\beta}_2^S = -0.817$ for logit model (6.5.4), and $\hat{\lambda}_{11}^{IS} + \hat{\lambda}_{22}^{IS} - \hat{\lambda}_{12}^{IS} - \hat{\lambda}_{21}^{IS} = -0.817$ for loglinear model (GLS, GI, LI, IS). We obtain the same results whether we use software for logit models or software for the equivalent loglinear model. Fitted values, goodness-of-fit statistics, residual df, and adjusted residuals for logit model (6.5.4) are identical to those in Tables 6.8 to 6.10 for loglinear model (GLS, GI, IL, IS).

Loglinear models are GLMs that treat the 16 cell counts in Table 6.8 as outcomes of 16 independent Poisson variates. Logit models are GLMs that treat the table as binomial counts. Logit models with I as the response treat the marginal G-L-S table $\{n_{g+ls}\}$ as fixed and regard $\{n_{g1ls}\}$ for $\{(g, l, s) = (1, 1, 1), (1, 1, 2), (1, 2, 1), (1, 2, 2), (2, 1, 1), (2, 1, 2), (2, 2, 1), (2, 2, 2)\}$ as eight independent binomial variates on that response. Though the sampling models differ, the results from fits of corresponding models are identical.

6.5.3 Correspondence Between Loglinear and Logit Models

Refer back to the derivation of logit model (6.5.2) from loglinear model (XY, XZ, YZ). The λ_{ik}^{XZ} term in model (XY, XZ, YZ) cancels when we form the logit. It might seem as if the model (XY, YZ) omitting this term is also equivalent to that logit model. Indeed, forming the logit on Y for loglinear model (XY, YZ) results in a logit model of the same form. However, the loglinear model that has the same fit as the logit model is the one containing a general interaction term for relationships among the explanatory variables. The logit model does not describe relationships among explanatory variables, so it allows an arbitrary interaction pattern for them.

Table 6.11 summarizes equivalent logit and loglinear models for three-way tables when Y is a binary response variable. The simple loglinear model (Y, XZ) states that Y is jointly independent of both X and Z, and is equivalent to the special case of logit model (6.5.2) with $\{\beta_i^X\}$ and $\{\beta_k^Z\}$ terms equal to zero. In each pairing of models in Table 6.11, the loglinear model contains the X-Z association term relating the variables that are explanatory in the logit models.

The saturated loglinear model (XYZ) contains the three-factor interaction term. When Y is a binary response, this model is equivalent to a logit model with an interaction between the predictors X and Z. For instance, the effect of X on Y depends on the level of Z, meaning that the X-Y odds ratio varies across levels of Z. That logit model is also saturated.

Analogous correspondences hold for higher-way tables, as we observed in the previous subsection. Logit model (6.5.4) for a four-way table contains main effect terms for the explanatory variables, but no interaction terms. This model corresponds to the loglinear model that contains the fullest interaction term among the explanatory variables, and associations between each explanatory variable and the response I, namely model (GLS, GI, LI, IS).

6.5.4 Strategies in Model Selection

We end this chapter by describing some strategies for selecting a loglinear model. In most applications, the potentially useful models are a small subset of all loglinear models. For instance, most studies distinguish between explanatory and response variables. The focus is then on modeling effects of explanatory variables on response variables and modeling associations among response variables, rather than modeling relationships among explanatory variables. When there is a single response and it is binary, relevant loglinear models correspond to logit models for that response. When the response has more than two categories, relevant loglinear models correspond to generalized logit models presented in Section 8.1. Of course, in such cases one can fit logit models directly, rather than approach the problem with loglinear models. Indeed, one can see by comparing equations (6.5.3) and (6.5.4) how much simpler the logit structure is.

An advantage of the loglinear approach is its generality. It applies when more than one response variable exists. The drug use example in Section 6.2.3, for instance,

Table 6.11 Equivalent Loglinear and Logit Models for a Three-Way Table With Binary Response Variable Y

Loglinear	Logit
(Y, XZ)	α
(XY, XZ)	$\alpha + \beta_i^X$
(YZ, XZ)	$\alpha + \beta_k^Z$
(XY, YZ, XZ)	$\alpha + \beta_i^X + \beta_k^Z$
(XYZ)	$\alpha + \beta_i^X + \beta_k^Z + \beta_{ik}^{XZ}$

used loglinear models to study association patterns among three response variables. In summary, loglinear models are most natural when at least two variables are response variables. When only one variable is a response, it is more sensible to use logit models directly.

For either model type, the selection process becomes more difficult as the number of variables increases, because of the increase in possible associations and interactions. Fitting all possible models is usually impractical, and one must balance two competing goals: The model should be complex enough to provide a good fit, but it should be simple to interpret, smoothing rather than overfitting the data. Often, the main questions posed by a study help to guide the choice of potential terms for a model and may suggest a restricted set of models. For instance, a study's hypotheses about associations might be tested by comparing fits of certain models.

For studies that are exploratory in nature, a search among potential models may provide clues about important facets of the dependence structure. Exploratory analyses attempt to unravel the complexity of association linkages among the variables. One approach first fits the loglinear model having only single-factor terms, the model having only two-factor and single-factor terms, the model having only three-factor and lower terms, and so forth, as shown in Section 6.4.2. Fitting such models often reveals a restricted range of good-fitting models. One can also use variable elimination procedures for this purpose, as shown in Section 5.5.5. Backward elimination starts with a complex model (such as the saturated model) and drops terms one at a time, according to what does the least damage to the working model fit.

Regardless of the strategy, when certain marginal totals are fixed by the sampling design or by the response–explanatory distinction, the model should contain terms that force those totals to be identical to the fitted totals. To illustrate, suppose one treats the counts $\{n_{g+l+}\}$ in Table 6.8 as fixed at each combination of levels of G = gender and L = location. It follows from a result presented in Section 7.5.1 that a loglinear model should contain the G-L two-factor term, which ensures that $\{\hat{\mu}_{g+l+} = n_{g+l+}\}$; that is, the model should be at least as complex as model (GL, S, I). If 20,629 women had accidents in urban locations, then the fitted counts have 20,629 women in urban locations.

Related to this point, models should recognize distinctions between response and explanatory variables. The modeling process should concentrate on terms linking responses and terms linking explanatory variables to responses. Allowing a general interaction term among the explanatory variables has the effect of fixing totals at combinations of their levels. If G and L are both explanatory variables, models assuming conditional independence between G and L are not of interest.

For Table 6.8, I is a response variable, and S might be treated either as a response or explanatory variable. If it is explanatory, we treat the $\{n_{g+ls}\}$ totals as fixed and consider logit models for the I response. If S is also a response, we consider the $\{n_{g+l+}\}$ totals as fixed and consider loglinear models at least as complex as (GL, S, I). Such models focus on the effects of G and L on S and I as well as the association between S and I.

Complicating factors arise as the number of dimensions increases. The number of cells increases dramatically. Unless the sample size is very large, many zero counts are likely. This can cause infinite parameter estimates and poor chi-squared approximations for goodness-of-fit statistics. In addition, some variables are likely to be ordinal, and the models should reflect this. The next chapter discusses these issues.

PROBLEMS

6.1. Fit the independence model to Table 2.2, and check the goodness of fit. Report $\{\hat{\lambda}_j^Y\}$ (a) using a form of constraints for which a baseline estimate equals 0, (b) using "sum to zero" constraints. Interpret $\hat{\lambda}_1^Y - \hat{\lambda}_2^Y$.

6.2. For the saturated model with Table 2.2, report $\{\hat{\lambda}_{ij}^{XY}\}$ (a) using "sum to zero" constraints, (b) using a form of constraints for which only $\hat{\lambda}_{11}^{XY}$ is nonzero. Show how to interpret these estimates using an odds ratio.

6.3. Refer to Table 3.1. Let D = defendant's race, V = victims' race, and P = death penalty verdict. Fit model (DV, DP, PV).

   a. Using the fitted values, calculate the odds ratio between D and P at each level of V. Note the "common odds ratio" property of this model, and interpret the value.

   b. Calculate the marginal odds ratio between D and P, (i) using the fitted values, (ii) using the sample data. (For loglinear models, all margins corresponding to terms in the model are identical for the observed and fitted data.) Contrast the fitted odds ratio with that in (**a**), and remark on Simpson's paradox.

   c. Test the goodness of fit of this model. Interpret.

   d. Fit the simpler model (DV, PV). Interpret associations, test the fit, and conduct a residual analysis.

   e. Test the D-P partial association by comparing the fits of models (DV, PV) and (DV, DP, PV). Interpret.

6.4. In the 1988 General Social Survey respondents were asked "Do you support or oppose the following measures to deal with AIDS? (1) Have the government pay all of the health care costs of AIDS patients; (2) Develop a government information program to promote safe sex practices, such as the use of condoms." Table 6.12 shows responses on these two items, classified also by the

Table 6.12

Gender	Information Opinion	Health Opinion	
		Support	Oppose
Male	Support	76	160
	Oppose	6	25
Female	Support	114	181
	Oppose	11	48

Source: 1988 General Social Survey.

respondent's gender. Denote the variables by G for Gender, H for opinion on Health care costs, and I for opinion on an Information program.

a. Fit the loglinear models (GH, GI), (GH, HI), (GI, HI), and (GH, GI, HI). Show that models that lack the H-I term fit poorly.

b. For model (GH, GI, HI), show that 95% confidence intervals equal $(0.55, 1.10)$ for the G-H conditional odds ratio and $(0.99, 2.55)$ for the G-I conditional odds ratio. Interpret, explaining why gender may have no effect on opinion for these issues. (Since the intervals contain values rather far from 1.0, however, these odds ratios could also be moderate. It is safest to describe these data using model (GH, GI, HI), even though simpler models fit adequately.)

6.5. Refer to Table 3.3.

a. Fit the loglinear model of homogeneous association. Report the estimated conditional odds ratio between smoking and lung cancer. Obtain a 99% confidence interval for the true odds ratio. Interpret.

b. Test goodness of fit of the model. Interpret.

c. Consider the simpler model of conditional independence between smoking and lung cancer, given city. Compare the fit to the homogeneous association model, and interpret.

d. Fit a logit model containing effects of smoking and city on lung cancer. Use the smoking effect to estimate the conditional odds ratio between smoking and lung cancer. How does this compare to the estimate for the loglinear model?

e. Use (**b**) to test the hypothesis of a common odds ratio between smoking and lung cancer for these eight studies. How does the result compare to the Breslow–Day test in Section 3.2.4?

6.6. Table 6.13 refers to applicants to graduate school at the University of California at Berkeley for the fall 1973 session. Admissions decisions are presented by gender of applicant, for the six largest graduate departments. Denote the three variables by A = whether admitted, G = gender, and D = department.

Table 6.13

Department	Whether admitted, male		Whether admitted, female	
	Yes	No	Yes	No
1	512	313	89	19
2	353	207	17	8
3	120	205	202	391
4	138	279	131	244
5	53	138	94	299
6	22	351	24	317

Source: P. Bickel et al., *Science, 187:* 398–403 (1975).

    a. Fit loglinear models (AD, DG) and (AD, AG, DG). Report G^2 and df values, and comment on the quality of fit.

    b. Conduct a residual analysis for (AD, DG). Describe the lack of fit.

    c. Deleting the data for Department 1, fit model (AD, DG). Interpret.

    d. Conduct equivalent analyses using logit models with A as the response.

6.7. Table 6.14 is from the 1991 General Social Survey. White subjects in the sample were asked: (B) Do you favor busing of (Negro/Black) and white school children from one school district to another?, (P) If your party nominated a (Negro/Black) for President, would you vote for him if he were qualified for the job?, (D) During the last few years, has anyone in your family brought a friend who was a (Negro/Black) home for dinner? The response scale for each item was (Yes, No, Don't know). Fit model (BD, BP, DP).

Table 6.14

President	Busing	Home 1	Home 2	Home 3
	1	41	65	0
1	2	71	157	1
	3	1	17	0
	1	2	5	0
2	2	3	44	0
	3	1	0	0
	1	0	3	1
3	2	0	10	0
	3	0	0	1

Source: 1991 General Social Survey, with categories 1 = yes, 2 = no, 3 = don't know.

    a. Using the "yes" and "no" categories of each response, estimate the conditional odds ratios for each pair of variables. Interpret.

    b. Analyze the model goodness of fit. Interpret.

    c. Test the significance of the B-P association. Interpret.

6.8. Refer to Problem 3.1.

    a. Analyze these data using loglinear models. For the model you choose to describe the data, interpret parameter estimates.

    b. For the model in (**a**), fit the corresponding logit model, treating death penalty as the response. Show the correspondence between parameter estimates for the two models.

6.9. Table 6.15 is based on automobile accident records in 1988, supplied by the state of Florida Department of Highway Safety and Motor Vehicles. Subjects were classified by whether they were wearing a seat belt, whether ejected, and whether killed.

Table 6.15

Safety Equipment In Use	Whether Ejected	Killed	
		No	Yes
Seat belt	Yes	1105	14
	No	411,111	483
None	Yes	4624	497
	No	157,342	1008

Source: Florida Department of Highway Safety and Motor Vehicles, Tallahassee, FL.

    a. Find a loglinear model that describes the data well. Interpret the associations.

    b. Treating whether killed as the response variable, fit an equivalent logit model. Interpret the effects on the response.

    c. Since the sample size is large, goodness-of-fit statistics are large unless the model fits very well. Calculate the dissimilarity index, and interpret.

6.10. Refer to the loglinear models for Table 6.8.

    a. Explain why the fitted odds ratios in Table 6.10 for model (GI, GL, GS, IL, IS, LS) suggest that the most likely case for injury is accidents for females not wearing seat belts in rural locations.

    b. Fit model (GLS, GI, IL, IS). Show, using model parameter estimates, that the fitted I-S conditional odds ratio equals 0.44. Show that, for each injury level, the estimated conditional L-S odds ratio is 1.17 for $(G = \text{female})$ and 1.03 for $(G = \text{male})$. How can you get these using the model parameter estimates? (*Hint:* Calculate the log odds ratio using the model formula.)

6.11. Consider the following two-stage model for Table 6.8. The first stage is a logit model with S as the response, for the three-way G-L-S table. The second stage is a logit model with these three variables as predictors for I in the four-way table. Explain why this composite model is sensible, fit the models, and interpret results.

6.12. Table 6.16, based on the 1991 General Social Survey, relates responses on four variables: How often you attend religious services ($R =$ At most a few times a year, At least several times a year); Political views ($P =$ Liberal, Moderate,

Table 6.16

		Premarital Sex							
		1				2			
	Religious Attendance	1		2		1		2	
	Birth control	1	2	1	2	1	2	1	2
Political	1	99	15	73	25	8	4	24	22
Views	2	73	20	87	37	20	13	50	60
	3	51	19	51	36	6	12	33	88

Source: 1991 General Social Survey.

Conservative); Methods of birth control should be available to teenagers between the ages of 14 and 16 (B = Agree, Disagree); Opinion about a man and woman having sex relations before marriage (S = Wrong only sometimes or not wrong at all, Always or almost always wrong).

a. Find a loglinear model that fits these data well.

b. Interpret this model by estimating conditional odds ratios for each pair of variables. Construct and interpret a 90% confidence interval for the true odds ratio relating B and S, controlling for P and R.

c. Consider the logit model predicting (S) using the other variables as main-effect predictors, without any interaction. Fit the corresponding loglinear model. Does it fit adequately? Interpret the effects of the predictors on the response, and compare to results from (b).

6.13. Table 6.17 is taken from the 1989 General Social Survey. Subjects were asked their opinions regarding government spending on the environment (E), health (H), assistance to big cities (C), and law enforcement (L). The common response scale was (too little, about right, too much). (Note that, despite the common public complaint about taxes, the highest counts occur in cells corresponding to preferring *increased* spending on most items!)

Table 6.17

Cities		1			2			3		
Law Enforcement		1	2	3	1	2	3	1	2	3
Environment	Health									
1	1	62	17	5	90	42	3	74	31	11
	2	11	7	0	22	18	1	19	14	3
	3	2	3	1	2	0	1	1	3	1
2	1	11	3	0	21	13	2	20	8	3
	2	1	4	0	6	9	0	6	5	2
	3	1	0	1	2	1	1	4	3	1
3	1	3	0	0	2	1	0	9	2	1
	2	1	0	0	2	1	0	4	2	0
	3	1	0	0	0	0	0	1	2	3

Source: 1989 General Social Survey.
Note: 1 = too little, 2 = about right, 3 = too much.

a. Fit the homogeneous association model. Test the fit, and interpret.

b. Show that the estimated conditional log odds ratio for the "too much" and "too little" categories of E and H equals

$$\hat{\lambda}_{11}^{EH} + \hat{\lambda}_{33}^{EH} - \hat{\lambda}_{13}^{EH} - \hat{\lambda}_{31}^{EH}.$$

Table 6.18 reports $\{\hat{\lambda}_{eh}^{EH}\}$ using constraints whereby parameters sum to zero within each row and within each column, and whereby parameters are zero in the first row and the first column. Calculate and interpret the odds ratio. Show how to obtain the same result using the corner fitted values for a partial table relating E and H at any combination of levels of the other two variables. The ASE of the estimated E-H log odds ratio is 0.523. Show that a 95% confidence interval for the true odds ratio equals $(3.1, 24.4)$. Interpret.

Table 6.18

E	Sum to Zero Constraints H			Zero for First Level H		
	1	2	3	1	2	3
1	0.509	0.166	-0.676	0	0	0
2	-0.065	-0.099	0.163	0	0.309	1.413
3	-0.445	-0.068	0.513	0	0.720	2.142

    c. Report the estimated conditional odds ratios using the "too much" and "too little" categories for each of the other pairs of variables. Summarize the association structure for this table.

6.14. Refer to Table 6.3. The survey also classified respondents by gender (G) and race (R). Table 7.1 shows the full data.

    a. Analyze these data using loglinear models, distinguishing between response and explanatory variables. Summarize your main conclusions based on studying the relevant estimated associations in the final model.

    b. Analyze these data using logit models, treating marijuana use as the response variable. Specify the loglinear model that is equivalent to your choice of logit model.

6.15. Refer to Problem 6.3. Using P as the response, fit logit models that give the same results as these two loglinear models. Show the relation between parameter estimates for the loglinear and logit models.

6.16. Refer to Problem 3.2. Analyze these data with loglinear models, using residuals to describe lack of fit. Show how logit models for the death penalty response provide the same results.

6.17. Refer to the logit model in Problem 5.23. Let A denote the response variable, opinion on abortion.

    a. Give the symbol for the loglinear model that is equivalent to this logit model. Which logit model corresponds to loglinear model (AR, AP, GRP)?

    b. State the equivalent loglinear and logit models for which: (i) A is jointly independent of G, R, and P, (ii) There are main effects of R on A, but A is conditionally independent of G and P, given R, (iii) There is interaction between P and R in their effects on A, and G has main effects.

6.18. Verify that logit model (6.5.4) follows from loglinear model (GLS, GI, LI, IS). Show that the conditional log odds ratio for the effect of S on I equals $\beta_1^S - \beta_2^S$ in the logit model and $\lambda_{11}^{IS} + \lambda_{22}^{IS} - \lambda_{12}^{IS} - \lambda_{21}^{IS}$ in the loglinear model.

6.19. For a multiway contingency table, when is a logit model more appropriate than a loglinear model? When is a loglinear model more appropriate than a logit model?

6.20. For four categorical variables W, X, Y, Z, explain why (WXZ, WYZ) is the most general loglinear model for which X and Y are conditionally independent.

CHAPTER 7

Building and Applying Logit and Loglinear Models

Chapter 5 presented the logistic regression model, which uses the logit link for a binomial response. Chapter 6 presented the loglinear model that uses the log link for Poisson cell counts in a contingency table. Chapter 4 showed that they are both generalized linear models (GLMs), and Section 6.5 discussed equivalences between them. This chapter discusses further topics relating to building and applying these two types of models.

Section 7.1 introduces graphical representations that portray a model's association and conditional independence patterns. They also provide simple ways of indicating when conditional odds ratios are identical to marginal odds ratios.

The loglinear models of Chapter 6 treat all variables as nominal. Section 7.2 presents a loglinear model of association between ordinal variables. Inferences utilizing the ordering are more powerful than those that ignore it.

Section 7.3 presents tests of the hypothesis of conditional independence in three-way tables for nominal and ordinal variables. One approach compares the fits of two loglinear or logit models. A related approach uses generalized versions of the Cochran–Mantel–Haenszel test for multicategory responses.

Most inferential analyses in this text use large-sample approximations. Section 7.4 discusses the effects on model parameter estimates and goodness-of-fit statistics of small samples or zero cell counts. Finally, Section 7.5 summarizes theory underlying the fitting of logit and loglinear models and their use for large-sample inference.

7.1 ASSOCIATION GRAPHS AND COLLAPSIBILITY

We begin by presenting a graphical representation for associations in loglinear models. For a model with a particular set of variables, the graph indicates which pairs of variables are independent and which pairs are associated, given the others. This representation is helpful for revealing implications of models, such as determining when marginal and conditional odds ratios are identical.

174

7.1.1 Association Graphs

An *association graph* for a loglinear model has a set of vertices, each vertex representing a variable. There are as many vertices as dimensions of the contingency table. An edge connecting two vertices represents a partial association between the corresponding two variables.

We illustrate for a four-way table with variables W, X, Y, Z. The loglinear model (WX, WY, WZ, YZ) lacks X-Y and X-Z association terms. It assumes that X and Y are independent and that X and Z are independent, conditional on the remaining two variables, but permits association between W and X and between each pair of variables in the set $\{W, Y, Z\}$. The association graph

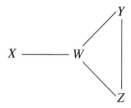

portrays this model. The four variables form the vertices of the graph. The four edges, connecting W and X, W and Y, W and Z, and Y and Z, represent pairwise partial associations. Edges do not connect X and Y or X and Z, since those pairs are conditionally independent, given the remaining variables.

Two loglinear models that have the same pairwise associations have the same association graph. For instance, the association graph just portrayed for model (WX, WY, WZ, YZ) is also the one for model (WX, WYZ) that also contains a three-factor WYZ interaction.

A *path* in an association graph is a sequence of edges leading from one variable to another. Two variables X and Y are said to be *separated* by a subset of variables if all paths connecting X and Y intersect that subset. For instance, in the above graph, W separates X and Y, since any path connecting X to Y goes through W. The subset $\{W, Z\}$ also separates X and Y. A fundamental result states that two variables are conditionally independent given *any* subset of variables that separates them. Thus, not only are X and Y conditionally independent given W and Z, but also given W alone. Similarly, X and Z are conditionally independent given W alone.

For another example, we consider loglinear model (WX, XY, YZ). Its association graph is

$$W \longrightarrow X \longrightarrow Y \longrightarrow Z.$$

Since W and Z are separated by X, by Y, and by X and Y, this graph reveals that W and Z are independent given X alone or given Y alone or given both X and Y. Also, W and Y are independent, given X alone or given X and Z; X and Z are independent, given Y alone or given Y and W.

7.1.2 Collapsibility in Three-Way Tables

Section 3.1.4 showed that associations in partial tables may differ from marginal associations. For instance, if X and Y are conditionally independent, given Z, they are not necessarily marginally independent. We next present conditions under which a model's odds ratios are identical in partial tables as in the marginal table. These *collapsibility conditions* imply that the association is unchanged when we combine the partial tables.

> For three-way tables, X-Y marginal and partial odds ratios are identical if either Z and X are conditionally independent or if Z and Y are conditionally independent.

The conditions state that the variable treated as the control (Z) is conditionally independent of X or Y, or both. These conditions correspond to loglinear models (XY, YZ) and (XY, XZ). That is, the X-Y association is identical in the partial tables and the marginal table for models with association graphs

$$X \text{———} Y \text{———} Z \quad \text{and} \quad Y \text{———} X \text{———} Z$$

or even simpler models, but not for the model with graph $X \text{———} Z \text{———} Y$ in which an edge connects Z to both X and Y.

We illustrate for the drug use data (Table 6.3) from Section 6.2.3, denoting $A =$ alcohol use, $C =$ cigarette use, and $M =$ marijuana use. Consider (AM, CM), the model of conditional independence of A and C, given M, which has association graph

$$A \text{———} M \text{———} C.$$

Consider the A-M association, controlling for C; that is, we identify C with Z in the collapsibility conditions. In this model, since C is conditionally independent of A, the A-M partial odds ratios are the same as the A-M marginal odds ratio collapsed over C. In fact, Table 6.5 showed that both the fitted marginal and partial A-M odds ratios equal 61.9. Similarly, the C-M association is collapsible. The A-C association is not, however. The collapsibility conditions are not satisfied, because M is conditionally dependent with both A and C in model (AM, CM). Thus, A and C may be marginally dependent, even though they are conditionally independent in this model. In fact, Table 6.5 showed that the fitted A-C marginal odds ratio for this model equals 2.7, not 1.0.

The model (AC, AM, CM) of homogeneous association has association terms for each pair of variables, so no pair is conditionally independent. No collapsibility conditions are fulfilled. In fact, Table 6.5 showed that each pair of variables has quite different fitted marginal and partial associations for this model. When a model contains all two-factor effects, collapsing over any variable may cause effects to change.

7.1.3 Collapsibility and Logit Models

The collapsibility conditions apply also to corresponding logit models. For instance, suppose a clinical trial studies the association between a treatment variable X and a binary response Y, using data from several centers regarded as levels of a control variable Z. The logit model

$$\text{logit}(\pi) = \alpha + \beta_i^X + \beta_k^Z \qquad (7.1.1)$$

for the probability π that Y is a "success" assumes that treatment effects are the same for each center. Since this logit model corresponds to loglinear model (XY, XZ, YZ), the estimated treatment effects may differ if we collapse the table over the center factor. That is, the estimated X-Y odds ratio for this model, $\exp(\hat{\beta}_1^X - \hat{\beta}_2^X)$, differs from the sample odds ratio in the marginal 2×2 table relating X and Y.

Next, consider the simpler model for this three-way table that lacks the center effects,

$$\text{logit}(\pi) = \alpha + \beta_i^X.$$

For each treatment, this model states that the success probability π is identical for each center. The partial and marginal treatment effects are identical for this model. It satisfies a collapsibility condition, because the model states that Z is conditionally independent of Y. This logit model is equivalent to loglinear model (XY, XZ) with association graph $Y \text{——} X \text{——} Z$, for which the X-Y association is collapsible. In practice, this suggests that when center effects seem negligible and the simpler model does not fit poorly compared to the full model (i.e., when center does not seem to be a "confounding" variable), one can collapse the table and estimate the treatment effect using the marginal odds ratio.

7.1.4 Collapsibility and Association Graphs for Multiway Tables

The next result provides collapsibility conditions for models for multiway tables.

Suppose that variables in a model for a multiway table partition into three mutually exclusive subsets, A, B, C, such that B separates A and C; thus, the model does not contain parameters linking variables from A with variables from C. When one collapses the table over the variables in C, model parameters relating variables in A and model parameters relating variables in A with variables in B are unchanged.

In other words, suppose that every path between a variable in A and a variable in C involves at least one variable in B. That is, the subsets of variables have the form

$$A \text{——} B \text{——} C.$$

When one collapses over the variables in C, the same parameter values relate the variables in A, and the same parameter values relate variables in A to variables in B.

It follows that the corresponding associations are unchanged, as described by odds ratios based on those parameters.

We illustrate using model (WX, WY, WZ, YZ) for a four-way table, which has association graph

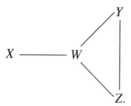

Let $A = \{X\}, B = \{W\}$, and $C = \{Y, Z\}$. Since the X-Y and X-Z association parameters do not appear in this model, all parameters linking set A with set C equal zero, and B separates A and C. If we collapse over Y and Z, the W-X association is unchanged. Next, identify $A = \{Y, Z\}, B = \{W\}, C = \{X\}$. Then, partial associations among W, Y, and Z remain the same when the table is collapsed over X.

When the set B contains more than one variable, the collapsibility conditions require a slight qualifier. Though the true parameter values are unchanged when one collapses over set C, the ML estimates of those parameters may differ slightly. (The estimates are also identical if the model contains the highest-order term relating variables in B to each other.)

7.1.5 Model Building for the Dayton Drug-Use Example

Sections 6.2 and 6.3 analyzed data on usage of alcohol (A), cigarettes (C), and marijuana (M) by a sample of high school seniors. When the students are also classified by the demographic factors gender (G) and race (R), the five-dimensional contingency table shown in Table 7.1 results. In selecting a model for these data, we treat A, C, and M as response variables and G and R as explanatory variables.

Table 7.1 Alcohol, Cigarette, and Marijuana Use for High School Seniors by Gender and Race

Race		White				Other			
Gender		Female		Male		Female		Male	
Alcohol	Cigarette				Marijuana Use				
Use	Use	Yes	No	Yes	No	Yes	No	Yes	No
Yes	Yes	405	268	453	228	23	23	30	19
	No	13	218	28	201	2	19	1	18
No	Yes	1	17	1	17	0	1	1	8
	No	1	117	1	133	0	12	0	17

Source: Prof. Harry Khamis, Wright State University.

Table 7.2   Goodness-of-Fit Tests for Loglinear Models Relating Alcohol (A), Cigarette (C), and Marijuana (M) Use, by Gender (G) and Race (R)

Model	G^2	df
1. Mutual Independence + GR	1325.1	25
2. Homogeneous Association	15.3	16
3. All Three-Factor Terms	5.3	6
4a. (2) $- AC$	201.2	17
4b. (2) $- AM$	107.0	17
4c. (2) $- CM$	513.5	17
4d. (2) $- AG$	18.7	17
4e. (2) $- AR$	20.3	17
4f. (2) $- CG$	16.3	17
4g. (2) $- CR$	15.8	17
4h. (2) $- GM$	25.2	17
4i. (2) $- MR$	18.9	17
5. $(AC, AM, CM, AG, AR, GM, GR, MR)$	16.7	18
6. $(AC, AM, CM, AG, AR, GM, GR)$	19.9	19
7. (AC, AM, CM, AG, AR, GR)	28.8	20

Since G and R are explanatory, it does not make sense to estimate association or assume conditional independence for that pair. It follows from remarks near the end of Section 6.5.4 that a model should contain the G-R term. Including this term forces the G-R fitted marginal totals to be the same as the corresponding sample marginal totals. The sample contained 1040 white females, for instance, so the model's fitted total of white females then equals 1040.

Table 7.2 displays results of goodness-of-fit tests for several loglinear models. Because many cell counts are small, the chi-squared approximation for G^2 may be poor. It is best not to take the G^2 values too seriously for a particular model, but this index is useful for comparing models. The first model listed in Table 7.2 contains only the G-R association and assumes conditional independence for the other nine pairs of associations. It fits horribly, which is no surprise. The homogeneous association model, on the other hand, seems to fit well. The only large adjusted residual results from a fitted value of 3.1 in the cell having a count of 8.

The model containing all the three-factor interaction terms also fits well, but the improvement in fit is not great (difference in G^2 of $15.3 - 5.3 = 10.0$ based on $df = 16 - 6 = 10$). Thus, we consider models without three-factor terms. Beginning with the homogeneous association model as the baseline, we eliminate two-factor associations that do not make significant contributions. We use a backward elimination process, sequentially taking out terms for which the resulting increase in G^2 is smallest, when refitting the model. However, we do not delete the G-R association term relating the explanatory variables.

Table 7.2 shows the start of this process. Nine pairwise associations are candidates for removal from model (2), shown in models numbered (4a)–(4i) in the table. The smallest increase in G^2, compared to model (2), occurs in removing the C-R term. The increase is $15.8 - 15.3 = 0.5$, based on $df = 17 - 16 = 1$, so this elimination seems reasonable. After removing the C-R term (model 4g), the smallest additional increase results from removing the C-G term (model 5), resulting in $G^2 = 16.7$ with $df = 18$, and a change in G^2 of 0.9 based on $df = 1$. Removing next the M-R term (model 6) yields $G^2 = 19.9$ with $df = 19$, a change in G^2 of 3.2 based on $df = 1$.

At this stage, the only large adjusted residual refers to a fitted value of 2.9 in the cell having a count of 8. Additional removals have a more severe effect. For instance, removing next the A-G term increases G^2 by 5.3, based on $df = 1$, for a P-value of .02. One cannot take such P-values too literally, since these tests are suggested by the data, but it seems safest not to drop additional terms. Model (6), denoted by $(AC, AM, CM, AG, AR, GM, GR)$, has association graph

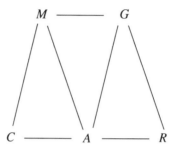

Consider the sets $\{C\}$, $\{A, M\}$, and $\{G, R\}$. For this model, every path between C and $\{G, R\}$ involves a variable in $\{A, M\}$. Given the outcome on alcohol use and marijuana use, the model states that cigarette use is independent of both gender and race. Collapsing over the explanatory variables race and gender, the partial associations between C and A and between C and M are the same as with the model (AC, AM, CM) fitted in Section 6.2.3.

Suppose we remove the G-M term from this model, yielding (AC, AM, CM, AG, AR, GR), model (7) in Table 7.2. Its association graph reveals that $\{G, R\}$ are separated from $\{C, M\}$ by A. It follows that all pairwise partial associations among A, C, and M in model (7) are identical to those in model (AC, AM, CM), collapsing over G and R. In fact, model (7) does not fit all that poorly ($G^2 = 28.8$ with $df = 20$, and only one adjusted residual exceeds 3), especially considering the large sample size. Its sample dissimilarity index equals $D = .036$. For practical purposes, one may be able to collapse over gender and race in studying associations among the drug-use variables. An advantage of the full five-variable model portrayed by the above graph is that one can also study the effects of gender and race on these responses, in particular the effects of race and gender on alcohol use and the effect of gender on marijuana use.

7.2 MODELING ORDINAL ASSOCIATIONS

The loglinear models presented so far have a serious limitation: they treat all classifications as nominal. If we change the order of a variable's categories in any way, we get the same fit. For ordinal data, these models ignore important information.

Table 7.3   Fit of Independence Model and Adjusted Residuals for Opinion about Premarital Sex and Availability of Teenage Birth Control

| Premarital Sex | Teenage Birth Control | | | |
	Strongly Disagree	Disagree	Agree	Strongly Agree
Always wrong	81	68	60	38
	(42.4)	(51.2)	(86.4)	(67.0)
	7.6	3.1	−4.1	−4.8
Almost always wrong	24	26	29	14
	(16.0)	(19.3)	(32.5)	(25.2)
	2.3	1.8	−0.8	−2.8
Wrong only sometimes	18	41	74	42
	(30.1)	(36.3)	(61.2)	(47.4)
	−2.7	1.0	2.2	−1.0
Not wrong at all	36	57	161	157
	(70.6)	(85.2)	(143.8)	(111.4)
	−6.1	−4.6	2.4	6.8

Source: 1991 General Social Survey.

Table 7.3, taken from the 1991 General Social Survey, illustrates the inadequacy of ordinary loglinear models for analyzing ordinal data. Subjects were asked their opinion about a man and woman having sex relations before marriage, with possible responses "always wrong," "almost always wrong," "wrong only sometimes," and "not wrong at all." They were also asked if they "strongly disagree," "disagree," "agree," or "strongly agree" that methods of birth control should be made available to teenagers between the ages of 14 and 16. Both classifications have ordered categories.

For these data, the loglinear model of independence (i.e., model (6.1.1)), which we denote by I, has goodness-of-fit statistics $G^2(I) = 127.6$ and $X^2(I) = 128.7$, based on $df = 9$. These tests of fit are simply the tests of independence presented in Section 2.4. The model fits poorly, providing strong evidence of dependence. Yet, adding the ordinary association term makes the model saturated (model (6.1.2)) and of little use.

Table 7.3 also contains fitted values and adjusted residuals (Section 2.4.5) for the independence model. The residuals in the corners of the table are very large. Observed counts are much larger than the independence model predicts in the corners where both responses are the most negative possible ("always wrong" with "strongly disagree") or the most positive possible ("not wrong at all" with "strongly agree"). By contrast, observed counts are much smaller than fitted counts in the other two corners, where one response is the most positive and the other is the most negative. Cross-classifications of ordinal variables often exhibit their greatest deviations from independence in the corner cells. This pattern for Table 7.3 indicates lack of fit in the form of a positive trend. Subjects who feel more favorable to making birth control available to teenagers also tend to feel more tolerant about premarital sex.

The independence model is too simple to fit most data well. Models for ordinal variables use association terms that permit negative or positive association trends. The models are more complex than the independence model yet simpler than the saturated model.

7.2.1 Linear-by-Linear Association

This section presents an ordinal loglinear model for two-way tables. It requires assigning scores $\{u_i\}$ to the I rows and $\{v_j\}$ to the J columns. To reflect category orderings, $u_1 \le u_2 \le \cdots \le u_I$ and $v_1 \le v_2 \le \cdots \le v_J$. The model is

$$\log \mu_{ij} = \lambda + \lambda_i^X + \lambda_j^Y + \beta u_i v_j. \tag{7.2.1}$$

The independence model is the special case $\beta = 0$.

Model (7.2.1) has form

$$\log \mu_{ij} = \text{independence} + \beta u_i v_j.$$

The final term represents the deviation of $\log \mu_{ij}$ from independence. The deviation is linear in the Y scores at a fixed level of X and linear in the X scores at a fixed level of Y. In column j, for instance, the deviation is a linear function of X, having form (slope) × (score for X), with slope βv_j. Because of this property, (7.2.1) is called the *linear-by-linear association model* (abbreviated, $L \times L$). This linear-by-linear deviation implies that the model has its greatest departures from independence in the corners of the table.

The parameter β in model (7.2.1) refers to the direction and strength of association. When $\beta > 0$, there is a tendency for Y to increase as X increases. Expected frequencies are larger than expected (under independence) in cells of the table where X and Y are both high or both low. When $\beta < 0$, there is a tendency for Y to decrease as X increases, and for expected frequencies to be relatively larger in cells where X is high and Y is low or where X is low and Y is high. When the data display a positive or negative trend, this model usually fits much better than the independence model.

We describe associations for this model using odds ratios for pairings of categories. For the 2×2 table using the cells intersecting rows a and c with columns b and d, the model has odds ratio equal to

$$\frac{\mu_{ab}\mu_{cd}}{\mu_{ad}\mu_{cb}} = \exp[\beta(u_c - u_a)(v_d - v_b)]. \tag{7.2.2}$$

The association is stronger as $|\beta|$ increases. For given β, pairs of categories that are farther apart have greater differences between their scores and odds ratios farther from 1.

In practice, the most common choice of scores is $\{u_i = i\}$ and $\{v_j = j\}$, simply the row and column numbers. These scores have equal spacings of 1 between each pair of adjacent row or column scores. The odds ratios formed using adjacent rows and adjacent columns are called *local odds ratios*. For these unit-spaced scores, (7.2.2) simplifies so that e^β is the common value of all the local odds ratios. Any set of equally-spaced row and column scores has the property of uniform local odds ratios. This special case of the model is called *uniform association*. Figure 7.1 portrays some of the local odds ratios that take uniform value in this model.

Fitting the linear-by-linear association model requires iterative methods. The model's fitted values, like those for the independence model, have the same row and column totals as the observed data. In addition, the correlation between the row scores for X and the column scores for Y is the same for the observed counts as it is for the joint distribution given by the fitted counts. Thus, the fitted counts display the same positive or negative trend as the observed data. Unlike the observed data, the fitted counts exactly satisfy the odds ratio pattern (7.2.2) implied by the model. Since the model has one more parameter (β) than the independence model, its residual $df = IJ - I - J$ are 1 less; it is unsaturated whenever $I > 2$ or $J > 2$.

7.2.2   Sex Opinions Example

Table 7.4 reports fitted values for the linear-by-linear $(L \times L)$ association model applied to the opinions about premarital sex and availability of teen birth control, using row scores $\{1, 2, 3, 4\}$ and column scores $\{1, 2, 3, 4\}$. The goodness-of-fit statistics for this uniform association version of the model are $G^2(L \times L) = 11.5$ and $X^2(L \times L) = 11.5$, with $df = 8$. Compared to the independence model, for which $G^2(I) = 127.6$ with $df = 9$, the $L \times L$ model provides a dramatic improvement in fit. This is especially noticeable in the corners of the table.

The ML estimate of the association parameter is $\hat{\beta} = 0.286$, with $ASE = 0.028$. The positive estimate suggests that subjects having more favorable attitudes about availability of teen birth control also tend to have more tolerant attitudes about premarital sex. The estimated local odds ratio is $\exp(\hat{\beta}) = \exp(0.286) = 1.33$. The strength of association seems weak. From (7.2.2), however, nonlocal odds ratios are

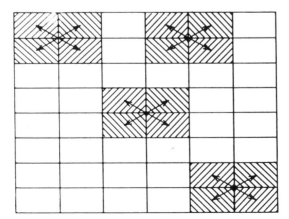

Figure 7.1   Constant local odds ratio implied by uniform association model. (*Note:* β = the constant log odds ratio for adjacent rows and adjacent columns.)

Table 7.4   Fit of Linear-by-Linear Association Model for Table 7.3

Premarital Sex	Teenage Birth Control			
	Strongly Disagree	Disagree	Agree	Strongly Agree
Always wrong	81	68	60	38
	(80.9)	(67.6)	(69.4)	(29.1)
Almost always wrong	24	26	29	14
	(20.8)	(23.1)	(31.5)	(17.6)
Wrong only sometimes	18	41	74	42
	(24.4)	(36.1)	(65.7)	(48.8)
Not wrong at all	36	57	161	157
	(33.0)	(65.1)	(157.4)	(155.5)

stronger. For instance, the estimated odds ratio for the four corner cells equals

$$\exp[\hat{\beta}(u_4 - u_1)(v_4 - v_1)] = \exp[0.286(4 - 1)(4 - 1)] = \exp(2.57) = 13.1.$$

One can also obtain this value directly using the fitted values for the corner cells in Table 7.4; that is, $(80.9)(155.5)/(29.1)(33.0) = 13.1$. For those who "strongly agree" with availability of teen birth control, the odds of response "not wrong at all" instead of "always wrong" on premarital sex are estimated to be 13.1 times the odds for those who "strongly disagree" with availability of teen birth control.

Two sets of scores having the same spacings yield the same estimate of β and the same fit. For instance, $\{u_1 = 1, u_2 = 2, u_3 = 3, u_4 = 4\}$ yields the same results as $\{u_1 = -1.5, u_2 = -0.5, u_3 = 0.5, u_4 = 1.5\}$. Any other sets of equally spaced scores yield the same fit but an appropriately rescaled estimate of β, so that the fitted odds ratios (7.2.2) do not change. For instance, the row scores $\{2, 4, 6, 8\}$ with $\{v_j = j\}$ also yield $G^2 = 11.5$, but have $\hat{\beta} = 0.143$ with standard error 0.014 (both half as large).

It is not necessary to use equally-spaced scores in the $L \times L$ model. For the classifications in Table 7.3, one might regard categories 2 and 3 as farther apart than categories 1 and 2 or categories 3 and 4. To recognize this, one could assign scores such as $\{1, 2, 4, 5\}$ to the row and column categories. The $L \times L$ model then has $G^2 = 8.8$. One need not, however, regard the model scores as approximations for distances between categories. They simply imply a certain pattern for the relative sizes of the odds ratios. From (7.2.2), fitted local odds ratios are stronger for pairs of adjacent categories having greater distances between scores. More general models, not discussed here, treat the row and/or column scores as parameters estimated using the data.

7.2.3   Ordinal Tests of Independence

For the linear-by-linear association model, the hypothesis of independence is $H_0 : \beta = 0$. The likelihood-ratio test statistic equals the reduction in G^2 goodness-of-fit statistics between the independence (I) and $L \times L$ models,

$$G^2(I \mid L \times L) = G^2(I) - G^2(L \times L). \tag{7.2.3}$$

This statistic refers to a single parameter (β), and is based on $df = 1$. For Table 7.3, the reduction is $127.6 - 11.5 = 116.1$. This has $P < .0001$, extremely strong evidence of an association.

The Wald statistic $z^2 = (\hat{\beta}/ASE)^2$ provides an alternative test for this hypothesis. It is also a chi-squared statistic with $df = 1$. For these data, $z^2 = (0.286/0.0282)^2 = 102.4$, also showing strong evidence of a positive trend. The correlation statistic (2.5.1) presented in Section 2.5.1 for testing independence with ordinal data is usually similar to the likelihood-ratio and Wald statistics for testing $\beta = 0$ in this model. (In fact, it is the efficient score statistic.) For Table 7.3, it equals 112.6, also based on $df = 1$.

Generalizations of the linear-by-linear association model exist for multi-way tables. We discuss one of these in the following section. Sections 8.2 and 8.3 present other ways of using ordinality, based on models that create logits for an ordinal response variable.

7.3 TESTS OF CONDITIONAL INDEPENDENCE

This section discusses ways of testing the hypothesis of conditional independence in three-way tables. Likelihood-ratio tests compare the fit of two loglinear or logit models. Alternatively, one can use generalizations of the Cochran–Mantel–Haenszel statistic.

7.3.1 Using Models to Test Conditional Independence

Section 6.3.4 showed how to test a partial association by comparing two loglinear models that contain or omit that association. The likelihood-ratio test compares the models by the difference of the G^2 goodness-of-fit statistics, which is identical to the difference of *deviances* (Section 4.5.3).

An important application of this test refers to the null hypothesis of X-Y conditional independence. One compares the model (XZ, YZ) of X-Y conditional independence to the more complex model (XY, XZ, YZ) that contains the X-Y association. The test statistic is $G^2[(XZ, YZ) \mid (XY, XZ, YZ)] = G^2(XZ, YZ) - G^2(XY, XZ, YZ)$. This test assumes that the homogeneous association model (XY, XZ, YZ) holds. It is a test of H_0 : all $\lambda_{ij}^{XY} = 0$ for this model.

When Y is binary, this test relates to logit models. The model for the logit of the probability π that $Y = 1$,

$$\text{logit}(\pi) = \alpha + \beta_i^X + \beta_k^Z,$$

corresponds to loglinear model (XY, XZ, YZ). The null hypothesis of X-Y conditional independence is H_0 : all $\beta_i^X = 0$ for this model. The likelihood-ratio test statistic is the difference between G^2 statistics for the reduced model $\text{logit}(\pi) = \alpha + \beta_k^Z$ and the full model for the three-way table.

For $2 \times 2 \times K$ tables, the test of conditional independence comparing two loglinear or logit models has the same purpose as the Cochran–Mantel–Haenszel (*CMH*) test (Section 3.2). The *CMH* test works well when the X-Y odds ratio is similar in each partial table. In this sense, it is also naturally directed toward the alternative of homogeneous association. For large samples, the model-based likelihood-ratio test usually gives similar results as the *CMH* test. In fact, the *CMH* procedure is an efficient score test (Section 4.5.2) of the hypothesis that the X-Y association parameters equal zero in the loglinear or logit model of homogeneous association.

7.3.2   Job Satisfaction and Income Example

Table 7.5, from the 1991 General Social Survey, refers to the relationship between job satisfaction (S) and income (I), stratified by gender (G). The test of the hypothesis of I-S conditional independence compares the conditional independence model (GI, GS) to model (IS, GI, GS). This analysis checks whether one can eliminate the I-S association term from model (IS, GI, GS), assuming that that model holds. The fit statistics are $G^2(GI, GS) = 19.4$, with $df = 18$, and $G^2(IS, GI, GS) = 7.1$, with $df = 9$. Comparing the models, $G^2[(GI, GS) \mid (IS, GI, GS)] = 19.4 - 7.1 = 12.3$, based on $df = 18 - 9 = 9$. This gives $P = .20$ and does not provide evidence of an association.

7.3.3   Direct Goodness-of-Fit Test

Another way to test the hypothesis of I-S conditional independence compares the model (GI, GS) directly to the saturated model. That is, one tests the hypothesis by performing a goodness-of-fit test of the model. For Table 7.5, $G^2(GI, GS) = 19.4$ with $df = 18$. This test of conditional independence has P-value of .37. Again, I-S conditional independence is plausible.

Table 7.5   Job Satisfaction and Income, Controlling for Gender

		Job Satisfaction			
Gender	Income	Very Dissatisfied	A Little Satisfied	Moderately Satisfied	Very Satisfied
Female	< 5000	1	3	11	2
	5000–15,000	2	3	17	3
	15,000–25,000	0	1	8	5
	> 25,000	0	2	4	2
Male	< 5000	1	1	2	1
	5000–15,000	0	3	5	1
	15,000–25,000	0	0	7	3
	> 25,000	0	1	9	6

Source: 1991 General Social Survey.

A statistic of form $G^2(XZ, YZ)$ does not require an assumption about homogeneous association. Since it is identical to $G^2[(XZ, YZ) \mid (XYZ)] = G^2(XZ, YZ) - G^2(XYZ)$, the null hypothesis for this test is H_0 : all $\lambda_{ij}^{XY} = 0$ and all $\lambda_{ijk}^{XYZ} = 0$ in the saturated loglinear model. The statistic could be large if there is three-factor interaction, or if there is no three-factor interaction but conditional dependence.

This test has the advantage of not assuming model structure, such as homogeneous association. A disadvantage is that it often has low power. If there truly is no (or little) three-factor interaction, the *CMH* test and the likelihood-ratio comparison statistic $G^2[(XY, YZ) \mid (XY, XZ, YZ)]$ are more likely to yield small P-values. In testing that the X-Y association parameters alone are zero, those chi-squared tests focus the analysis on fewer degrees of freedom. Capturing an effect with a smaller df value yields a test with greater power (Section 2.5.3). Unless the degree of heterogeneity in X-Y partial associations is severe, it is better to use the test having an unsaturated baseline model.

The baseline models in this test and the test using $G^2[(GI, GS) \mid (IS, GI, GS)]$ treat income and job satisfaction as nominal, but they are ordinal. More powerful tests of conditional independence exploit the ordinality. We next construct a test using an ordinal loglinear model.

7.3.4   Detecting Ordinal Conditional Association

Generalizations of the linear-by-linear association model (7.2.1) apply to modeling association between ordinal variables X and Y while controlling for a third variable that may be nominal or ordinal. A useful model,

$$\log \mu_{ijk} = \lambda + \lambda_i^X + \lambda_j^Y + \lambda_k^Z + \beta u_i v_j + \lambda_{ik}^{XZ} + \lambda_{jk}^{YZ}, \qquad (7.3.1)$$

is the special case of model (XY, XZ, YZ) that replaces the general X-Y association term λ_{ij}^{XY} by the simpler linear-by-linear (abbreviated, $L \times L$) term $\beta u_i v_j$ based on ordered row and column scores. The form of the $L \times L$ association is the same in each partial table, since the term $\beta u_i v_j$ does not depend on k. This model is appropriate when the X-Y association has the same positive or negative trend at each level of Z. The model is called a *homogeneous linear-by-linear association model*.

The conditional independence model (XZ, YZ) is the special case of this model with $\beta = 0$. One can use the ordinality of X and Y in testing conditional independence by comparing G^2 for model (7.3.1) to G^2 for model (XZ, YZ), or by forming the Wald statistic $z^2 = (\hat{\beta}/ASE)^2$. These statistics concentrate evidence about the association on a single degree of freedom. Unless model (7.3.1) fits very poorly, these tests are more powerful than tests that ignore the ordering.

For Table 7.5 with scores $\{1, 2, 3, 4\}$ for income and job satisfaction, the homogeneous $L \times L$ model (7.3.1) has $G^2 = 12.3$, with $df = 17$; by comparison, $G^2(GI, GS) = 19.4$ with $df = 18$. The difference equals 7.1, with $df = 1$, and has $P = .01$ for testing conditional independence. This contrasts with the lack of evidence provided by models that ignore the ordinality.

The positive ML estimate $\hat{\beta} = 0.388$, based on $ASE = 0.155$, reveals a tendency for job satisfaction to be greater at higher levels of income. The estimated local odds ratio for each gender is $\exp(0.388) = 1.47$. The Wald statistic $z^2 = (0.388/0.155)^2 = 6.3$, with $df = 1$, also provides strong evidence of an association ($P = .01$).

7.3.5 Generalized Cochran–Mantel–Haenszel Tests

Alternative tests of conditional independence generalize the Cochran–Mantel–Haenszel (CMH) statistic (3.2.1) to $I \times J \times K$ tables. Like the CMH statistic and the model-based statistic $G^2[(XZ, YZ) \mid (XY, XZ, YZ)]$, these statistics perform well when the X-Y association is similar at each level of Z. There are three versions, according to whether both, one, or neither of X and Y are treated as ordinal.

When X and Y are ordinal, the test statistic generalizes the correlation statistic (2.5.1) for two-way tables. It is designed to detect a linear trend in the X-Y association that has the same direction in each partial table. The statistic takes larger value as the correlation increases in magnitude and as the sample size grows in each table.

The generalized correlation statistic has approximately a chi-squared distribution with $df = 1$. Its formula is complex (Agresti (1990), p. 284), and we omit computational details since it is available in standard software. In fact, it is the efficient score statistic for testing that $\beta = 0$ in model (7.3.1). For large samples, it usually gives similar results as the likelihood-ratio statistic or the Wald statistic for that hypothesis.

For Table 7.5 with the row and column numbers as the scores, the sample correlation between income and job satisfaction equals .171 for females and .381 for males. The generalized correlation statistic equals 6.6 with $df = 1$ ($P = .01$), giving the same conclusion as the likelihood-ratio and Wald tests of the previous subsection.

Often scores other than the row and column numbers are more sensible. For instance, with grouped continuous variables, one might use scores that are midpoints of the class intervals. For Table 7.5, the row scores (3, 10, 20, 35) use midpoints of the middle two categories, in thousands of dollars. Alternatively, midrank scores are based on the relative numbers of observations in the various response categories. When in doubt about scoring, perform a sensitivity analysis by using a few different choices that seem sensible. Unless the categories exhibit severe imbalance in their totals, the choice of scores usually has little impact on the conclusion. Different choices yield similar results for Table 7.5. For the scores (3, 10, 20, 35) for income and (1, 3, 4, 5) for satisfaction, for instance, the generalized correlation statistic equals 6.2 with $df = 1$ ($P = .01$).

7.3.6 Detecting Nominal–Ordinal Conditional Association

When X is nominal and Y is ordinal, scores are relevant only for levels of Y. We summarize the responses of subjects within a given row by the mean of their scores on the ordinal variable and then average this row-wise mean information across the K strata. The test of conditional independence compares the I rows using a statistic based on the variation in those I averaged row mean responses that is designed to detect differences among their true values. It has a large-sample chi-squared distribution with $df = (I - 1)$. The power is strong when the differences among the row means are similar in each partial table.

The formula for the test statistic is complex (Agresti (1990), pp. 286–287). We present it only for the special case of a two-way table (i.e., $K = 1$), to illustrate the basic idea. For column scores $\{v_j\}$, let $\bar{y}_i = \sum_j v_j n_{ij}/n_{i+}$. The numerator sums the scores on Y for all subjects in row i, and the denominator is the sample size for that row. The measure $\bar{y}_i$ is the sample mean response on the ordinal variable Y in row i, for the chosen scores. Let $\bar{y} = \sum_j v_j n_{+j}/n$ denote the mean response on Y for the overall sample, using the column totals. The test statistic equals

$$(n - 1)\frac{\sum_i n_{i+}(\bar{y}_i - \bar{y})^2}{\sum_j v_j^2 n_{+j} - n\bar{y}^2}. \tag{7.3.2}$$

If Y were normally distributed, a one-way ANOVA would compare means of Y for I levels of a nominal variable X. Thus, statistic (7.3.2) is an analog of a one-way ANOVA statistic when Y is ordinal categorical rather than continuous. In fact, with midrank scores for $\{v_j\}$, it is the *Kruskal–Wallis* statistic for comparing mean ranks for I groups. When $I = 2$, it is identical to the correlation statistic (2.5.1), which then compares two row means. For K partial tables, the formula for the test of whether the row mean scores differ is complex, but it is available in standard software. It is the efficient score test for a generalization of model (7.3.1) in which the row scores are parameters.

For Table 7.5, this test treats job satisfaction as ordinal and income (the row variable) as nominal. The test searches for differences among the four income levels in their mean job satisfaction. Using scores $\{1, 2, 3, 4\}$, the mean job satisfaction at the four levels of income equal $(2.82, 2.84, 3.29, 3.00)$ for females and $(2.60, 2.78, 3.30, 3.31)$ for males. For instance, the mean for the 17 females with income < 5000 equals $[1(1) + 2(3) + 3(11) + 4(2)]/17 = 2.82$. The pattern of means is similar for each gender, roughly increasing as income increases. The generalized *CMH* statistic for testing whether the true row mean scores differ equals 9.2 with $df = 3$ $(P = .03)$.

Unlike this statistic, the correlation statistic of the previous subsection also treats the rows as ordinal. It detects a linear trend across rows in the row mean scores, and it utilizes the approximate increase in mean satisfaction as income increases. One can use the nominal-ordinal statistic based on variability in row means when X and Y are ordinal, but such a linear trend may not occur. For instance, one might expect responses on Y to tend to be higher in some rows than in others, without the mean of Y increasing consistently or decreasing consistently as the level of X increases. An analogous test treats X as ordinal and Y as nominal and compares J mean scores on X computed within columns. It has $df = J - 1$.

7.3.7 Detecting Nominal–Nominal Conditional Association

Another *CMH*-type statistic, based on $df = (I - 1)(J - 1)$, provides a "general association" test. It is designed to detect *any* type of association that is similar in

Table 7.6 Summary of Generalized Cochran–Mantel–Haenszel Tests of Conditional Independence for Table 7.5

Alternative Hypothesis	Statistic	df	P-value
General association	10.2	9	.34
Row mean scores differ	9.2	3	.03
Nonzero correlation	6.6	1	.01

each partial table. It treats both X and Y as nominal, so does not require category scores. Because of its complexity (Agresti (1990), pp. 234–235), we do not present its formula, but it is available in software. It is an efficient score test of X-Y conditional independence for loglinear model (XY, XZ, YZ). For a single table $(K = 1)$, the general association statistic is similar to the Pearson chi-squared statistic, equaling $[n/(n-1)]X^2$.

For Table 7.5, the general association statistic equals 10.2, with $df = 9$ ($P = .34$). We pay a price for ignoring the ordinality of job satisfaction and income. For ordinal variables, the general association test is usually not as powerful as narrower tests with smaller df values that use the ordinality.

Table 7.6 summarizes results of the three generalized CMH tests for $I \times J \times K$ tables applied to Table 7.5. (The format is similar to that used by SAS with the CMH option in PROC FREQ.) The *general association* alternative treats both X and Y as nominal, and has $df = (I - 1)(J - 1) = 9$. It is sensitive to any departure that is similar in each level of Z. The *row mean scores differ* alternative treats the rows of X as nominal and the columns of Y as ordinal and has $df = I - 1 = 3$. It is sensitive to variation among the I mean scores on Y computed within levels of X, when the nature of that variation is similar in each level of Z. Finally, the *nonzero correlation* alternative treats both X and Y as ordinal and has $df = 1$. Its test statistic is sensitive to a linear trend between X and Y that is similar in each level of Z. When $I = J = 2$, all three test statistics have $df = 1$ and simplify to the CMH statistic (3.2.1).

7.4 EFFECTS OF SPARSE DATA

This section discusses the effects of small cell counts on the fitting of loglinear models and logit models to contingency tables. Tables having many cells with small counts are said to be *sparse*. Sparse tables occur when the sample size is small. They also occur when the sample size is large but so is the number of cells. Sparseness is common in tables with many variables or with classifications having several categories.

7.4.1 Empty Cells

Sparse tables usually contain cells with zero counts. Such cells are called *empty cells* and are of two types: *sampling zeroes* and *structural zeroes*.

In most cases, even though a cell is empty, its true probability is positive. That is, it is theoretically possible to have observations in the cell, and a positive count would occur if the sample size were sufficiently large. This type of empty cell is called a *sampling zero*. The empty cells in Table 7.1 on drug use and in Table 7.5 on job satisfaction are sampling zeroes.

An empty cell in which observations are theoretically impossible is called a *structural zero*. Such cells have true probabilities equal to zero, and the cell count is zero regardless of the sample size. To illustrate, suppose that professors employed in a given department at the University of Rochester for at least five years were cross-classified on their current rank (assistant professor, associate professor, professor) and their rank five years ago. Professors cannot be demoted in rank, so three of the nine cells in the table contain structural zeroes. One of these is the cell corresponding to the rank of professor five years ago and assistant professor now; it cannot contain any observations. Contingency tables containing structural zeroes are called *incomplete tables*.

Sampling zeroes are part of the observed data set. For instance, a count of 0 is a possible outcome for a Poisson variate. It contributes to the likelihood function and the model-fitting process. A structural zero, on the other hand, is not an observation and is not part of the data. Sampling zeroes are much more common than structural zeroes, and the remaining discussion refers to them.

Sampling zeroes can affect the existence of ML estimates of loglinear and logit model parameters. When all cell counts are positive, parameter estimates are necessarily finite. When any marginal counts corresponding to qualitative terms in a model equal zero, infinite estimates occur for that term. For instance, when any X-Y marginal totals equal zero, infinite estimates occur among $\{\hat{\lambda}_{ij}^{XY}\}$ for loglinear models such as (XY, XZ, YZ) and (XY, XZ), and infinite estimates occur among $\{\hat{\beta}_i^X\}$ for the effect of X on Y in logit models.

A value of ∞ (or $-\infty$) for a parameter estimate means that the likelihood function keeps increasing as the parameter moves toward ∞ $(-\infty)$. Such results imply that ML fitted values equal 0 in some cells, and some odds ratio estimates have values of ∞ or 0. Most software cannot distinguish and does not indicate when infinite estimates occur. One potential sign is when the iterative process for fitting the model does not converge, typically because a parameter estimate keeps getting larger from cycle to cycle. Another sign is when the software reports large estimates, in relative terms, with very large estimated standard errors. Slight changes in the data then often cause dramatic changes in the estimates and their standard errors. A danger with sparse data is that one might not realize that a true estimated effect is infinite and, as a consequence, report estimated effects and results of statistical inferences that are invalid and highly unstable.

Empty cells and sparse tables can cause severe bias in estimators of odds ratios and poor chi-squared approximations for goodness-of-fit statistics. One remedy to estimation bias is to add a small constant to cell counts before conducting an analysis. For saturated models, adding $\frac{1}{2}$ to each cell reduces the bias in sample odds ratio estimators (Section 2.3.3). For instance, this shrinks infinite (or zero) estimates of odds ratios to finite values corresponding to positive probabilities in all the cells. For unsaturated models, though, adding $\frac{1}{2}$ to each cell before fitting the model smooths the data too much, causing havoc with sampling distributions. This operation has too conservative an influence on fitted odds ratios and test statistics.

Many ML analyses for unsaturated models are unharmed by empty cells. For instance, when a single cell is empty, finite estimates exist for all parameters in unsaturated models presented so far. In fact, they usually exist when all the marginal totals corresponding to terms in the model are positive. Even when a parameter estimate is infinite, this is not fatal to data analysis. Though an infinite estimate for an odds ratio is rather unsatisfactory, one can construct a confidence interval for the true odds ratio for which one bound is finite (Section 5.7.3).

When iterative fitting processes fail to converge because of infinite estimates, adding a very small constant (such as 10^{-8}) is adequate for ensuring convergence. One can then estimate parameters for which the true estimates are finite and are not affected by the empty cells, as the example in the following subsection shows. When in doubt about the effect of empty cells, one should perform a sensitivity analysis. Repeat the analysis by adding constants of various sizes, (say .00000001, .0001, .01, .1) in order to gauge their effect on parameter estimates and goodness-of-fit statistics. The total count added should be only a tiny percentage of the total sample size. Also, for each possibly influential observation, delete it or move it to another cell to see how much the results vary with small perturbations to the data. Often, some associations are not affected by the empty cells and give stable results for the various analyses, whereas others that are affected are highly unstable. Use caution in making conclusions about an association if small changes in the data are highly influential. In some cases, it makes sense to fit the model by excluding part of the data containing empty cells or by combining that part with other parts of the data.

An alternative to ML estimation, using Bayesian methods, provides a way of smoothing data in a less *ad hoc* manner than adding arbitrary constants to cells. Bayesian methods are beyond the scope of this text, but Bishop et al. ((1975), Ch. 12) and Agresti ((1990), Sec. 13.4) describe their use for dealing with sparse data.

7.4.2   Clinical Trials Example

Table 7.7 shows results of a clinical trial conducted at five centers. The purpose was to compare an active drug to placebo in terms of a binary (success, failure) response for treating fungal infections. For these data, let C = Center, T = Treatment (Active drug or Placebo), and R = Response.

Centers 1 and 3 had no successes. Thus, the 6×2 marginal table relating center to response, collapsed over treatment, contains zero counts. This marginal table is shown in the last two columns of Table 7.7. Infinite ML estimates occur for terms in loglinear or logit models containing the C-R association. An example is the logit model containing main effects for C and T in their effects on R. The likelihood function continually increases as the parameters for Centers 1 and 3 decrease toward $-\infty$; that is, as the logit decreases toward $-\infty$, so the fitted probability of success decreases toward 0 for those centers. Most software reports estimates that

Table 7.7 Clinical Trial Relating Treatment (T) to Response (R) for Six Centers (C), with T-R and C-R Marginal Tables

| Center | Treatment | Response | | C-R Marginal | |
		Success	Failure	Success	Failure
1	Active Drug	0	5	0	14
	Placebo	0	9		
2	Active Drug	1	12	1	22
	Placebo	0	10		
3	Active Drug	0	7	0	12
	Placebo	0	5		
4	Active Drug	6	3	8	9
	Placebo	2	6		
5	Active Drug	5	9	7	21
	Placebo	2	12		
T-R	Active Drug	12	36		
Marginal	Placebo	4	42		

Source: Diane Connell, Sandoz Pharmaceuticals Corp.

are truly infinite as large numbers with large standard errors. For instance, when SAS (GENMOD) fits the logit model (setting the center estimate to be 0 for Center 5), the reported center estimates for Centers 1 and 3 are both about -26 with standard errors of about 200,000.

The counts in the 2×2 marginal table relating treatment to response, shown in the bottom panel of Table 7.7, are all positive. The empty cells in Table 7.7 affect the center estimates, but not the treatment estimates, for this logit model. For instance, if we add any positive constant to each cell, the fitting process converges, all center parameter estimates being finite; moreover, the treatment effects and goodness of fit are stable, as the addition of any such constant less than 0.001 yields an estimated log odds ratio equal to 1.55 for the treatment effect ($ASE = 0.70$) and a G^2 goodness-of-fit statistic equal to 0.50.

This treatment estimate also results from deleting Centers 1 and 3 from the analysis. When a center contains responses of only one type, it provides no information about the association between treatment and response. In fact, such tables also make no contribution to the Cochran–Mantel–Haenszel test (Section 3.2.1) or to the exact test of conditional independence between treatment and response (Section 3.3.1).

An alternative strategy in multi-center analyses combines centers of a similar type. Then, if each resulting partial table has responses with both outcomes, the inferences use all data. This, however, affects somewhat the interpretations and conclusions made from those inferences. For Table 7.7, perhaps Centers 1 and 3 are similar to Center 2, since the success rate is very low for that center. Combining these three centers and re-fitting the model to this table and the tables for the other two centers yields an estimated treatment effect of 1.56 ($ASE = 0.70$), with $G^2 = 0.56$.

7.4.3   Effect of Small Samples on X^2 and G^2

The true sampling distributions of goodness-of-fit statistics converge to chi-squared as the sample size $n \to \infty$, for a fixed number of cells N. The adequacy of the chi-squared approximation depends both on n and N. It tends to improve as n/N, the average number of observations per cell, increases.

The quality of the approximation has been studied carefully for the Pearson X^2 test of independence for two-way tables. Most guidelines refer to the fitted values. When $df > 1$, a minimum fitted value of about 1 is permissible as long as no more than about 20% of the cells have fitted values below 5. The size of permissible fitted values decreases as N increases. However, the chi-squared approximation can be poor for sparse tables containing both very small and very large fitted values. Unfortunately, a single rule cannot cover all cases.

The X^2 statistic tends to be valid with smaller samples and sparser tables than G^2. The distribution of G^2 is usually poorly approximated by chi-squared when n/N is less than 5. Depending on the sparseness, P-values based on referring G^2 to a chi-squared distribution can be too large or too small. When most fitted values are smaller than 0.5, treating G^2 as chi-squared gives a highly conservative test; that is, when H_0 is true, reported P-values tend to be much larger than true ones. When most fitted values are between about .5 and 5, G^2 tends to be too liberal; the reported P-value tends to be too small.

For fixed values of n and N, the chi-squared approximation is better for tests with smaller values of df. For instance, consider tests of conditional independence in $I \times J \times K$ tables. The statistic $G^2[(XZ, YZ) \mid (XY, XZ, YZ)]$, which has $df = (I - 1)(J - 1)$, is closer to chi-squared than $G^2(XZ, YZ)$, which has $df = K(I - 1)(J - 1)$. The ordinal test based on the homogeneous $L \times L$ association model (7.3.1) has $df = 1$, and behaves even better. It is difficult to provide general guidelines about how large n must be. The adequacy of model-comparison tests depends more on the two-way marginal totals than on cell counts. Cell counts can be small (which often happens when K is large) as long as most totals in the two-way marginal tables exceed about 5.

When cell counts are so small that chi-squared approximations may be inadequate, one could combine categories of variables to obtain larger counts. This is usually not advisable unless there is a natural way to combine them and little information loss in defining the variable more crudely. In any case, poor sparse-data performance of chi-squared tests is becoming less problematic because of the development of exact small-sample methods. This text has presented several exact tests, such as Fisher's exact test for 2 × 2 tables and analogous exact tests for $I \times J$ tables (Section 2.6), an exact test of conditional independence for 2 × 2 × K tables (Sec. 3.3), and exact tests for logistic regression (Section 5.7).

Exact analyses are now feasible due to recent improvements in computer power and sophistication of algorithms. For instance, the StatXact software conducts many exact inferences for two-way and three-way tables and LogXact handles exact inference in logistic regression. In principle, exact inferences about parameters or about goodness of fit exist for any loglinear or logit model, and software should soon enable us to conduct exact analyses for general situations.

7.5 SOME MODEL-FITTING DETAILS*

Most nonstatisticians will not have the interest or the prerequisite theoretical background to understand all the technical details underlying loglinear and logit model-fitting and inference. This is not a handicap for applying the methods. Thus, this text has omitted theoretical derivations in favor of emphasizing application and interpretation of models. With available software, one can fit models and analyze data sets without understanding how one maximizes likelihood functions, derives standard error formulas, proves large-sample chi-squared distributions for X^2 and G^2, and so forth. This section provides a brief introduction to these topics by giving a heuristic discussion of some loglinear and logit model-fitting details.

7.5.1 Sufficient Statistics and Likelihood Equations

Loglinear and logit models have fitted values and ML estimates of model parameters that depend on the data only through certain *sufficient statistics*. One can replace the data by these summary statistics without losing any information needed to fit the model.

For models with qualitative factors, the sufficient summaries of the data are marginal counts for terms in the model. For loglinear model (XZ, YZ), for instance, the sufficient statistics are the X-Z and Y-Z two-way marginal tables $\{n_{i+k}\}$ and $\{n_{+jk}\}$. Every three-way table having the same entries in these two marginal tables has the same fit.

The ML parameter estimates provide fitted values that satisfy the model and maximize the likelihood function. The estimates are solutions to a set of *likelihood equations*, which equate the fitted values to the sufficient statistics. For instance, model (XZ, YZ) has likelihood equations

$$\hat{\mu}_{i+k} = n_{i+k}, \qquad \hat{\mu}_{+jk} = n_{+jk}.$$

The fitted values satisfy the model but have the same X-Z and Y-Z marginal totals as the observed data. The formula for the fitted values for model (XZ, YZ) is $\hat{\mu}_{ijk} = n_{i+k}n_{+jk}/n_{++k}$. One can verify from this formula that the likelihood equations hold; that is, the X-Z and Y-Z marginals of $\{\hat{\mu}_{ijk}\}$ equal the corresponding observed totals.

Fitted values for loglinear models have similarities to the sample data, since certain marginal totals are the same for each. The fitted values are smoothed versions of the sample counts that match them in those margins, but which have associations and interactions satisfying the model. For instance, though the fitted values for model (XZ, YZ) match the data in the X-Z and Y-Z margins, they have X-Y conditional odds ratios equal to 1.0 at each level of Z. Analogous results apply to logistic regression models. For either model type, a parameter estimate is infinite when its sufficient statistic takes its maximum or minimum possible value. This happens, for instance, when a sufficient marginal total for a loglinear model equals zero.

Many loglinear and logit models do not have direct ML estimates. Unlike model (XZ, YZ), they require iterative algorithms for solving likelihood equations to produce fitted values and parameter estimates. The most popular iterative procedure is the *Newton–Raphson* method. This method, described in Section 4.5.1, maximizes successive parabolic approximations for the log likelihood function. Calculations for doing this involve solving a system of linear equations at each step. The parameter values that maximize the parabolic functions serve as successive approximations for the ML estimates.

7.5.2 Asymptotic Chi-Squared Distributions

We next sketch the reason that X^2 and G^2 statistics for testing model fit have large-sample chi-squared distributions. Consider the sampling scheme by which N cell counts $\{n_i\}$ are Poisson variates with means $\{\mu_i\}$. For simplicity, we use a single subscript, though the table may have any dimension.

The X^2 statistic has form $X^2 = \sum e_i^2$, where $e_i = (n_i - \hat{\mu}_i)/\sqrt{\hat{\mu}_i}$ is the Pearson residual for cell i. This residual estimates $(n_i - \mu_i)/\sqrt{\mu_i}$, which has a large-sample standard normal distribution, since the Poisson distribution is approximately normal when its mean, which is also its variance, is large. Squaring standard normal variates produces chi-squared variates with $df = 1$. Adding N independent chi-squared variates with $df = 1$ yields a chi-squared variate with $df = N$. Thus, $\sum (n_i - \mu_i)^2/\mu_i$ has an approximate chi-squared distribution with df equal to the number of cells, N.

The df for X^2 do not equal N, however, because substituting $\{\hat{\mu}_i\}$ for $\{\mu_i\}$ in the Pearson residuals $\{e_i\}$ reduces their variance and yields correlated values. The df equal the rank of the covariance matrix of these residuals. This depends on the complexity of the model, equaling the number of cells minus the number of nonredundant model parameters.

Expanding G^2 in a Taylor series approximation, one obtains X^2 from the first two terms. Under the null hypothesis that the model holds, the higher-order terms in the expansion are negligible as the sample size increases. In other words, X^2 is then a quadratic approximation for G^2. The two statistics have the same large-sample chi-squared distribution as the sample size n increases. When the null hypothesis is false, X^2 and G^2 tend to increase as n increases, but their limiting *noncentral* chi-squared distributions can be quite different.

7.5.3 Comparing Nested Models

Many model-building procedures involve comparing the fits of two models, when one is a special case of the other. Our discussion here pertains to a GLM of any type for categorical data. For an arbitrary model, denoted by M, let $G^2(M)$ denote the value of G^2 for testing the fit of the model. In GLM terminology, this is the model's *deviance* (Section 4.5.3).

Consider two models, M_0 and M_1, such that M_0 is simpler than M_1. For instance, M_1 could be loglinear model (XY, XZ, YZ), and M_0 could be model (XZ, YZ). Model M_0 is *nested* within M_1, being a special case in which certain parameters equal zero. Since M_1 is more complex than M_0, it has a larger set of parameter values to search over in maximizing the likelihood function. Thus, the fit of M_1 is better, in the sense that necessarily

$$G^2(M_1) \leq G^2(M_0).$$

A test comparing the models checks whether the more complex model M_1 gives a better fit than the simpler model M_0. The test of ($H_0 : M_0$ holds) against ($H_a : M_1$ holds) analyzes whether the extra terms in M_1 that are not in M_0 equal zero. Assuming that model M_1 holds, the likelihood-ratio approach for testing that M_0 holds uses test statistic

$$G^2(M_0 \mid M_1) = G^2(M_0) - G^2(M_1).$$

We used statistics of form $G^2(M_0 \mid M_1)$ for loglinear models in Sections 6.3.4, 7.1.5, 7.2.3, and 7.3.1 to test whether a model term equals zero. The simpler model M_0 omits the term, and the more complex model M_1 contains it. We also used this test in Sections 5.3.2 and 5.5.2 to compare nested logistic regression models.

Theory states that $G^2(M_0 \mid M_1)$ is a large-sample chi-squared statistic when the parameter spaces are fixed for M_0 and M_1 as n increases. The df value measures the difference between the number of parameters for the two models. This large-sample theory is not always appropriate in practice. An example is when M_0 is a logistic regression model with continuous predictors and M_1 is the saturated model, so the test refers to the goodness of fit of M_0. In that case, one usually observes data at additional levels of the predictors as the sample size increases. The saturated model has a separate parameter at each combination of predictor levels, so its number of parameters increases with the sample size. Thus, the parameter space is not fixed for M_1, and the df value comparing its size to the number of parameters for the simpler model is not fixed, invalidating the chi-squared theory. One can, however, compare the fits of two unsaturated logistic regression models. Their numbers of parameters and the difference df between them stays fixed as n increases. For a given sample size, the chi-squared approximation tends to be better for smaller values of df, such as when the two models differ by just one term.

Let $\{\hat{\mu}_{0i}\}$ and $\{\hat{\mu}_{1i}\}$ denote fitted values for models M_0 and M_1. One can show that the likelihood-ratio statistic for comparing the models also equals

$$G^2(M_0 \mid M_1) = 2 \sum \hat{\mu}_{1i} \log \left(\frac{\hat{\mu}_{1i}}{\hat{\mu}_{0i}} \right). \tag{7.5.1}$$

This statistic has the form of the usual G^2 statistic, but with $\{\hat{\mu}_{1i}\}$ in place of the observed cell counts. In fact, $G^2(M_0)$ is the special case of $G^2(M_0 \mid M_1)$ with M_1 being the saturated model, in which case the fitted values $\{\hat{\mu}_{1i}\}$ for M_1 are simply the cell counts $\{n_i\}$. For the Pearson statistic X^2, the difference $X^2(M_0) - X^2(M_1)$ for nested models is not necessarily nonnegative. A more appropriate Pearson statistic for comparing nested models is

$$X^2(M_0 \mid M_1) = \sum \frac{(\hat{\mu}_{1i} - \hat{\mu}_{0i})^2}{\hat{\mu}_{0i}}. \tag{7.5.2}$$

The Pearson X^2 for testing the fit of a model has this form with M_1 as the saturated model.

These remarks relate to the results in Section 2.4.6 on the partitioning of chi-squared. The likelihood-ratio statistic for the simpler model partitions into

$$G^2(M_0) = G^2(M_1) + [G^2(M_0) - G^2(M_1)] = G^2(M_1) + G^2(M_0 \mid M_1),$$

a statistic for testing M_1 and a statistic for testing M_0 given that M_1 holds. By contrast, $X^2(M_0)$ does not partition exactly into $X^2(M_1) + X^2(M_0 \mid M_1)$.

7.5.4 Distribution of Parameter Estimators

Finally, we discuss standard errors for ML estimates of Poisson loglinear model parameters and binomial logit parameters. The standard errors are square roots of variances, which are the diagonal elements of the covariance matrix of the parameter estimators. The estimated covariance matrix is the inverse of a matrix called the *information matrix*, which is a by-product of the Newton–Raphson fitting procedure. The information matrix measures the curvature of the log likelihood function at the ML estimates. More highly curved log likelihood functions yield greater information about the parameter values; this results in smaller elements of the inverse of the information matrix and smaller standard errors.

We first illustrate this for loglinear models for Poisson counts. Let $\boldsymbol{\mu}$ denote a column vector of expected frequencies. Loglinear models for Poisson cell means have form

$$\log \boldsymbol{\mu} = \mathbf{X}\boldsymbol{\beta}. \qquad (7.5.3)$$

The matrix $\mathbf{X}$, called a *model matrix* or *design matrix*, contains known constants. The column vector $\boldsymbol{\beta}$ contains the parameters. The log means are linearly related to the parameters, so the model is "loglinear." To illustrate, consider the independence model, $\log \mu_{ij} = \lambda + \lambda_i^X + \lambda_j^Y$, for a 2×2 table. The model has three nonredundant parameters. For the constraints $\lambda_2^X = \lambda_2^Y = 0$, these are $\boldsymbol{\beta} = (\lambda, \lambda_1^X, \lambda_1^Y)$. Expression (7.5.3) is then

$$\begin{pmatrix} \log \mu_{11} \\ \log \mu_{12} \\ \log \mu_{21} \\ \log \mu_{22} \end{pmatrix} = \begin{pmatrix} 1 & 1 & 1 \\ 1 & 1 & 0 \\ 1 & 0 & 1 \\ 1 & 0 & 0 \end{pmatrix} \begin{pmatrix} \lambda \\ \lambda_1^X \\ \lambda_1^Y \end{pmatrix}$$

For instance, $\log \mu_{12} = \lambda + \lambda_1^X + \lambda_2^Y = \lambda + \lambda_1^X$, so the second row of $\mathbf{X}$ is $(1, 1, 0)$.

The model matrix $\mathbf{X}$ occurs in the expression for the covariance matrix of parameter estimators. As the sample size n increases, the ML estimator $\hat{\boldsymbol{\beta}}$ has distribution approaching the normal. The estimated covariance matrix of $\hat{\boldsymbol{\beta}}$ is

$$\hat{\text{Cov}}(\hat{\boldsymbol{\beta}}) = [\mathbf{X}'\mathbf{Diag}(\hat{\boldsymbol{\mu}})\mathbf{X}]^{-1}, \qquad (7.5.4)$$

where the diagonal matrix has the fitted values on the main diagonal. The elements on the main diagonal of the covariance matrix are the large-sample variances, and their square roots are the standard errors. As the fitted values increase, the standard errors decrease.

Logistic regression models have a similar formula. Let π denote the vector of "success" probabilities at the various settings of the explanatory variables. The model has form

$$\text{logit}(\pi) = X\beta.$$

Let $\hat{\pi}$ denote the estimated probabilities for the model fit, with value $\hat{\pi}_i$ for the n_i observations at the ith setting of explanatory variables. The large-sample estimated covariance matrix of the ML estimator $\hat{\beta}$ equals

$$\hat{\text{Cov}}(\hat{\beta}) = [X'\text{Diag}X]^{-1}, \tag{7.5.5}$$

where **Diag** denotes a diagonal matrix having elements $n_i \hat{\pi}_i (1 - \hat{\pi}_i)$ on the main diagonal. The standard errors decrease as the sample size increases. This can happen by $\{n_i\}$ increasing or by more observations occurring at additional settings of the explanatory variables.

PROBLEMS

7.1. Draw the association graph for loglinear model (WXZ, WYZ). Which, if any, variables are conditionally independent in this model?

7.2. For a four-way table, are X and Y independent, given Z alone, for model (i) (WX, XZ, YZ, WZ), (ii) (WX, XZ, YZ, WY)?

7.3. Refer to Problem 6.3 with Table 3.1. Show the association graph for model (DV, PV), and fit the model. Using the fitted values, compute the estimated P-V odds ratios at the two levels of D, and compute the marginal P-V odds ratio. Compare the values. Are the collapsibility conditions satisfied?

7.4. Refer to the clinical trial in Problem 3.10 with Table 3.6.

   a. Fit logit model (7.1.1). Using the fitted values, compute the estimated odds ratio between group and response (i) for each center, (ii) for the marginal table. Why do they differ?

   b. Fit the simpler logit model deleting the center effects to this $2 \times 2 \times 3$ table. Using the fitted values, compute the estimated odds ratio between group and response (i) for each center, (ii) for the marginal table. When this model fits well, can we collapse over centers? Compare the fit to the model in (a). Is the simpler model adequate?

7.5. Refer to Problem 6.13 with Table 6.17.

   a. Show that model (CE, CH, CL, EH, EL, HL) fits well. Show that model (CEH, CEL, CHL, EHL) also fits well but does not provide a significant improvement. Beginning with (CE, CH, CL, EH, EL, HL), show that backward elimination yields (CE, CL, EH, HL). Interpret its fit.

b. Based on the association graph for (CE, CL, EH, HL), (i) show that every path between C and H involves a variable in $\{E, L\}$; (ii) explain why, collapsing over H, one obtains the same associations between C and E and between C and L, and collapsing over C, one obtains the same associations between H and E and between H and L; (iii) explain why the conditional independence patterns between C and H and between E and L are not collapsible.

7.6. Refer to Problem 6.12 with Table 6.16 and the loglinear model you selected for those data. Draw the association graph for the model. Remark on conditional independence patterns. For each pair of variables, indicate whether the fitted marginal and partial associations are identical.

7.7. Fit the model to Table 7.1 having association graph portrayed in Section 7.1.5.

   a. Explain why the A-M conditional odds ratio is unchanged by collapsing over race, but it is not unchanged by collapsing over gender.

   b. Examine and interpret the associations among the drug-use variables and the effects of gender and race on those responses.

7.8. Consider model (WX, YZ). Applying the collapsibility conditions in Section 7.1.4 with sets $A = \{W, X\}$, $C = \{Y, Z\}$, and with B empty, explain why the W-X partial association is the same as the W-X marginal association, collapsed over Y and Z.

7.9. Show that model (WX, WY, WZ, XY, XZ, YZ) or any more complex model for four variables has the same association graph. Explain why each pair of variables may have differing partial and marginal associations.

7.10. Consider loglinear model (WX, WY, WZ, XZ, YZ). Though X and Y are conditionally independent, explain why they may be marginally dependent, collapsing over W and Z. Explain why the W-X and X-Z associations are unchanged when one collapses over Y, and the W-Y and Y-Z associations are unchanged when one collapses over X.

7.11. Show that when any variable is independent of all others, collapsing over that variable does not affect other model terms. Illustrate using model (W, XY, XZ, YZ), showing that associations among X, Y, and Z are the same as in model (XY, XZ, YZ).

7.12. Consider logit models for a four-way table in which X_1, X_2, and X_3 are predictors of Y. When the table is collapsed over X_3, indicate whether the association between X_1 and Y remains unchanged, for the model (i) that has main effects of all predictors, (ii) that has main effects of X_1 and X_2 but assumes no effect for X_3.

7.13. Table 7.8 is taken from the 1991 General Social Survey. Subjects were asked whether methods of birth control should be available to teenagers between the ages of 14 and 16, and how often they attend religious services.

   a. Fit the independence model, and use residuals to describe lack of fit.

   b. Using equally spaced scores, fit the linear-by-linear association model. Describe the association. Test goodness of fit, test independence in a way that uses the ordinality, and interpret.

    c. Repeat **(b)** using row and column scores that reflect your perceived distances between categories. Are the results substantively any different?

Table 7.8

	Teenage Birth Control			
Religious Attendance	Strongly Disagree	Disagree	Agree	Strongly Agree
Never	49	49	19	9
Less than once a year	31	27	11	11
Once or twice a year	46	55	25	8
Several times a year	34	37	19	7
About once a month	21	22	14	16
2–3 times a month	26	36	16	16
Nearly every week	8	16	15	11
Every week	32	65	57	61
Several times a week	4	17	16	20

Source: 1991 General Social Survey.

7.14. Refer to Table 2.5. Treat party identification as ordinal by fitting the linear-by-linear association model. Conduct a likelihood-ratio test of independence for that model. Interpret, and compare results to the ordinary likelihood-ratio test of independence.

7.15. Refer to Problem 2.21 with Table 2.15. Test the hypothesis of independence by testing the fit of the independence model. Using an ordinal model, describe the association and conduct an alternative test that may be more powerful. Interpret.

7.16. For Table 7.3, fit the linear-by-linear association model using scores $\{1, 2, 4, 5\}$. Explain why the fitted local log odds ratio using levels 2 and 3 of each classification equals four times the fitted local log odds ratio using levels 1 and 2 of each or 3 and 4 of each. What is the relation between the odds ratios?

7.17. Political ideology is often measured using the categories (Liberal, Moderate, Conservative). For the 1991 General Social Survey, Democrats had counts $(161, 171, 96)$ in these categories, whereas Republicans had counts $(76, 148, 183)$. Analyze these data.

7.18. Refer to Problem 3.10 with Table 3.6. Test conditional independence of group and response, given center, by comparing two logit or loglinear models. Using a relevant model, estimate the conditional odds ratio between group and response, given center. Interpret.

7.19. Refer to Table 3.3 in Section 3.2. Using a model-based procedure, test conditional independence and estimate the strength of conditional association between smoking and lung cancer. Compare results with the *CMH* approach to testing and estimation.

7.20. Refer to Problem 6.6 with Table 6.13.

   a. Fit model (AD, AG, DG). Using the fitted values, estimate the A-G conditional odds ratio and the A-G marginal odds ratio. Explain why they give different indications of the A-G association.

   b. Is it proper to use $G^2[(AD, DG) \mid (AD, AG, DG)]$ to test conditional independence of A and G? Explain.

   c. From Problem 6.6, the adjusted residuals for model (AD, DG) suggest that Department 1 differs from the others. Using only Departments 2 through 6, fit models (AD, DG) and (AD, AG, DG) and test A-G conditional independence. Interpret.

   d. Explain how results in (a)–(c) apply to corresponding logit models, with A as the response.

7.21. Refer to Table 7.5. Perform a sensitivity analysis to check the dependence of the analyses in Section 7.3 on the choice of scores.

7.22. The sample in Table 7.5 consists of 104 black Americans. A similar table relating income and job satisfaction for white subjects in the 1991 General Social Survey had counts (by row) of $(3, 10, 30, 27/7, 8, 45, 39/8, 7, 46, 51/4, 2, 28, 47)$ for females and $(1, 4, 9, 9/1, 2, 37, 29/0, 10, 35, 39/7, 14, 69, 109)$ for males. Analyze these data.

7.23. Refer to the data in the previous problem. Test the hypothesis of conditional independence, (a) using ordinary loglinear models, (b) using the category orderings. Interpret, and compare results.

7.24. Refer to the data in Problem 7.22.

   a. Test conditional independence against the alternative that the mean job satisfaction varies by level of income, controlling for gender. Report and interpret a relevant test statistic.

   b. Conduct the test and interpret results for the alternative that mean income varies by level of job satisfaction, controlling for gender.

   c. Conduct the test using a general association statistic not designed to detect any particular pattern of association. Interpret, and indicate the extent to which your conclusions suffer from not using the ordinality.

7.25. For $K = 1$, the generalized *CMH* correlation statistic equals (2.5.1). When there truly is a trend, Section 2.5.3 noted that this test is more powerful than the X^2 and G^2 tests of Section 2.4. To illustrate, for Table 7.5, construct the marginal 4×4 table relating job satisfaction and income. Show that the Pearson test (2.4.3) has $X^2 = 11.5$ with $df = 9$ $(P = .24)$. Show that the correlation statistic with equally-spaced scores equals 7.6 based on $df = 1$ $(P = .006)$.

7.26. Table 7.9, taken from the 1991 General Social Survey, shows the relation between political party affiliation and political ideology, stratified by gender. Analyze these data.

Table 7.9

		Political Ideology				
Gender	Political Party	Very Liberal	Slightly Liberal	Moderate	Slightly Conservative	Very Conservative
Female	Democratic	44	47	118	23	32
	Republican	18	28	86	39	48
Male	Democratic	36	34	53	18	23
	Republican	12	18	62	45	51

Source: 1991 General Social Survey.

7.27. Refer to the fit of the homogeneous $L \times L$ model in Section 7.3.4. Describe the conditional association using the fitted odds ratio for the corner cells at the extreme categories of income and job satisfaction. Interpret, and compare to the local odds ratios. Show that the estimated odds ratio for the corner cells using the fitted values for model (GI, GS, IS) equals ∞. (For each gender the fitted value for the cell at the highest income and lowest satisfaction is zero). For each gender, the sample value is also ∞. Thus, an advantage of using the simple $L \times L$ model is that the model estimate shrinks sample estimates of ∞ for odds ratios to more realistic values.

7.28. Explain why replacing β in model (7.3.1) by β_k gives a *heterogeneous $L \times L$ model*. The fit is equivalent to fitting the $L \times L$ association model for two-way tables separately for each gender. Fit the heterogeneous model to Table 7.5, and test whether it fits significantly better than the homogeneous model (7.3.1).

7.29. Provide an example of contingency tables in which certain cells contain (a) structural zeroes, (b) sampling zeroes.

7.30. Show that a single cell containing any positive count makes a large contribution to X^2 if the fitted value is close to 0. To illustrate, calculate the contribution of (a) a count of 1 in a cell having a fitted value of .01, (b) the count of 2 in the cell having a fitted value of 0.24 for model (AM, CM) fitted to Table 6.3.

7.31. Suppose that all row and column marginal totals of a two-way table are positive, but some cells are empty. Show that all fitted values for the loglinear model of independence (6.1.1) are positive.

7.32. Consider Table 7.10, from a study of nonmetastatic osteosarcoma described in A. M. Goorin, *J. Clinical Oncology, 5:* 1178–1184 (1987) and the *LogXact Turbo User Manual* (1993, p. 5–22). The response is whether the subject achieved a three-year disease-free interval.

   a. Show that each predictor has a significant effect when it is used individually without the other predictors.

   b. Try to fit a main-effects logistic regression model containing all three predictors. Explain why the ML estimate for the effect of lymphocytic infiltration is infinite.

    c. Using conditional logistic regression, (i) conduct an exact test of the hypothesis of no effect of lymphocytic infiltration, controlling for the other variables; (ii) compute a 95% confidence interval for the effect. Interpret results.

Table 7.10

Lymphocytic Infiltration	Sex	Osteoblastic Pathology	Disease-Free Yes	No
High	Female	No	3	0
		Yes	2	0
	Male	No	4	0
		Yes	1	0
Low	Female	No	5	0
		Yes	3	2
	Male	No	5	4
		Yes	6	11

Source: LogXact-Turbo User Manual, Cambridge, MA: Cytel Software (1993), p. 5–23.

7.33. For the model $\text{logit}(\pi) = \alpha + \beta x$, let (x_i, y_i) denote the x and y values for subject i, $i = 1, \ldots, N$. Suppose $y_i = 0$ for all x below some point and $y_i = 1$ for all x above that point. Explain intuitively why $\hat{\beta} = \infty$. (*Note:* In technical terms, the sufficient statistics are $\sum y_i$ for α and $\sum x_i y_i$ for β. For a given $\sum y_i$, $\sum x_i y_i$ takes its maximum possible value for this assignment of $\{y_i\}$ to fixed $\{x_i\}$.)

7.34. Refer to $\{\hat{\mu}_{ijk} = n_{i+k}n_{+jk}/n_{++k}\}$ for model (XZ, YZ). Show that these fitted values have the same X-Z and Y-Z marginal totals as the observed data. For $2 \times 2 \times K$ tables, show that $\hat{\theta}_{XY(k)} = 1$. Illustrate these results for the fit of model (AM, CM) to Table 6.3.

7.35. Refer to formula (7.5.3) applied to the independence model for a 2×2 table. Show that for the constraints $\lambda_1^X = \lambda_1^Y = 0$, for which $\boldsymbol{\beta} = (\lambda, \lambda_2^X, \lambda_2^Y)$, $\mathbf{X}$ has rows $(1, 0, 0)$, $(1, 0, 1)$, $(1, 1, 0)$, $(1, 1, 1)$.

7.36. For Table 6.3, show the matrix representation of loglinear model (AC, AM, CM). Specify the parameter constraints used in your model matrix. Show the matrix representation of the corresponding logit model, when M is a response.

7.37. A GLM has form $g(\boldsymbol{\mu}) = \mathbf{X}\boldsymbol{\beta}$, for a monotone function g. Explain what each symbol in this formula represents for fitting the ordinary linear regression model to n observations on a normally distributed response and a single predictor.

CHAPTER 8

Multicategory Logit Models

This chapter presents generalizations of logistic regression models that handle multicategory responses. Section 8.1 presents models for *nominal* response variables, and Sections 8.2 and 8.3 present models for *ordinal* response variables. As in ordinary logistic regression modeling, explanatory variables can be continuous and/or categorical.

At each combination of levels of the explanatory variables, the models assume that the response counts for the categories of Y have a *multinomial* distribution. This generalization of the binomial applies when the number of response categories exceeds two. Logistic regression models are a special case of these models for binary responses.

Like logistic regression models and unlike loglinear models, multicategory logit models treat one classification as a response and the other variables as explanatory. Nevertheless, when explanatory variables are categorical, some of these models are equivalent to loglinear models for contingency tables.

8.1 LOGIT MODELS FOR NOMINAL RESPONSES

Suppose Y is a nominal variable with J categories. The order of listing the categories is irrelevant. Let $\{\pi_1, \ldots, \pi_J\}$ denote the response probabilities, satisfying $\sum_j \pi_j = 1$. When one takes n independent observations based on these probabilities, the probability distribution for the number of outcomes that occur of each of the J types is the multinomial. It specifies the probability for each possible way of allocating the n observations to the J categories. Here, we simply mention this as a sampling model for the counts, and we will not need to calculate such probabilities. For the case $J = 2$, this is the binomial distribution (1.2.2).

Multicategory (also called *polychotomous*) logit models simultaneously refer to all pairs of categories, and describe the odds of response in one category instead of another. Once the model specifies logits for a certain $(J - 1)$ pairs of categories, the rest are redundant.

8.1.1 Baseline-Category Logits

Logit models for nominal responses pair each response category with a baseline category, the choice of which is arbitrary. When the last category (J) is the baseline, the *baseline-category logits* are

$$\log\left(\frac{\pi_j}{\pi_J}\right), \qquad j = 1, \ldots, J - 1.$$

Given that the response falls in category j or category J, this is the log odds that the response is j. For $J = 3$, for instance, the logit model uses $\log(\pi_1/\pi_3)$ and $\log(\pi_2/\pi_3)$.

The logit model using the baseline-category logits with a predictor x has form

$$\log\left(\frac{\pi_j}{\pi_J}\right) = \alpha_j + \beta_j x, \qquad j = 1, \ldots, J - 1. \tag{8.1.1}$$

The model consists of $J - 1$ logit equations, with separate parameters for each. That is, the effects vary according to the response category paired with the baseline. When $J = 2$, this model simplifies to a single equation for $\log(\pi_1/\pi_2) = \mathrm{logit}(\pi_1)$, resulting in the ordinary logistic regression model for binary responses.

Parameters in the $J - 1$ equations (8.1.1) determine parameters for logits using all other pairs of response categories. For instance, for an arbitrary pair of categories a and b,

$$\log\left(\frac{\pi_a}{\pi_b}\right) = \log\left(\frac{\pi_a/\pi_J}{\pi_b/\pi_J}\right) = \log\left(\frac{\pi_a}{\pi_J}\right) - \log\left(\frac{\pi_b}{\pi_J}\right)$$

$$= (\alpha_a + \beta_a x) - (\alpha_b + \beta_b x)$$

$$= (\alpha_a - \alpha_b) + (\beta_a - \beta_b)x. \tag{8.1.2}$$

Thus, the logit equation for categories a and b has intercept parameter ($\alpha_a - \alpha_b$) and slope parameter ($\beta_a - \beta_b$).

For optimal efficiency, one should use software that fits the $J - 1$ logit equations (8.1.1) *simultaneously*. Estimates of the model parameters then have smaller standard errors than when binary logistic regression software fits each component equation in (8.1.1) separately. For simultaneous fitting, the same parameter estimates occur for a pair of categories no matter which category is the baseline.

This logit model is an important tool in marketing research for modeling how subjects choose one of a discrete set of options. A generalization of model (8.1.1) allows the explanatory variables to take different values for different response categories. For instance, the choice of a brand of car would likely depend on price, which would vary among the brand options. This generalized model, often called a *multinomial logit model*, is beyond the scope of this text (see Agresti (1990), Sec. 9.2). The following example deals with discrete choice in a somewhat different context: for alligators rather than humans.

Table 8.1 Alligator Size (Meters) and Primary Food Choice,[a] for 59 Florida Alligators

1.24 I	1.30 I	1.30 I	1.32 F	1.32 F	1.40 F	1.42 I	1.42 F
1.45 I	1.45 O	1.47 I	1.47 F	1.50 I	1.52 I	1.55 I	1.60 I
1.63 I	1.65 O	1.65 I	1.65 F	1.65 F	1.68 F	1.70 I	1.73 O
1.78 I	1.78 I	1.78 O	1.80 I	1.80 F	1.85 F	1.88 I	1.93 I
1.98 I	2.03 F	2.03 F	2.16 F	2.26 F	2.31 F	2.31 F	2.36 F
2.36 F	2.39 F	2.41 F	2.44 F	2.46 F	2.56 O	2.67 F	2.72 I
2.79 F	2.84 F	3.25 O	3.28 O	3.33 F	3.56 F	3.58 F	3.66 F
3.68 O	3.71 F	3.89 F					

[a] F = Fish, I = Invertebrates, O = Other.

Source: Thanks to M. F. Delany and Clint T. Moore for providing these data.

8.1.2 Alligator Food Choice Example

Table 8.1 is taken from a study by the Florida Game and Fresh Water Fish Commission of factors influencing the primary food choice of alligators. For 59 alligators sampled in Lake George, Florida, Table 8.1 shows the alligator length (in meters) and the primary food type, in volume, found in the alligator's stomach. Primary food type has three categories: Fish, Invertebrate, and Other. The invertebrates were primarily apple snails, aquatic insects, and crayfish. The "other" category includes reptiles (primarily turtles, though one stomach contained tags of 23 baby alligators that had been released in the lake during the previous year!), amphibean, mammal, plant material, and stones or other debris.

We apply logit model (8.1.1) with $J = 3$ to these data, using Y = "primary food choice" as the response and X = "length of alligator" as predictor. Table 8.2 shows ML parameter estimates and standard errors for the two logits using "other" as the baseline category. From Table 8.2, $\log(\hat{\pi}_1 / \hat{\pi}_3) = 1.490 - 0.070x$, and $\log(\hat{\pi}_2 / \hat{\pi}_3) = 5.716 - 2.473x$. By (8.1.2), the estimated log odds that the response is "fish" rather than "invertebrate" equals $\log(\hat{\pi}_1 / \hat{\pi}_2) = (1.490 - 5.716) + [-0.070 - (-2.473)]x = -4.227 + 2.403x$.

For each logit, one interprets the estimates just as in ordinary binary logistic regression models, conditional on the event that the response outcome was one of those two categories. For instance, given that the primary food type is fish or invertebrate, the estimated probability that it is fish increases in length x according to an S-shaped curve. For alligators of length $x + 1$ meters, the estimated odds that

Table 8.2 Parameter Estimates and Standard Errors (in parentheses) for Generalized Logit Model Fitted to Table 8.1

	Food Choice Categories for Logit	
Parameter	(Fish/Other)	(Invertebrate/Other)
Intercept	1.490	5.716
Length	−0.070 (.521)	−2.473 (.901)

primary food type is "fish" rather than "invertebrate" equal $\exp(2.403) = 11.1$ times the estimated odds for alligators of length x meters.

To test the hypothesis that primary food choice is independent of alligator length, we test $H_0 : \beta_j = 0$ for $j = 1, 2$ in model (8.1.1). The likelihood-ratio test takes twice the difference in maximized log likelihoods between this model and the simpler one having response independent of length. The test statistic equals 9.2, with $df = 2$, giving a P-value of .01 and strong evidence of a length effect.

8.1.3   Estimating Response Probabilities

One can alternatively express the multicategory logit model directly in terms of the response probabilities, as

$$\pi_j = \frac{\exp(\alpha_j + \beta_j x)}{\sum_h \exp(\alpha_h + \beta_h x)}, \qquad j = 1, \ldots, J - 1. \tag{8.1.3}$$

The denominator is the same for each probability, and the numerators for various j sum to the denominator, so $\sum_j \pi_j = 1$. The parameters equal zero in (8.1.3) for whichever category is the baseline in the logit expressions. For instance, the model that defines parameters with category J as the baseline for the denominator of each logit (as in (8.1.1)) has $\alpha_J = \beta_J = 0$. For category j, (α_j, β_j) then refers to the logit equation for $\log[\pi_j / \pi_J]$.

The estimates in Table 8.2 contrast "fish" and "invertebrate" to "other" as the baseline category. The estimated probabilities (8.1.3) of the outcomes (Fish, Invertebrate, Other) equal

$$\hat{\pi}_1 = \frac{\exp(1.49 - 0.07x)}{1 + \exp(1.49 - 0.07x) + \exp(5.72 - 2.47x)}$$

$$\hat{\pi}_2 = \frac{\exp(5.72 - 2.47x)}{1 + \exp(1.49 - 0.07x) + \exp(5.72 - 2.47x)}$$

$$\hat{\pi}_3 = \frac{1}{1 + \exp(1.49 - 0.07x) + \exp(5.72 - 2.47x)}$$

The 1 term in each denominator and in the numerator of $\hat{\pi}_3$ represents $\exp(\hat{\alpha}_3 + \hat{\beta}_3 x)$ using $\hat{\alpha}_3 = \hat{\beta}_3 = 0$. The three probabilities sum to 1, since the numerators sum to the common denominator. The logs of the ratios of $[\hat{\pi}_1 / \hat{\pi}_3]$ and $[\hat{\pi}_2 / \hat{\pi}_3]$ produce the results in Table 8.2.

For an alligator of length $x = 3.9$ meters, for instance, the estimated probability that primary food choice is "other" equals

$$\hat{\pi}_3 = \frac{1}{1 + \exp[1.49 - 0.07(3.9)] + \exp[5.72 - 2.47(3.9)]} = 0.23.$$

Figure 8.1 displays the three response probabilities as a function of length.

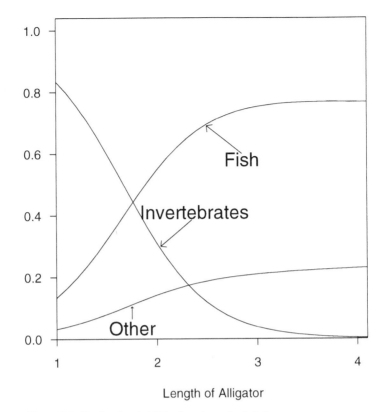

Figure 8.1 Predicted probabilities for primary food choice.

8.1.4 Belief in Afterlife Example

Logit models for nominal responses also apply when explanatory variables are categorical, or a mixture of categorical and continuous. When predictors are entirely categorical, one can display the data as a contingency table. If the data are not sparse, one can test model goodness of fit using X^2 or G^2 statistics. In addition, the models then have equivalent loglinear models.

To illustrate, Table 8.3, taken from the 1991 General Social Survey, has response variable Y = belief in life after death, with categories (Yes, Undecided, No), and explanatory variables X_1 = gender and X_2 = race. We use dummy variables for the predictors, with $x_1 = 1$ for females and 0 for males, and $x_2 = 1$ for whites and 0 for blacks. Using "no" as the baseline category for belief in life after death, we form the model

$$\log\left(\frac{\pi_j}{\pi_3}\right) = \alpha_j + \beta_j^G x_1 + \beta_j^R x_2, \qquad j = 1, 2, \qquad (8.1.4)$$

where G and R superscripts identify the gender and race parameters. The model assumes a lack of interaction between gender and race in their effects on belief in

Table 8.3   Belief in Afterlife, by Race and Gender, with Fitted Values for Generalized Logit Model

		Belief in Afterlife		
Race	Gender	Yes	Undecided	No
White	Female	371	49	74
		(372.8)	(49.2)	(72.1)
	Male	250	45	71
		(248.2)	(44.8)	(72.9)
Black	Female	64	9	15
		(62.2)	(8.8)	(16.9)
	Male	25	5	13
		(26.8)	(5.2)	(11.1)

Source: 1991 General Social Survey.

life after death. The effect parameters represent log odds ratios with the baseline category. For instance, β_1^G is the conditional log odds ratio between gender and response categories 1 and 3 (yes and no), given race; β_2^G is the conditional log odds ratio between gender and response categories 2 and 3 (undecided and no).

Table 8.3 also shows ML fitted values for this model. The goodness-of-fit statistics are $G^2 = 0.9$ and $X^2 = 0.9$. The sample has two non-redundant logits at each of four gender–race combinations, for a total of eight logits. Model (8.1.4), considered for $j = 1$ and 2, contains six parameters. Thus, the model has residual $df = 8 - 6 = 2$. The model fits well. Table 8.4 shows the parameter estimates and their standard errors. The estimated odds of a "yes" rather than a "no" response for females are $\exp(0.419) = 1.5$ times those for males, controlling for race; for whites, they are $\exp(0.342) = 1.4$ times those for blacks, controlling for gender.

To test the effect of gender, we test $H_0 : \beta_j^G = 0$ for $j = 1, 2$. The likelihood-ratio test compares $G^2 = 0.8$ ($df = 2$) to $G^2 = 8.0$ ($df = 4$) obtained by dropping gender from the model. The difference of 7.2, based on $df = 2$, has a P-value of .03 and shows evidence of a gender effect. By contrast, the effect of race is not significant, the model deleting race having $G^2 = 2.8$ ($df = 4$). This partly reflects the larger standard errors that the effects of race have, due to a much greater imbalance between sample sizes in the two race categories than occurs with gender.

Table 8.4   Parameter Estimates and Standard Errors (in parentheses) for Generalized Logit Model Fitted to Table 8.3

	Belief Categories for Logit	
Parameter	(Yes/No)	(Undecided/No)
Intercept	0.883 (.243)	−0.758 (.361)
Gender ($F = 1$)	0.419 (.171)	0.105 (.246)
Race ($W = 1$)	0.342 (.237)	0.271 (.354)

Table 8.5   Predicted Probabilities for Belief in Afterlife

		Belief in Afterlife		
Race	Gender	Yes	Undecided	No
White	Female	0.76	0.10	0.15
	Male	0.68	0.12	0.20
Black	Female	0.71	0.10	0.19
	Male	0.62	0.12	0.26

Table 8.5 displays predicted probabilities for the three response categories. To illustrate, for white females ($x_1 = x_2 = 1$), the estimated probability of response 1 ("yes") equals

$$\frac{\exp[0.883 + 0.419(1) + 0.342(1)]}{1 + \exp[0.883 + 0.419(1) + 0.342(1)] + \exp[-0.758 + 0.105(1) + 0.271(1)]} = .76.$$

This also follows directly from the fitted values, as $372.8/(372.8 + 72.1 + 49.2) = .76$.

8.1.5   Connection with Loglinear Models

When all explanatory variables are categorical, logit models have corresponding loglinear models. The loglinear model has the most general interaction among the explanatory variables from the logit model. It has the same association and interaction structure relating the explanatory variables to the response.

To illustrate, the model fitted to Table 8.3 assumes main effects of gender (G) and race (R) on belief (B) in afterlife, with no interaction. It corresponds to the loglinear model (GR, BG, BR) of homogeneous association. The same estimated effect of G on B results from logit or loglinear parameters. For instance, model (GR, BG, BR) has estimated conditional odds ratio relating B categories $1 =$ "yes" and $3 =$ "no" to G equal to $\exp(\hat{\lambda}_{11}^{BG} + \hat{\lambda}_{32}^{BG} - \hat{\lambda}_{12}^{BG} - \hat{\lambda}_{31}^{BG}) = 1.5$, as obtained above for the logit model.

The simpler logit model that deletes the race effect on belief corresponds to loglinear model (GR, BG). For this logit model and for the one having both G and R as predictors of belief, the corresponding loglinear model contains the G-R term. This term represents the interaction among variables that are explanatory in the logit model.

8.2   CUMULATIVE LOGIT MODELS FOR ORDINAL RESPONSES

When response categories are ordered, logits can directly incorporate the ordering. This results in models having simpler interpretations and potentially greater power than ordinary multicategory logit models.

The *cumulative probabilities* are the probabilities that the response Y falls in category j or below, for each possible j. The jth cumulative probability is

$$P(Y \le j) = \pi_1 + \cdots + \pi_j, \qquad j = 1, \ldots, J.$$

The cumulative probabilities reflect the ordering, with $P(Y \leq 1) \leq P(Y \leq 2) \leq \cdots \leq P(Y \leq J) = 1$. Models for cumulative probabilities do not use the final one, $P(Y \leq J)$, since it necessarily equals 1. The logits of the first $J - 1$ cumulative probabilities are

$$\text{logit}[P(Y \leq j)] = \log\left(\frac{P(Y \leq j)}{1 - P(Y \leq j)}\right)$$

$$= \log\left(\frac{\pi_1 + \cdots + \pi_j}{\pi_{j+1} + \cdots + \pi_J}\right), \qquad j = 1, \ldots, J - 1.$$

These are called *cumulative logits*.

Each cumulative logit uses all J response categories. A model for the jth cumulative logit looks like an ordinary logit model for a binary response in which categories 1 to j combine to form a single category, and categories $j + 1$ to J form a second category. In other words, the response collapses into two categories. Ordinal models simultaneously provide a structure for all $J - 1$ cumulative logits. For $J = 3$, for instance, models refer both to $\log[\pi_1/(\pi_2 + \pi_3)]$ and $\log[(\pi_1 + \pi_2)/\pi_3]$.

8.2.1 Proportional Odds Model

For a predictor X, the model

$$\text{logit}[P(Y \leq j)] = \alpha_j + \beta x, \qquad j = 1, \ldots, J - 1, \tag{8.2.1}$$

has parameter β describing the effect of X on the log odds of response in category j or below. In this formula, β does not have a j subscript, so the model assumes an identical effect of X for all $J - 1$ collapsings of the response into binary outcomes. When this model fits well, it requires a single parameter rather than $J - 1$ parameters to describe the effect of X.

Interpretations for this model refer to odds ratios for the collapsed response scale, for any fixed j. For two values x_1 and x_2 of X, the odds ratio utilizes cumulative probabilities and their complements,

$$\frac{P(Y \leq j \mid X = x_2)/P(Y > j \mid X = x_2)}{P(Y \leq j \mid X = x_1)/P(Y > j \mid X = x_1)}.$$

The log of this odds ratio is the difference between the cumulative logits at those two values of x. This equals $\beta(x_2 - x_1)$, proportional to the distance between the x values. The same proportionality constant (β) applies for each possible point j for the collapsing. Because of this property, model (8.2.1) is called a *proportional odds model*. In particular, for $x_2 - x_1 = 1$, the odds of response below any given category multiply by e^β for each unit increase in X. When the model holds with $\beta = 0$, X and Y are statistically independent.

Explanatory variables in cumulative logit models can be continuous, categorical, or of both types. The ML fitting process uses an iterative algorithm simultaneously

$P(Y \le j)$

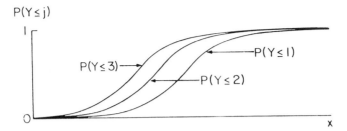

Figure 8.2 Depiction of cumulative probabilities in proportional odds model.

for all j. When the categories are reversed in order, one gets the same fit, but the sign of $\hat{\beta}$ reverses.

Figure 8.2 depicts the proportional odds model for a four category response and a single continuous x predictor. A separate curve applies to each cumulative probability, describing its change as a function of x. The curve for $P(Y \le j)$ looks like a logistic regression curve for a binary response with pair of outcomes $(Y \le j)$ and $(Y > j)$. The common effect β for each j implies that the three response curves have the same shape. Any one curve is identical to any of the others simply shifted to the right or shifted to the left. As in logistic regression, the size of $|\beta|$ determines how quickly the curves climb or drop. At any fixed x value, the curves have the same ordering as the cumulative probabilities, the one for $P(Y \le 1)$ being lowest.

Figure 8.2 has $\beta > 0$. Figure 8.3 shows corresponding curves for the category probabilities, $P(Y = j) = P(Y \le j) - P(Y \le j - 1)$. As x increases, the response on Y is more likely to fall at the low end of the ordinal scale. When $\beta < 0$, the

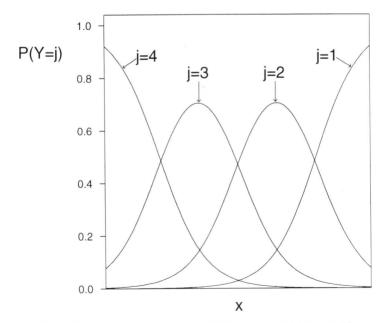

Figure 8.3 Depiction of category probabilities in proportional odds model.

curves in Figure 8.2 descend rather than ascend, and the labels in Figure 8.3 reverse order. Then, as x increases, Y is more likely to fall at the high end of the scale. (The model is sometimes written instead as $\text{logit}[P(Y \le j)] = \alpha_j - \beta x$, so that $\beta > 0$ corresponds to Y being more likely to fall at the high end of the scale as x increases.)

8.2.2 Political Ideology Example

Table 8.6, from the 1991 General Social Survey, relates political ideology to political party affiliation. Political ideology uses a five-point ordinal scale, ranging from very liberal to very conservative. Let x be a dummy variable for political party, with $x = 1$ for Democrats and $x = 0$ for Republicans. The ML fit of the proportional odds model (8.2.1) has estimated effect $\hat{\beta} = 0.975$ ($ASE = 0.129$). For any fixed j, the estimated odds that a Democrat's response is in the liberal direction rather than the conservative direction (i.e., $Y \le j$ rather than $Y > j$) equal $\exp(0.975) = 2.65$ times the estimated odds for Republicans. A 95% confidence interval for this odds ratio equals $\exp(0.975 \pm 1.96 \times 0.129)$, or $(2.1, 3.4)$. A fairly substantial association exists, Democrats tending to be more liberal than Republicans.

Table 8.6 also displays the fitted values for the model. These have an odds ratio of $\exp(\hat{\beta}) = 2.65$ for each of the four collapsings of the data to a 2×2 table. For instance, for the estimated odds of a very liberal response,

$$\frac{(78.4)(44.0 + 151.7 + 75.5 + 104.0)}{(83.2 + 168.2 + 49.1 + 49.1)(31.8)} = 2.65.$$

One can use the fitted values or parameter estimates to calculate estimated probabilities. The cumulative probabilities equal

$$P(Y \le j) = \frac{\exp(\alpha_j + \beta x)}{1 + \exp(\alpha_j + \beta x)}.$$

For instance, $\hat{\alpha}_1 = -2.469$, so the first estimated cumulative probability for Democrats ($x = 1$) equals

$$\frac{\exp[-2.469 + .975(1)]}{1 + \exp[-2.469 + .975(1)]} = .18.$$

From the fitted values for the 428 Democrats, this also equals $78.4/428$.

Table 8.6 Political Ideology by Party Affiliation, with Fitted Values for Cumulative Logit Model

| Party Affiliation | Political Ideology | | | | | Total |
	Very Liberal	Slightly Liberal	Moderate	Slightly Conservative	Very Conservative	
Democratic	80	81	171	41	55	428
	(78.4)	(83.2)	(168.2)	(49.1)	(49.1)	
Republican	30	46	148	84	99	407
	(31.8)	(44.0)	(151.7)	(75.5)	(104.0)	

Source: 1991 General Social Survey.

The model fits Table 8.6 well, with $G^2 = 3.7$ and $X^2 = 3.7$, based on $df = 3$. The model with $\beta = 0$ specifies independence between ideology and party affiliation, and is equivalent to the loglinear model of independence. The G^2 test of fit of that special case is simply the G^2 test of independence (2.4.3) for two-way contingency tables, which equals $G^2 = 62.3$ based on $df = 4$. The model permitting an effect fits considerably better than the independence model.

The likelihood-ratio statistic for an ordinal test of independence ($H_0 : \beta = 0$) is the difference between G^2 values for the independence model and the proportional odds model. The difference in G^2 values of $62.3 - 3.7 = 58.6$, based on $df = 4 - 3 = 1$, gives extremely strong evidence of an association ($P < .0001$). Like the test for the ordinal loglinear model presented in Section 7.2.3, this test uses the ordering of the response categories. When the model fits well, it is more powerful than the tests of Section 2.4 based on $df = (I - 1)(J - 1)$, since it focuses on a restricted alternative and has only a single degree of freedom. Similar strong evidence results from the Wald test, using $z^2 = (\hat{\beta}/ASE)^2 = (0.975/0.129)^2 = 57.1$.

8.2.3 Invariance to Choice of Response Categories

When the proportional odds model holds for a given response scale, it also holds with the same effects for any collapsing of the response categories. For instance, if a model for categories (Very liberal, Slightly liberal, Moderate, Slightly conservative, Very conservative) fits well, approximately the same estimated effects result from collapsing the response scale to (Liberal, Moderate, Conservative). This *invariance* to the choice of response categories is a nice feature of the model. Two researchers who use different response categories in studying an association should reach similar conclusions.

To illustrate, we collapse Table 8.6 to a three-category response, combining the two liberal categories and combining the two conservative categories. Then, the ML estimated effect of party affiliation changes only from 0.975 ($ASE = .129$) to 1.006 ($ASE = 0.132$). Interpretations are unchanged. Some loss of efficiency occurs in collapsing ordinal scales, resulting in larger standard errors. In practice, when observations are spread fairly evenly among the categories, the efficiency loss is minor unless one collapses to a binary response. It is usually inadvisable to collapse ordinal data to binary.

The proportional odds model implies trends upward or downward among distributions of Y at different values of explanatory variables. When X refers to two groups, as in Table 8.6, the model fits well when subjects in one group tend to make higher responses on the ordinal scale than subjects in the other group. The model does not fit well when the response distributions differ in their dispersion rather than their average. If Democrats tended to be primarily moderate in ideology, while Republicans tended to be both very conservative and very liberal, then the Republicans' responses would show greater dispersion than the Democrats'. The two ideology distributions would be quite different, but the proportional odds model would not detect this if the average responses were similar.

When a proportional odds model does not fit well, an improved fit may result from a more complex model having quadratic or interaction terms, or from a different link that allows $P(Y \le j)$ to approach 1 at a different rate than it approaches 0 and permits a nonsymmetric appearance for the category probability curves of Figure 8.3 (Agresti (1990), Section 9.5). When simplified ordinal models do not fit well, one can fit ordinary multicategory logit models of form (8.1.1) and use the ordinality in an informal way in interpreting the associations.

8.3 PAIRED-CATEGORY LOGITS FOR ORDINAL RESPONSES*

Cumulative logit models for ordinal responses use the entire response scale in forming each logit. This section presents alternative logits for ordered categories that, like baseline-category logits for nominal responses, use pairs of categories.

8.3.1 Adjacent-Categories Logits

One alternative forms $J - 1$ logits using all pairs of adjacent categories. The *adjacent-categories logits* are

$$\log \left(\frac{\pi_{j+1}}{\pi_j} \right), \qquad j = 1, \ldots, J - 1.$$

For $J = 3$, for instance, the logits are $\log(\pi_2/\pi_1)$ and $\log(\pi_3/\pi_2)$. A model using these logits with a predictor x has form

$$\log \left(\frac{\pi_{j+1}}{\pi_j} \right) = \alpha_j + \beta_j x, \qquad j = 1, \ldots, J - 1. \tag{8.3.1}$$

These logits, like the baseline-category logits, determine logits for all pairs of response categories. Model (8.3.1) is equivalent to logit model (8.1.1), with β_j in (8.3.1) being equivalent to $\beta_{j+1} - \beta_j$ in (8.1.1).

A simpler version of model (8.3.1),

$$\log \left(\frac{\pi_{j+1}}{\pi_j} \right) = \alpha_j + \beta x, \qquad j = 1, \ldots, J - 1. \tag{8.3.2}$$

has identical effects $\{\beta_j = \beta\}$ for each pair of adjacent categories. For this model, the effect of X on the odds of making the higher instead of the lower response is the same for all pairs of adjacent categories. This model and proportional odds model (8.2.1) use a single parameter, rather than $J - 1$ parameters, for the effect of X. When the model holds, independence is equivalent to $\beta = 0$.

The simpler adjacent-categories logit model (8.3.2) implies that the coefficient of x for the logit based on arbitrary response categories a and b equals $\beta(a - b)$. The effect depends on the distance between categories, so this model recognizes the ordering of the response scale. One can check whether this simplification is justified by comparing the model fit to that of (8.3.1), or equivalently to logit model (8.1.1).

8.3.2 Political Ideology Example Revisited

To illustrate the adjacent-categories logit model, we return to Table 8.6 and model political ideology using the simple model (8.3.2) having a common effect for each logit. Let $x = 0$ for Democrats and $x = 1$ for Republicans.

The ML estimate of the party affiliation effect is $\hat{\beta} = 0.435$. The estimated odds that a Republican's ideology classification is in category $j + 1$ instead of j are $\exp(\hat{\beta}) = 1.54$ times the estimated odds for Democrats. This is the estimated odds ratio for each of the four 2×2 tables consisting of a pair of adjacent columns of Table 8.6. For instance, the estimated odds of "slightly conservative" instead of "moderate" ideology are 54% higher for Republicans than for Democrats. The estimated odds ratio for an arbitrary pair of columns $a > b$ equals $\exp[\hat{\beta}(a - b)]$. The estimated odds that a Republican's ideology is "very conservative" (category 5) instead of "very liberal" (category 1) are $\exp[0.435(5 - 1)] = (1.54)^4 = 5.7$ times those for Democrats. Republicans tend to be much more conservative than Democrats.

The model fit has $G^2 = 5.5$ with $df = 3$, a reasonably good fit. The special case of the model with $\beta = 0$ specifies independence of ideology and party affiliation and is equivalent to the loglinear model of independence. The G^2 fit of that model is simply the G^2 statistic (2.4.3) for testing independence, which equals $G^2 = 62.3$ with $df = 4$. The model permitting a party affiliation effect fits much better than the independence model.

The likelihood-ratio test statistic for the hypothesis that party affiliation has no effect on ideology ($H_0 : \beta = 0$) is based on the difference between the G^2 values for the two models. It equals $62.3 - 5.5 = 56.8$ with $df = 4 - 3 = 1$. There is very strong evidence of an association ($P < .0001$). Results are similar to those for the cumulative-logit analysis in Section 8.2.2.

For two-way contingency tables, the more general model (8.3.1) and the equivalent logit model (8.1.1) are saturated. An advantage of using the single effect parameter is that the simpler model is unsaturated.

8.3.3 Connection with Loglinear Models

Many loglinear models for ordinal variables have simple representations as adjacent-category logit models. For instance, Section 7.2.1 introduced the linear-by-linear association model,

$$\log \mu_{ij} = \lambda + \lambda_i^X + \lambda_j^Y + \beta u_i v_j, \tag{8.3.3}$$

for which the association term uses monotone scores for the rows and columns. For this model with column scores $\{v_j = j\}$, the adjacent-category logits within row i are

$$\log \left(\frac{\pi_{j+1}}{\pi_j} \right) = \log \left(\frac{\mu_{i,j+1}}{\mu_{ij}} \right)$$

$$= \log(\mu_{i,j+1}) - \log(\mu_{ij})$$

$$= (\lambda + \lambda_i^X + \lambda_{j+1}^Y + \beta u_i v_{j+1}) - (\lambda + \lambda_i^X + \lambda_j^Y + \beta u_i v_j)$$

$$= (\lambda_{j+1}^Y - \lambda_j^Y) + \beta u_i.$$

This has form (8.3.2), identifying α_j with $(\lambda_{j+1}^Y - \lambda_j^Y)$ and the row scores $\{u_i\}$ with the levels of x. In fact, the two models are equivalent. The logit representation (8.3.2) provides an interpretion for model (8.3.3). The conclusions presented above for fitting logit model (8.3.2) to Table 8.6 also result from fitting the loglinear model (8.3.3) with equally-spaced column scores.

Analogous remarks apply to multi-way tables. For instance, consider Table 7.5 on job satisfaction (S), income (I), and gender (G), analyzed in Section 7.3. Unlike the loglinear models described there, logit models treat job satisfaction as a response variable and income and gender as explanatory variables. Let g denote a dummy variable for gender, and let $\{x_i\}$ denote scores assigned to the levels of income. Consider the model for job satisfaction response probabilities,

$$\log\left(\frac{\pi_{j+1}}{\pi_j}\right) = \alpha_j + \beta_1 x_i + \beta_2 g, \qquad j = 1, 2, 3. \tag{8.3.4}$$

The model assumes identical effects of explanatory variables for each adjacent pair of response categories.

This model is equivalent to a special case of loglinear model (GI, GS, IS) in which the G-S and I-S associations each have linear-by-linear $(L \times L)$ form, with equally-spaced scores for S and the scores $\{x_i\}$ for I and g for G. That special case is also a special case of the homogeneous $L \times L$ model (7.3.1) (which assumed $L \times L$ structure for the I-S association) that also provides linear-by-linear structure for the G-S association. Using income categories $\{1, 2, 3, 4\}$ as scores, model (8.3.4) fits well, with $G^2 = 12.6$ having $df = 19$. It provides estimates $\hat{\beta}_1 = 0.389$ ($ASE = 0.155$) and $\hat{\beta}_2 = 0.045$ ($ASE = 0.314$). Satisfaction is positively associated with income, given gender, but shows no evidence of association with gender, given income.

8.3.4 Continuation-Ratio Logits

Another approach forms logits for ordered response categories in a sequential manner. One constructs models simultaneously for

$$\log\left(\frac{\pi_1}{\pi_2}\right), \log\left(\frac{\pi_1 + \pi_2}{\pi_3}\right), \dots, \log\left(\frac{\pi_1 + \cdots + \pi_{J-1}}{\pi_J}\right).$$

These are called *continuation-ratio logits*. They refer to a binary response that contrasts each category with a grouping of categories from lower levels of the response scale. A second type of continuation-ratio logit contrasts each category with a grouping of categories from higher levels of the response scale; that is,

$$\log\left(\frac{\pi_1}{\pi_2 + \cdots + \pi_J}\right), \log\left(\frac{\pi_2}{\pi_3 + \cdots + \pi_J}\right), \dots, \log\left(\frac{\pi_{J-1}}{\pi_J}\right).$$

Models using these logits have different parameter estimates and goodness-of-fit statistics than models using the other continuation-ratio logits.

Table 8.7   Outcomes for Pregnant Mice in Developmental Toxicity Study

Concentration (mg/kg per day)	Response		
	Dead	Malformation	Normal
0 (controls)	15	1	281
62.5	17	0	225
125	22	7	283
250	38	59	202
500	144	132	9

Source: Based on results in C. J. Price et al., *Fund. Appl. Toxicol., 8:* 115–126 (1987).
I thank Dr. Louise Ryan for showing me these data.

We illustrate continuation-ratio logits using Table 8.7, which refers to a developmental toxicity study. Rodent studies are commonly used to test and regulate substances posing potential danger to developing fetuses. This study administered diethylene glycol dimethyl ether, an industrial solvent used in the manufacture of protective coatings, to pregnant mice. Each mouse was exposed to one of five concentration levels for ten days early in the pregnancy. Two days later, the uterine contents of the pregnant mice were examined for defects. Each fetus had the three possible outcomes (Dead, Malformation, Normal). The outcomes are ordered, normal being the most desirable result.

We use continuation-ratio logits to model the probability of a dead fetus, using $\log[\pi_1/(\pi_2 + \pi_3)]$, and the conditional probability of a malformed fetus, given that the fetus was live, using $\log(\pi_2/\pi_3)$. We used scores $\{0, 62.5, 125, 250, 500\}$ for concentration level. The two models are ordinary logistic regression models in which the responses are column 1 and columns 2–3 combined for one fit and column 2 and column 3 for the second fit. The estimated effect of concentration level is 0.0064 ($ASE = 0.0004$) for the first logit, and 0.0174 ($ASE = 0.0012$) for the second logit. In each case, the less desirable outcome is more likely as concentration level increases. For instance, given that a fetus was live, the estimated odds that it was malformed rather than normal changes by a multiplicative factor of $\exp(100 \times 0.0174) = 5.7$ for every 100-unit increase in concentration level.

When models for different continuation-ratio logits have separate parameters, as in this example, separate fitting of models for different logits gives the same results as simultaneous fitting. The sum of the separate G^2 statistics is an overall goodness-of-fit statistic pertaining to the simultaneous fitting of the models. For Table 8.7, the G^2 values are 5.8 for the first logit and 6.1 for the second, each based on $df = 3$. We summarize the fit by their sum, $G^2 = 11.8$, based on $df = 6$.

This analysis treats pregnancy outcomes for different fetuses as independent observations. In fact, each pregnant mouse had a litter of fetuses, and statistical dependence may exist among different fetuses in the same litter. The model also treats different fetuses at a given concentration level as having the same response probabilities. Heterogeneity of various types among the litters (for instance, due to different physical conditions of different pregnant mice) would usually cause these probabilities to vary somewhat among litters. Either statistical dependence or heterogeneous probabilities violates the binomial assumption and typically causes *overdispersion*—greater variation than the binomial model predicts. For example, at a fixed concentration level, the number of mice that die in a litter may vary among pregnant mice to a greater degree than if the counts were independent and identical binomial variates.

The total G^2 for testing the continuation-ratio model shows evidence of lack of fit. The structural form chosen for the model may be incorrect. The lack of fit may partly or entirely, however, reflect overdispersion caused by dependence within litters or heterogeneity among litters. Both factors are common in developmental toxicity studies. See Collett ((1992), Ch. 6) and Morgan ((1992), Ch. 6) for ways of handling overdispersion.

PROBLEMS

8.1. Refer to the example in Section 8.1.2.

   a. Using the model fit, calculate an odds ratio that describes the estimated effect of length on primary food choice being either "invertebrate" or "other."

   b. Estimate the probability that food choice is "invertebrate," for an alligator of length 3.9 meters.

   c. Find the length at which the outcomes "invertebrate" and "other" are equally likely.

8.2. Table 8.8 displays primary food choice for a sample of alligators, classified by length (≤ 2.3 meters, > 2.3 meters) and by the lake in Florida in which they were caught.

Table 8.8

Lake	Size	Fish	Invertebrate	Reptile	Bird	Other
			Primary Food Choice			
Hancock	≤ 2.3	23	4	2	2	8
	> 2.3	7	0	1	3	5
Oklawaha	≤ 2.3	5	11	1	0	3
	> 2.3	13	8	6	1	0
Trafford	≤ 2.3	5	11	2	1	5
	> 2.3	8	7	6	3	5
George	≤ 2.3	16	19	1	2	3
	> 2.3	17	1	0	1	3

Source: Wildlife Research Laboratory, Florida Game and Fresh Water Fish Commission.

   a. Use a logit model to describe effects of length and lake on primary food choice.

   b. Using your model, estimate the probability that the primary food choice is "fish," for the various length and lake combinations. Interpret.

   c. Which loglinear model is equivalent to this logit model?

8.3. Refer to the example in Section 8.1.4.
  a. Using the fitted model, estimate the probability of response "no" for black females, (i) using parameter estimates, (ii) using fitted values.
  b. Fit the logit model for which race has no effect on belief in an afterlife, controlling for gender. Test the goodness of fit, and interpret. Which loglinear model is equivalent to this model?
  c. Describe the gender effect for the model in (**b**) by reporting the estimated odds ratio for each pair of response categories. Interpret, and explain how these relate to the marginal gender–belief odds ratios.
  d. Does the model in (**b**) imply that race and belief in an afterlife are marginally independent, collapsing over gender? (*Hint:* Consider collapsibility for the corresponding loglinear model.)

8.4. Refer to Problem 6.7. Treating vote for President (P) as the response variable and B and D as explanatory variables, find a logit model that fits these data well. Interpret parameter estimates for the effects of B on P. Which loglinear model is equivalent to this model? Test the significance of the B-P conditional association, and interpret.

8.5. Refer to Problem 2.21 with Table 2.15. Using modeling methods, describe and make inferences about the effect of mammography experience on women's attitudes.

8.6. Table 8.9 shows results of a survey of women, relating frequency of breast self-examination and age. Fit a proportional odds model. Analyze goodness of fit, and conduct descriptive and inferential analyses about the association.

Table 8.9

	Frequency of Breast Self-Examination		
Age	Monthly	Occasionally	Never
< 45	91	90	51
45–59	150	200	155
60+	109	198	172

Source: R. T. Senie et al., *Am. J. Public Health, 71:* 583–590 (1981).
Reprinted with permission of the American Public Health Association.

8.7. For a sample of 40 subjects, Table 8.10 relates Y = mental impairment (1 = None , 2 = Mild, 3 = Moderate, 4 = Impaired) to socioeconomic status (SES = 1, high; SES = 0, low) and a life events index, which is a composite measure of both the number and severity of important life events (such as loss of job, divorce, death in family) that the subject experienced within the past three years.
  a. Fit a proportional odds model, and interpret effects.
  b. Add an interaction term between the predictors. Interpret this model by describing the impact of life events on mental impairment separately at each level of SES. Check whether this model provides an improved fit.

Table 8.10 Responses on (Mental impairment, SES, Life events)

Subject	Responses	Subject	Responses	Subject	Responses	Subject	Responses
1	(1, 1, 1)	11	(1, 0, 1)	21	(2, 1, 9)	31	(3, 0, 3)
2	(1, 1, 9)	12	(1, 0, 2)	22	(2, 0, 3)	32	(4, 1, 8)
3	(1, 1, 4)	13	(2, 1, 5)	23	(2, 1, 3)	33	(4, 1, 2)
4	(1, 1, 3)	14	(2, 0, 6)	24	(2, 1, 1)	34	(4, 1, 7)
5	(1, 0, 2)	15	(2, 1, 3)	25	(3, 0, 0)	35	(4, 0, 5)
6	(1, 1, 0)	16	(2, 0, 1)	26	(3, 1, 4)	36	(4, 0, 4)
7	(1, 0, 1)	17	(2, 1, 8)	27	(3, 0, 3)	37	(4, 0, 4)
8	(1, 1, 3)	18	(2, 1, 2)	28	(3, 0, 9)	38	(4, 1, 8)
9	(1, 1, 3)	19	(2, 0, 5)	29	(3, 1, 6)	39	(4, 0, 8)
10	(1, 1, 7)	20	(2, 1, 5)	30	(3, 0, 4)	40	(4, 0, 9)

8.8. The Human Suffering Index was created by the Population Crisis Committee (Washington, D.C.) to compare living conditions among countries. Table 8.11 shows results of this index in 1987 for countries classified by their continent. Though inference is not relevant here, one could use a model to describe how suffering varies by region. Treating region as a nominal predictor, attempt to fit a proportional odds model, using North America (NA) as the baseline category for the dummy variables. The region estimates are infinite, because all observations for NA fall in an extreme category of the response. To obtain finite estimates, add (a) .001, (b) .01, (c) .1, to the moderate category for NA. Note that the estimates for odds ratio comparisons of other regions with NA depend strongly on the choice of added constant, but the comparisons for other pairs of regions are similar for each such choice and similar to those obtained by fitting the model to all but the NA data. Use the parameter estimates from (a) to rank the regions on degree of human suffering and to compare Africa and Asia on this index.

Table 8.11

Region	Human Suffering Index			
	Minimal	Moderate	High	Extreme
Africa	0	1	16	24
Asia	2	10	16	6
Europe	20	7	0	0
Latin America	1	10	12	0
North America	2	0	0	0

Source: H. F. Spirer and L. Spirer, *Data Analysis for Monitoring Human Rights.* © 1993, American Association for the Advancement of Science. Reprinted with permission.

8.9. Table 8.12 refers to results of a clinical trial for the treatment of small-cell lung cancer. Patients were randomly assigned to two treatment groups. In the sequential therapy, the same combination of chemotherapeutic agents was administered in each treatment cycle; in the alternating therapy, three different combinations were given, alternating from cycle to cycle.

Table 8.12

| | | Response to Chemotherapy | | | |
| | | Progressive | No | Partial | Complete |
Therapy	Gender	Disease	Change	Remission	Remission
Sequential	Male	28	45	29	26
	Female	4	12	5	2
Alternating	Male	41	44	20	20
	Female	12	7	3	1

Source: W. Holtbrugge and M. Schumacher, *Appl. Statist., 40:* 249–259 (1991).

a. Fit a proportional odds model with main effects for treatment and gender. Interpret estimated effects.

b. Fit the model that also contains an interaction term between treatment and gender. Interpret. Does this model give a better fit? Show that this model is equivalent to using the four gender–treatment combinations as levels of a single explanatory factor.

8.10. Show that the generalization of model (8.2.1) that has a separate slope β_j for each cumulative logit has a structural problem, cumulative probabilities possibly being misordered for some values of x.

8.11. Refer to Problem 7.26 with Table 7.9.

a. Analyze these data using a proportional odds model.

b. Analyze the data using adjacent-category logits.

c. Specify the loglinear model that corresponds to the model in (b).

8.12. Refer to Table 7.5. Treating job satisfaction as the response, analyze the data using a model with cumulative logits.

a. Test the fit, and describe the effect of income.

b. Conduct a likelihood-ratio test for the effect of income, controlling for gender, and interpret. Compare results to those for the test about this effect in Sections 7.3.4 and 7.3.5. Does it help to utilize the ordinality of income and satisfaction?

c. Compare the estimated effect of income to the estimate obtained by combining categories "Very dissatisfied" and "A little satisfied." What property of the proportional odds model does this reflect?

d. Can gender be dropped from the model? Interpret the effect of income on satisfaction for this simpler model.

8.13. Fit the adjacent-categories logit model (8.3.4) to Table 7.5, using scores $\{1, 2, 3, 4\}$ for income.

a. Test goodness of fit, and interpret the estimated effects of income and gender.

   b. Perform a likelihood-ratio test of no income effect, controlling for gender, and interpret.

   c. Fit the multicategory logit model that corresponds to loglinear model (GI, GS, IS). Does this model give a significantly better fit? Test the income-satisfaction association in this model, and compare results to **(b)**.

   d. Show that the logit model corresponding to loglinear model (S, IG), by which satisfaction is jointly independent of income and gender, seems to fit well. Argue that this is misleading because of sparseness and because it ignores ordering information. Show that an analysis that utilizes ordinality for the *I-S* association provides different conclusions.

   e. Redo **(a)**–**(d)** using scores $\{4, 10, 20, 30\}$ that represent approximate income in thousands of dollars. Compare results.

8.14. Show that the homogeneous $L \times L$ model (7.3.1) with equally spaced scores for response Y corresponds to an adjacent categories logit model with a linear trend effect for X and a factor effect for Z. Illustrate for Table 7.5.

8.15. Analyze Table 7.5 using continuation-ratio logits. Interpret.

8.16. Refer to Table 8.3. Treating belief in an afterlife as ordinal, fit a model using adjacent-category logits to these data. Test the goodness of fit, and interpret the estimated effects of gender and race. Test whether this model gives a poorer fit than the baseline-category logit model.

8.17. Refer to the previous problem. Analyze the data using an alternative ordinal model. Interpret results.

8.18. Table 8.13 refers to a study in which subjects were randomly assigned to a control group or a treatment group. Daily during the study, treatment subjects ate cereal containing psyllium. The purpose of the study was to analyze whether this had a desirable effect in lowering LDL cholesterol.

Table 8.13

	Ending LDL Cholesterol Level							
	Control				Treatment			
Beginning	≤ 3.4	3.4–4.1	4.1–4.9	> 4.9	≤ 3.4	3.4–4.1	4.1–4.9	> 4.9
≤ 3.4	18	8	0	0	21	4	2	0
3.4–4.1	16	30	13	2	17	25	6	0
4.1–4.9	0	14	28	7	11	35	36	6
> 4.9	0	2	15	22	1	5	14	12

Source: Dr. Sallee Anderson, Kellogg Co.

   a. Model the ending cholesterol level as a function of treatment, using the beginning level as a covariate. Analyze the treatment effect, and interpret.

   b. Repeat the analysis in **(a)**, treating the beginning level as a qualitative control variable. Compare results.

   c. An alternative approach to **(b)** uses a generalized Cochran–Mantel–Haenszel test (Section 7.3.5) with 2×4 tables relating treatment to the ending response for four partial tables based on beginning cholesterol level. Apply such a test, taking into account the ordering of the response, to compare the treatments. Interpret, and compare results to **(b)**.

8.19. Refer to Table 7.8. Treating opinion on birth control as the response variable, analyze these data.

8.20. Table 8.14 is an expanded version of the data set discussed in Section 6.4. The response categories are (1) not injured, (2) injured but not transported by emergency medical services, (3) injured and transported by emergency medical services but not hospitalized, (4) injured and hospitalized but did not die, (5) injured and died. Analyze these data.

Table 8.14

Gender	Location	Seat-Belt	Response				
			1	2	3	4	5
Female	Urban	No	7287	175	720	91	10
		Yes	11,587	126	577	48	8
	Rural	No	3246	73	710	159	31
		Yes	6134	94	564	82	17
Male	Urban	No	10,381	136	566	96	14
		Yes	10,969	83	259	37	1
	Rural	No	6123	141	710	188	45
		Yes	6693	74	353	74	12

Source: Dr. Cristanna Cook, Medical Care Development, Augusta, Maine.

8.21. Suppose that a response scale has the categories (strongly agree, mildly agree, mildly disagree, strongly disagree, don't know). How might one model this response? (*Hint:* One approach uses two models, handling the ordered categories in one model, and combining them and modeling the "don't know" response in the other.)

CHAPTER 9

Models for Matched Pairs

This chapter introduces methods for comparing categorical responses for two samples when each sample has the same subjects or when a natural pairing exists between each subject in one sample and a subject from the other sample. The responses in the two samples are then statistically *dependent*. Methods that treat the two sets of observations as independent samples are inappropriate.

Table 9.1 illustrates dependent-samples data. For a poll of a random sample of 1600 voting-age Canadian citizens, 944 indicated approval of the Prime Minister's performance in office. Six months later, of these same 1600 people, 880 indicated approval. The rows of Table 9.1 are the response categories for the first survey. The columns are the same categories for the second survey. The two cells for which the row response is the same as the column response form the "main diagonal" of the table. The 794 + 570 subjects in these cells had the same opinion at both surveys. They compose most of the sample, since relatively few people changed their opinion. A strong association exists between opinions six months apart, the sample odds ratio being $(794 \times 570)/(150 \times 86) = 35.1$.

For two dependent samples, each observation in one sample pairs with an observation in the other sample. The pairs of observations are called *matched pairs*. A two-way table having the same categories for both classifications summarizes such data. The table is *square*, having the same number of rows and columns. The marginal counts display the frequencies for the response outcomes for the two samples. In Table 9.1, the row marginal counts (944, 656) are the (approve, disapprove) totals for the first survey, and the column marginal counts (880, 720) summarize results for the second survey.

This chapter presents analyses of square contingency tables with matched-pairs data. Section 9.1 presents ways of comparing proportions from dependent samples. Section 9.2 discusses logistic regression analyses of matched-pairs data for binary responses. For matched-pairs data on multicategory responses, Section 9.3 introduces loglinear models and Section 9.4 uses them for comparisons of marginal distributions of square tables. The final two sections discuss two applications that yield matched-pairs data: measuring agreement between two observers who rate a common set of subjects (Section 9.5), and evaluating preferences between pairs of treatments (Section 9.6).

Table 9.1   Rating of Performance of Prime Minister

First	Second Survey		Total
Survey	Approve	Disapprove	Total
Approve	794	150	944
Disapprove	86	570	656
Total	880	720	1600

9.1   COMPARING DEPENDENT PROPORTIONS

We first discuss methods of comparing dependent proportions for binary response variables. For Table 9.1, we compare the probabilities of approval for the prime minister's performance at the times of the two surveys.

Let n_{ij} denote the number of subjects making response i at the first survey and response j at the second. In Table 9.1, $n_{1+} = n_{11} + n_{12} = 944$ is the number of "approve" responses at the first survey, and $n_{+1} = n_{11} + n_{21} = 880$ is the number approving at the second survey. The sample proportions approving are $944/1600 = .59$ and $880/1600 = .55$. These marginal proportions are correlated, and statistical analyses must recognize this.

Let π_{ij} denote the probability that a subject makes response i at survey 1 and response j at survey 2. The probabilities of approval at the two surveys are π_{1+} and π_{+1}, the first row and first column totals. When these are identical, the probabilities of disapproval are also identical, and there is *marginal homogeneity*. Since

$$\pi_{1+} - \pi_{+1} = (\pi_{11} + \pi_{12}) - (\pi_{11} + \pi_{21}) = \pi_{12} - \pi_{21},$$

marginal homogeneity in 2×2 tables is equivalent to equality of "off-main-diagonal" probabilities; that is, $\pi_{12} = \pi_{21}$. The table then shows *symmetry* across the main diagonal.

9.1.1   McNemar Test

A test of marginal homogeneity for matched binary responses has null hypothesis $H_0 : \pi_{1+} = \pi_{+1}$, or equivalently $H_0 : \pi_{12} = \pi_{21}$. When the null hypothesis is true, we expect about the same frequency for the counts n_{12} and n_{21}. Let $n^* = n_{12} + n_{21}$ denote the total count in the two off-main-diagonal cells. Their allocation to those two cells are outcomes of a binomial variate with n^* trials. Under H_0, each of these n^* observations has a $\frac{1}{2}$ chance of contributing to n_{12} and a $\frac{1}{2}$ chance of contributing to n_{21}. So, n_{12} and n_{21} are numbers of "successes" and "failures" for a binomial distribution having n^* trials and success probability $\frac{1}{2}$.

The P-value for the test is a binomial tail probability. For instance, for Table 9.1, consider the alternative hypothesis that the probability of approval was greater at the first survey; that is, $H_a : \pi_{1+} > \pi_{+1}$, or equivalently $H_a : \pi_{12} > \pi_{21}$. In Table 9.1, $n_{12} = 150$, $n_{21} = 86$, and $n^* = 150 + 86 = 236$. The reference distribution is the binomial with 236 trials and success probability $\frac{1}{2}$. The P-value for the one-sided alternative is the binomial probability of at least 150 successes out of 236 trials, which equals .00002. The P-value for the two-sided alternative $H_a : \pi_{1+} \neq \pi_{+1}$ doubles this single-tail probability.

When $n^* > 10$, the reference binomial distribution has similar shape as the normal distribution with mean $\frac{1}{2}n^*$ and variance $n^*(\frac{1}{2})(\frac{1}{2})$. The standardized normal test statistic equals

$$z = \frac{n_{12} - (\frac{1}{2})n^*}{[n^*(\frac{1}{2})(\frac{1}{2})]^{1/2}} = \frac{n_{12} - n_{21}}{(n_{12} + n_{21})^{1/2}}. \tag{9.1.1}$$

The square of this statistic has a chi-squared distribution with $df = 1$. This test for a comparison of two dependent proportions is called *McNemar's test*.

One can apply the large-sample normal or chi-squared approximation to obtain the P-value whenever n^* is too large to use a binomial table. For Table 9.1, the normal test statistic equals $z = (150 - 86)/(150 + 86)^{1/2} = 4.2$. There is very strong evidence that the probability of approval was higher at the time of the first survey.

If the two surveys sampled *separate* random samples of subjects rather than the same subjects for each survey, the samples would be *independent* rather than dependent. One would enter the data in a different two-way table, the two rows representing the two surveys and the two columns representing the (approve, disapprove) response categories. One could use the methods of Chapter 2 to analyze the data, comparing the rows of the table rather than the marginal distributions. For instance, one could use the tests of independence from Section 2.4.3 to analyze whether the probability of approval was the same for each survey. Those chi-squared tests are not relevant for dependent-samples data of form Table 9.1. We naturally expect an association between the row and column classifications, because of the matched-pairs connection. The more relevant question concerns whether the marginal distributions differ.

9.1.2 Estimating Differences of Proportions

A confidence interval for the difference of proportions is more informative than a hypothesis test. Let $\{p_{ij} = n_{ij}/n\}$ denote the sample cell proportions. The difference of sample marginal proportions $p_{1+} - p_{+1}$ estimates the true difference $\pi_{1+} - \pi_{+1}$. The estimated variance of the sample difference equals

$$\frac{p_{1+}(1 - p_{1+}) + p_{+1}(1 - p_{+1}) - 2(p_{11}p_{22} - p_{12}p_{21})}{n}. \tag{9.1.2}$$

The first two terms in this expression provide the variance of a difference between uncorrelated proportions. The parenthetical part of the last term represents the effect of the dependence of the marginal proportions through their covariance. Matched-pairs data usually exhibit a strong positive association, responses for most subjects being the same for the column and the row classification. A sample odds ratio exceeding 1.0 corresponds to $p_{11}p_{22} > p_{12}p_{21}$ and hence a negative contribution from the third term in this variance expression. Thus, an advantage of using dependent samples rather than independent samples is a smaller variance for the estimated difference in proportions.

The square root of the estimated variance provides a standard error, used in constructing a confidence interval for the true difference of proportions. To illustrate, for Table 9.1, the difference of sample proportions equals $.59 - .55 = .04$. Its estimated variance equals

$$\frac{(.59)(.41) + (.55)(.45) - 2[(794/1600)(570/1600) - (150/1600)(86/1600)]}{1600}$$

$$= \frac{.1459}{1600} = .000091,$$

and the standard error is $(.000091)^{1/2} = .0095$. A 95% confidence interval equals $.04 \pm 1.96(.0095)$, or $(.021, .059)$. The probability of approval was between about .02 and .06 higher at the first survey than the second survey. A decrease in popularity occurred, though apparently a small one.

9.2 LOGISTIC REGRESSION FOR MATCHED PAIRS

Logistic regression models for binary responses extend to handle binary matched-pairs data. This section shows a connection between a matched-pairs logit model and McNemar's test and discusses logit modeling for case-control studies. First, we present a three-way representation of matched-pairs data for which a test of conditional independence is equivalent to McNemar's test.

9.2.1 Connection between McNemar and Cochran–Mantel–Haenszel Tests

Table 9.1 cross-classified results of two surveys having the same 1600 subjects. An alternative representation of binary responses for n matched pairs presents the data as n separate 2×2 partial tables, one for each pair. The kth partial table shows the responses for the kth matched pair. It has columns that are the two possible outcomes for each measurement; it contains the outcome of the first measurement in row 1 and the outcome of the second measurement in row 2. For instance, Table 9.2 shows a partial table for a subject classified as "approval" at both surveys. The full three-way table corresponding to Table 9.1 has 1600 partial tables; 794 of them look like Table 9.2, 150 of them have first row (1, 0) and second row (0, 1), representing approval

Table 9.2 Representation of Matched Pair Contributing to Count n_{11} in Table 9.1

Survey	Response	
	Approve	Disapprove
First	1	0
Second	1	0

at the first survey and disapproval at the second, 86 of them have first row (0, 1) and second row (1, 0), and 570 have (0, 1) in each row.

Each subject has a partial table, displaying the two matched responses. The 1600 subjects provide 3200 observations in a $2 \times 2 \times 1600$ contingency table. Collapsing this table over the 1600 partial tables yields a 2×2 table with first row equal to (944, 656) and second row equal to (880, 720). These are the total number of "approve" and "disapprove" responses for the two surveys and form the marginal counts in Table 9.1.

Suppose that for each subject the probability of an "approve" response is identical in each survey. Then conditional independence exists between the opinion response and the survey, controlling for subject. The probability of approval is also the same for each survey in the marginal table collapsed over the subjects. But this implies that the true probabilities for Table 9.1 satisfy marginal homogeneity. Thus, a test of conditional independence in the $2 \times 2 \times 1600$ table provides a test of marginal homogeneity for Table 9.1.

To test conditional independence in such a three-way table, one can use the Cochran–Mantel–Haenszel (*CMH*) statistic (3.2.1). In fact, the result of that chi-squared statistic is algebraically identical to the chi-squared version of McNemar's statistic, namely $(n_{12} - n_{21})^2/(n_{12} + n_{21})$ for tables of form 9.1. That is, McNemar's test is a special case of the *CMH* test applied to the binary responses of n matched pairs displayed in n partial tables. This connection is not helpful for computational purposes, since McNemar's statistic is so simple. But it does suggest ways of constructing statistics to handle more complex types of matched data, as discussed at the end of this section.

9.2.2 A Logit Model for Matched-Pairs Data

McNemar's test and other analyses for matched-pairs data also result from logistic regression modeling. The model refers to the n partial tables of form Table 9.2. It differs from other models in this text by permitting subjects to have their own probability distributions. Refer to outcome 1 as "success" and outcome 2 as "failure," and to the matched responses as measurement 1 and measurement 2. Let ϕ_i denote the probability of a success for subject i's response. The model has form

$$\text{logit}(\phi_i) = \alpha_i + \beta x, \tag{9.2.1}$$

where $x = 1$ for measurement 1 and $x = 0$ for measurement 2.

The probability of success ϕ_i for subject i equals $\exp(\alpha_i + \beta)/[1 + \exp(\alpha_i + \beta)]$ for measurement 1 and $\exp(\alpha_i)/[1 + \exp(\alpha_i)]$ for measurement 2. The $\{\alpha_i\}$ parameters permit the probabilities to vary among subjects. A subject with a large positive α_i has a high probability of success for each measurement and is likely to have a success each time; a subject with a large negative α_i has a low probability of success for each measurement and is likely to have a failure each time. The greater the variability in these parameters, the greater the overall positive association between responses, successes (failures) at time 1 tending to occur with successes (failures) at time 2.

This model assumes that, for each subject, the odds of success for measurement 1 are $\exp(\beta)$ times the odds of success for measurement 2. Since each partial table refers to a single subject, this conditional association is called a *subject-specific effect*. The value $\beta = 0$ implies marginal homogeneity. In that case, for each subject, the probability of success is the same for both measurements.

Inference for this model focuses on comparing the response distributions through the parameter β. The large number of subject parameters $\{\alpha_i\}$ causes difficulties with the fitting process and with the properties of ordinary ML estimators. One remedy, *conditional maximum likelihood*, maximizes the likelihood function for a conditional distribution that eliminates the subject parameters. In terms of tables of form Table 9.1, the conditional ML estimate of the odds ratio $\exp(\beta)$ for model (9.2.1) equals n_{12}/n_{21}. For Table 9.1, for instance, $\exp(\hat{\beta}) = 150/86 = 1.74$. Assuming the model holds, a subject's estimated odds of approval of the prime minister's performance were 74% higher for the first survey than the second survey.

An alternative remedy treats $\{\alpha_i\}$ as *random effects*. With this approach, $\{\alpha_i\}$ are an unobserved sample from a particular form of probability distribution, usually the normal distribution with unknown mean and standard deviation. Model (9.2.1) is then an example of a *mixed model*, containing both random effects and the fixed effect $\{\beta\}$. It is called the *logistic-normal model*. The analysis for this model eliminates $\{\alpha_i\}$ by averaging with respect to their distribution, yielding a marginal distribution. The resulting marginal likelihood function, which depends on β as well as the mean and standard deviation of the distribution for $\{\alpha_i\}$, has a much smaller number of parameters and is more manageable. The estimate of β that results from maximizing the marginal likelihood function is called the *marginal ML estimate*. For model (9.2.1), in most cases this approach yields exactly the same estimate of $\exp(\beta)$ as the conditional ML approach.

Model (9.2.1) implies that the true odds ratio for each of the n partial tables equals $\exp(\beta)$. Section 3.2.3 presented the Mantel–Haenszel estimate of a common odds ratio for several 2×2 tables. In fact, the Mantel–Haenszel estimate for matched-pairs partial tables of form Table 9.2 is algebraically identical to n_{12}/n_{21} for tables of form Table 9.1.

In summary, the Mantel–Haenszel estimate, the conditional ML estimate, and usually the marginal ML estimate for logit model (9.2.1) require only n_{12} and n_{21} in Table 9.1 to estimate β. This model provides justification for ignoring the main-diagonal counts in testing $H_0 : \beta = 0$. That is, this model suggests that McNemar's test is adequate for comparing the marginal distributions of matched-pairs data. If the model seems plausible, then McNemar's procedure is reasonable.

9.2.3 Logistic Regression for Matched Case-Control Studies*

Case-control studies that match a single control with each case are an important application having matched-pairs data. For a binary response Y, each case ($Y = 1$) is matched with a control ($Y = 0$) according to certain criteria that could affect the response. Subjects in the matched pairs are measured on the predictor variable of interest, X, and the X-Y association is analyzed.

Table 9.3   Previous Diagnoses of Diabetes for Myocardial Infarction (MI) Case-Control Pairs

	MI Controls		
MI Cases	Diabetes	No Diabetes	Total
Diabetes	9	37	46
No Diabetes	16	82	98
Total	25	119	144

Source: J. L. Coulehan et al., *Amer. J. Public Health, 76:* 412–414 (1986). Reprinted with permission of the American Public Health Association. See also M. Pagano and K. Gauvreau, *Principles of Biostatistics* (Duxbury Press 1993, p. 319).

Table 9.3 illustrates results of a matched case-control study. A study of acute myocardial infarction (MI) among Navajo Indians matched 144 victims of MI according to age and gender with 144 individuals free of heart disease. Subjects were then asked whether they had ever been diagnosed as having diabetes ($x = 0$, no; $x = 1$, yes). Table 9.3 has the same form as Table 9.1, except that the levels of X rather than the levels of Y form the two rows and the two columns.

One can display the data for each matched case-control pair using a partial table of form similar to Table 9.2, but reversing the roles of X and Y. Each matched pair has one subject with $Y = 1$ (the case) and one subject with $Y = 0$ (the control). There are four possible patterns of X values, shown in Table 9.4. There are 37 partial tables of type 9.4a, since for 37 pairs the case had diabetes and the control did not, 16 partial tables of type 9.4b, 8 of type 9.4c, and 82 of type 9.4d.

Now, for matched pair i, consider the model for the probability ϕ_i that $Y = 1$ having form

$$\text{logit}(\phi_i) = \alpha_i + \beta x.$$

The odds that a subject with diabetes ($x = 1$) is a MI case equal $\exp(\beta)$ times the odds that a subject without diabetes ($x = 0$) is a MI case. The probabilities $\{\phi_i\}$ refer to the distribution of Y given X, but these retrospective data provide information only about the distribution of X given Y. One can estimate $\exp(\beta)$, however, since it refers to the X-Y odds ratio, which relates to both types of conditional distribution (Section 2.3.4). Even though the roles of X and Y are reversed in terms of which is fixed and which is random in the study, the conditional ML estimate of the odds ratio $\exp(\beta)$ for Table 9.3 is simply $n_{12}/n_{21} = \frac{37}{16} = 2.3$.

Table 9.4   Possible Case-Control Pairs

	a		b		c		d	
	Case	Control	Case	Control	Case	Control	Case	Control
Yes ($x = 1$)	1	0	0	1	1	1	0	0
No ($x = 0$)	0	1	1	0	0	0	1	1

When the binary case-control classification has k predictors, this model generalizes to

$$\text{logit}(\phi_i) = \alpha_i + \beta_1 x_1 + \beta_2 x_2 + \cdots + \beta_k x_k.$$

Typically, one x_j is an explanatory variable of interest, such as diabetes status. The others are covariates one would like to control, in addition to those already controlled by virtue of using them to form the matched pairs. The standard approach uses conditional ML to estimate $\{\beta_j\}$.

Software exists for conditional ML estimation (e.g., LogXact). For modeling matched case-control pairs, a simple way exists to obtain conditional ML estimates of $\{\beta_j\}$ using software for ordinary ML logistic regression. One calculates the difference between the values for the case and the control on Y and on $\{x_1, \ldots, x_k\}$; one enters those differences into the ordinary software and fits the model that forces the intercept parameter α to equal zero.

We illustrate using Table 9.3. Denote the difference values by Y^* and x^*. For each pair, if we form the difference using the observation for the case minus the observation for the control, then always $Y^* = 1$. Since $x = 1$ represents "yes" for diabetes and $x = 0$ represents "no," the differences equal ($Y^* = 1$, $x^* = -1$) for 16 observations, ($Y^* = 1$, $x^* = 0$) for $9 + 82 = 91$ observations, and ($Y^* = 1$, $x^* = +1$) for 37 observations. The logit model that forces $\hat{\alpha} = 0$ has $\hat{\beta} = 0.84$. For this case of a single binary predictor, the estimate is identical to $\log(n_{12}/n_{21})$.

9.2.4 Extensions to Multiple Measurements and Multiple Categories*

The logit model (9.2.1) for matched pairs extends to a matched set of arbitrary size T, such as responses on the same subjects at T times. Let β_t denote a parameter for measurement t in the matched set, $t = 1, \ldots, T$. In the model

$$\text{logit}(\phi_i) = \alpha_i + \beta_t, \qquad (9.2.2)$$

the $\{\beta_t\}$ permit the probabilities to vary by measurement.

This model is often called the *Rasch model*, in honor of the Danish statistician who studied and promoted it in the 1960s. The Rasch model has a substantial literature in psychometric journals. It is commonly used to describe performance of students on a battery of questions. Formula (9.2.2) describes the probability that student i with ability α_i correctly answers question t, which has "easiness" parameter β_t. The data for each subject has the form of a generalization of Table 9.2 with T rows and two columns. In row t, entries $(1, 0)$ denote a correct response and $(0, 1)$ an incorrect response for question t.

For $T = 2$, Section 9.2.1 noted that the *CMH* test applied to the $2 \times 2 \times n$ table with strata of form Table 9.2 yields McNemar's test. For arbitrary T, a generalized *CMH* test of conditional independence (Sections 7.3.5–7.3.7) applied to the $T \times 2 \times n$ table is a test of identical $\{\beta_t\}$ in model (9.2.2). The test statistic for that case is sometimes called *Cochran's Q*.

The *CMH* representation also suggests ways to test marginal homogeneity for larger tables having possibly several response categories as well as several measurements. Consider a matched set of T measurements and a response scale having I categories. One can display the data in a $T \times I \times n$ table. A partial table displays the T measurements for a given subject or matched set, one measurement in each row. A generalized *CMH* test of conditional independence provides a test of homogeneity of the T marginal distributions. When response categories are ordered, one can use a version of that test having ordered scores for the columns (Section 7.3.6). This builds power for detecting marginal differences in true mean responses for the T measurements.

The *CMH* representation is also helpful for more complex forms of matched-pairs data. For instance, suppose that a case-control study matches each case with several controls. With n matched sets, one displays each matched set as a stratum of a $2 \times 2 \times n$ table. Each stratum has one observation in column 1 (the case) and several observations in column 2 (the controls). McNemar's test no longer applies, but one can use the ordinary *CMH* test to perform the analysis.

The next section presents models for matched-pairs data when the response scale has more than two categories.

9.3 SYMMETRY AND QUASI-SYMMETRY MODELS FOR SQUARE TABLES

The definition of marginal homogeneity for binary matched pairs extends directly to matched pairs with multicategory responses. For an I-category response, an $I \times I$ table displays results. The cell probabilities $\{\pi_{ij}\}$ satisfy *marginal homogeneity* if

$$\pi_{i+} = \pi_{+i}, \qquad i = 1, \ldots, I; \tag{9.3.1}$$

that is, for each category, its probability for the row classification equals its probability for the column classification.

The probabilities in the square table satisfy *symmetry* if

$$\pi_{ij} = \pi_{ji} \tag{9.3.2}$$

for all pairs of cells. Cell probabilities on one side of the main diagonal are a mirror image of those on the other side. When symmetry holds, necessarily marginal homogeneity also holds. When $I > 2$, though, marginal homogeneity can occur without symmetry.

9.3.1 Symmetry as Logit and Loglinear Models

The symmetry condition has the simple logit form

$$\log(\pi_{ij}/\pi_{ji}) = 0 \qquad \text{for all } i \text{ and } j.$$

For expected frequencies $\{\mu_{ij}\}$, the symmetry model also has a loglinear model representation,

$$\log \mu_{ij} = \lambda + \lambda_i + \lambda_j + \lambda_{ij} \tag{9.3.3}$$

where all $\lambda_{ij} = \lambda_{ji}$. This is the special case of the saturated loglinear model (6.1.2) in which the association terms are symmetric and the X and Y main-effect terms are identical; that is, $\lambda_{ij}^{XY} = \lambda_{ji}^{XY}$ and $\lambda_i^X = \lambda_i^Y$. For (9.3.3), $\log \mu_{ij} = \log \mu_{ji}$, so that $\mu_{ij} = \mu_{ji}$.

The ML fit of the symmetry model is

$$\hat{\mu}_{ij} = \frac{n_{ij} + n_{ji}}{2}.$$

The fit satisfies $\hat{\mu}_{ij} = \hat{\mu}_{ji}$. It has $\hat{\mu}_{ii} = n_{ii}$, a perfect fit on the main diagonal. The residual df for chi-squared goodness-of-fit tests equal $I(I-1)/2$. The adjusted residuals equal

$$r_{ij} = \frac{n_{ij} - n_{ji}}{\sqrt{(n_{ij} + n_{ji})}}.$$

Only one residual for each pair of categories is nonredundant, since $r_{ij} = -r_{ji}$. The sum of squared adjusted residuals for the pairs of categories equals X^2 for testing the model fit.

9.3.2 Coffee Market Share Example

A survey recorded the brand choice for a sample of buyers of instant coffee. At a later coffee purchase by these subjects, the brand choice was again recorded. Table 9.5 shows results for five brands of decaffinated coffee. The cell counts on the main diagonal are relatively large, indicating that most buyers did not change their brand choice.

The symmetry model fitted to these data has $G^2 = 22.5$ and $X^2 = 20.4$, with $df = 10$. The lack of fit results primarily from the discrepancy between n_{13} and n_{31}. For that pair, the adjusted residual equals $(44 - 17)/(44 + 17)^{1/2} = 3.5$; consumers of High Point changed to Sanka more often than the reverse. Otherwise, the symmetry model fits most of the table fairly well.

9.3.3 Quasi Symmetry

The symmetry model is so simple that it rarely fits well. For instance, when the marginal distributions differ substantially, the model fits poorly. For square tables with matched-pairs data, the questions of main interest usually refer to comparing the marginal distributions rather than pairs of cells. One can accommodate marginal heterogeneity by permitting the loglinear main-effect terms in (9.3.3) to differ. The

Table 9.5 Choice of Decaffinated Coffee at Two Purchase Dates, with Fit of Quasi-Symmetry Model

First Purchase	Second Purchase					Total
	High Point	Taster's	Sanka	Nescafe	Brim	
High Point	93	17	44	7	10	171
	(93)	(17.0)	(41.0)	(8.0)	(12.0)	
Taster's Choice	9	46	11	0	9	75
	(9.0)	(46)	(11.0)	(0.0)	(9.0)	
Sanka	17	11	155	9	12	204
	(20.0)	(11.0)	(155)	(7.9)	(10.1)	
Nescafe	6	4	9	15	2	36
	(5.0)	(4.0)	(10.1)	(15)	(1.9)	
Brim	10	4	12	2	27	55
	(8.0)	(4.0)	(13.9)	(2.1)	(27)	
Total	135	82	231	33	60	541

Source: Based on data from R. Grover and V. Srinivasan, *J. Marketing Research, 24:* 139–153 (1987). Reprinted with permission of the American Marketing Association.

resulting model, called the *quasi-symmetry model*, is

$$\log \mu_{ij} = \lambda + \lambda_i^X + \lambda_j^Y + \lambda_{ij}, \qquad (9.3.4)$$

where $\lambda_{ij} = \lambda_{ji}$ for all i and j.

This model is more complex than symmetry, because of its differing main-effect terms. The fitted marginal totals equal the observed totals,

$$\hat{\mu}_{i+} = n_{i+} \quad \text{and} \quad \hat{\mu}_{+i} = n_{+i}, \qquad i = 1, \ldots, I.$$

The symmetry model is the special case of (9.3.4) in which all $\lambda_i^X = \lambda_i^Y$. The model is also more complex than the independence model, since it has an association term. The independence model is the special case in which all $\lambda_{ij} = 0$. The quasi-symmetry model is useful partly because it contains these two models as special cases.

Fitting the quasi-symmetry model requires iterative procedures for loglinear models. For the symmetry model, $\hat{\mu}_{ij}$ is the average of n_{ij} and n_{ji}. For quasi symmetry, the average of $\hat{\mu}_{ij}$ and $\hat{\mu}_{ji}$ equals the average of n_{ij} and n_{ji}; thus,

$$\hat{\mu}_{ij} + \hat{\mu}_{ji} = n_{ij} + n_{ji}.$$

Like the symmetry model, the quasi-symmetry model fits perfectly on the main diagonal. Its residual df equal $(I - 1)(I - 2)/2$. When $I > 2$, it is unsaturated because it assumes a symmetric structure ($\lambda_{ij} = \lambda_{ji}$) for the association term. This association symmetry implies that odds ratios on one side of the main diagonal are identical to corresponding odds ratios on the other side. When $I > 2$, for instance,

$$\frac{\mu_{12}\mu_{23}}{\mu_{13}\mu_{22}} = \frac{\mu_{21}\mu_{32}}{\mu_{31}\mu_{22}}.$$

For Table 9.5 on coffee market share, the quasi-symmetry model has $G^2 = 10.0$ and $X^2 = 8.5$, with $df = 6$. Table 9.5 exhibits the fit. Permitting the marginal distributions to differ yields a better fit than the symmetry model provides. It is plausible that, adjusting for market shares, brand switching has a symmetric pattern. Section 9.4.1 utilizes this model to test marginal homogeneity. Section 9.5.1 presents a simpler model, *quasi independence*, that fits these data well.

9.3.4 An Ordinal Quasi-Symmetry Model

The symmetry and quasi-symmetry models treat the classifications as nominal. A special case of quasi symmetry often is useful when the categories are ordinal. Let $u_1 \le u_2 \le \cdots \le u_I$ denote ordered scores for both the row and column categories. The *ordinal quasi-symmetry model* has form

$$\log \mu_{ij} = \lambda + \lambda_i + \lambda_j + \beta u_j + \lambda_{ij}, \tag{9.3.5}$$

where $\lambda_{ij} = \lambda_{ji}$ for all i and j. It is the special case of the quasi-symmetry model (9.3.4) in which

$$\lambda_j^Y - \lambda_j^X = \beta u_j.$$

That is, the difference in main-effect terms has a linear trend across the response categories. The symmetry model is the special case $\beta = 0$.

For this model, the fitted marginal counts need not equal the observed marginal counts, but they do have the same means. For the chosen category scores $\{u_i\}$, the sample mean for the row classification is $\sum_i u_i p_{i+}$. This equals the row mean $\sum_i u_i \hat{\pi}_{i+}$ for the fitted values. A similar equality holds for the column means. When responses in one margin tend to be higher on the ordinal scale than those in the other margin, the fit of model (9.3.5) exhibits this same ordering.

When $\beta \ne 0$, this model implies ordered margins, the cumulative proportions in one margin exceeding those in the other margin. When $\beta > 0$, then $\pi_{1+} > \pi_{+1}$, $\pi_{1+} + \pi_{2+} > \pi_{+1} + \pi_{+2}$, and so forth. That is, responses are more likely to be at the low end of the ordinal scale for the row variable than for the column variable. It then follows that $\sum_i u_i \pi_{i+} < \sum_i u_i \pi_{+i}$; the mean response is higher for the column classification. When $\beta < 0$, the mean response is higher for the row classification.

The ordinal quasi-symmetry model is equivalent to the model of logit form,

$$\log(\pi_{ij}/\pi_{ji}) = \beta(u_j - u_i). \tag{9.3.6}$$

This is the special case of the usual logit model, $\mathrm{logit}(\pi) = \alpha + \beta x$, with $\alpha = 0$, $x = u_j - u_i$ and π equal to the conditional probability for cell (i, j), given response in cell (i, j) or (j, i). The greater the value of $|\beta|$, the greater the difference between π_{ij} and π_{ji} and between the marginal distributions. The probability that the response on the column variable is x units higher than the response on the row variable equals $\exp(x\beta)$ times the probability that the response on the row variable is x units higher than the response on the column variable.

One can estimate β in the ordinal quasi-symmetry model by fitting model (9.3.6) using logit model software. One identifies (n_{ij}, n_{ji}) as binomial numbers of successes and failures in $n_{ij} + n_{ji}$ trials, and fits a logit model with no intercept and with value of the predictor x equal to $u_j - u_i$. One can also easily fit the ordinal or the ordinary quasi-symmetry models with software for loglinear models. The Appendix shows details.

9.3.5 Premarital and Extramarital Sex Example

Table 9.6 is taken from the 1989 General Social Survey. Subjects were asked their opinion about a man and a woman having sex relations before marriage and a married person having sexual relations with someone other than the marriage partner. The response categories are 1 = always wrong, 2 = almost always wrong, 3 = wrong only sometimes, 4 = not wrong at all.

A cursory glance at the data reveals that the symmetry model is doomed. Indeed, $G^2 = 402.2$ and $X^2 = 297.6$ for testing its fit, with $df = 6$. By comparison, the quasi-symmetry model fits well, having $G^2 = 1.4$ and $X^2 = 0.9$, with $df = 3$. The simpler ordinal quasi-symmetry model also fits well. For the scores $\{1, 2, 3, 4\}$, $G^2 = 2.1$ and $X^2 = 2.1$, with $df = 5$. Table 9.6 displays its fitted values.

The cumulative sample marginal proportions are $p_{1+} = 146/475 = .307$, $p_{1+} + p_{2+} = (146 + 39)/475 = .389$, $p_{1+} + p_{2+} + p_{3+} = (146 + 39 + 105)/475 = .611$ for premarital sex, and $p_{+1} = 387/475 = .815$, $p_{+1} + p_{+2} = (387 + 49)/475 = .918$, $p_{+1} + p_{+2} + p_{+3} = (387 + 49 + 33)/475 = .987$ for extramarital sex. (The cumulative proportion necessarily equals 1.0 in the final category.) Each cumulative proportion for extramarital sex is much larger than the corresponding one for premarital sex, suggesting that responses on extramarital sex tend to be lower on the ordinal scale (i.e., more negative) than those on premarital sex. The mean responses also show this ordering. The mean for extramarital sex is $[387 + 2(49) + 3(33) + 4(6)]/475 = 1.28$,

Table 9.6 Opinions on Premarital Sex and Extramarital Sex, with Fit of Ordinal Quasi-Symmetry Model

Premarital Sex	Extramarital Sex				
	1	2	3	4	Total
1	144	2	0	0	146
	(144)	(1.9)	(0.3)	(0.0)	
2	33	4	2	0	39
	(33.1)	(4)	(0.9)	(0.1)	
3	84	14	6	1	105
	(83.7)	(15.1)	(6)	(1.4)	
4	126	29	25	5	185
	(126.0)	(28.9)	(24.6)	(5)	
Total	387	49	33	6	475

Source: 1989 General Social Survey.

close to the "always wrong" score, and the mean for premarital sex is $[146 + 2(39) + 3(105) + 4(185)]/475 = 2.69$, close to the "wrong only sometimes" score.

The ML estimate $\hat{\beta} = -2.86$ for the ordinal quasi-symmetry model also shows this tendency. From (9.3.6), the estimated probability that response on premarital sex is x categories higher than the response on extramarital sex equals $\exp(2.86x)$ times the reverse probability. For instance, the estimated probability that the response on premarital sex is "almost always wrong" and the response on extramarital sex is "always wrong" equals $\exp(2.86) = 17.4$ times the estimated probability that the response on premarital sex is "always wrong" and the response on extramarital sex is "almost always wrong."

9.4 COMPARING MARGINAL DISTRIBUTIONS

Having introduced models for square contingency tables, we next study ways of using them to compare marginal distributions. Quasi-symmetric models imply marginal homogeneity when their parameters take certain values. Tests of the hypothesis of marginal homogeneity compare the fit of the model to the fit of the special case specifying marginal homogeneity.

9.4.1 Testing Marginal Homogeneity

For quasi-symmetry model (9.3.4), marginal homogeneity is the special case in which the two sets of main-effect parameters are identical; that is, all $\lambda_i^X = \lambda_i^Y$. But this special case is simply the symmetry model. In other words, for the quasi-symmetry model, marginal homogeneity is equivalent to symmetry. A test of marginal homogeneity tests the null hypothesis that the symmetry (S) model holds against the alternative hypothesis of quasi symmetry (QS). The likelihood-ratio test compares the G^2 goodness-of-fit statistics,

$$G^2(S \mid QS) = G^2(S) - G^2(QS).$$

For $I \times I$ tables, the test has $df = I - 1$.

We illustrate using Table 9.5, the 5×5 table on choice of coffee brand at two purchases. For these data, $G^2(S) = 22.5$ and $G^2(QS) = 10.0$. The difference $G^2(S \mid QS) = 12.5$, based on $df = 4$, provides evidence of differing marginal distributions $(P = .014)$.

The sample marginal proportions for brands (High Point, Taster's Choice, Sanka, Nescafe, Brim) were $(.316, .139, .377, .067, .102)$ originally and $(.250, .152, .427, .061, .111)$ later. One can compare any matched (row, column) pair of these proportions by combining the other categories and using the methods of Section 9.1 for 2×2 tables. To illustrate, to compare the proportions selecting High Point at the two times, we construct the table with row and column categories (High Point, Others). This table has counts, by row, of $(93, 78/42, 328)$. The McNemar z statistic (9.1.1) equals $(78 - 42)/(78 + 42)^{1/2} = 3.3$. There is strong evidence of a change in the proportion choosing this brand $(P = .001)$.

From formula (9.1.2), the standard error equals .020 for the difference .316 − .250 = .067 between the sample proportions choosing High Point at the two times. A 95% confidence interval for the true decrease in proportions equals .067 ± .039. Further investigation reveals that the small P-value for testing marginal homogeneity reflects this decrease in the proportion choosing High Point and an increase in the proportion choosing Sanka, with no evidence of change for the other coffees.

When the quasi-symmetry model fits poorly, other tests of marginal homogeneity are available (Agresti (1990), pp. 359–360 and 499–500), including generalized *CMH* procedures as discussed in Section 9.2.4. A simple test uses the differences in the I row and column marginal proportions, $p_{i+} - p_{+i}, i = 1, \ldots, I$. A statistic based on these differences and their covariance matrix also has a large-sample chi-squared distribution with $df = I - 1$. Larger differences in marginal proportions yield a larger statistic. Though complex computationally, this statistic is available in standard software (e.g., PROC CATMOD in SAS). In practice, for nominal classifications the statistic $G^2(S \mid QS)$ usually captures most of the information about marginal heterogeneity even if the quasi-symmetry model shows lack of fit. In fact, the alternative test based on pairwise differences of marginal proportions has a connection to that model, being the model's "efficient score" statistic for testing marginal homogeneity.

9.4.2 Marginal Homogeneity and Ordered Categories

Tests of marginal homogeneity based on the quasi-symmetry model treat the classifications as nominal. When response categories are ordered, ordinal tests of marginal homogeneity analyze whether responses tend to be higher on the ordinal scale in one margin than the other.

One can utilize the ordering by choosing as baseline model the ordinal quasi-symmetry model (9.3.5). For that model, symmetry and thus marginal homogeneity is the special case $\beta = 0$. A likelihood-ratio test of marginal homogeneity uses the difference between the G^2 values for the symmetry and ordinal quasi-symmetry models, with $df = 1$. Or, a Wald statistic treats $(\hat{\beta}/ASE)^2$ as chi-squared with $df = 1$.

A third option for an ordinal chi-squared test does not require fitting this model, but is related to it, being its "efficient score" test of marginal homogeneity. It compares the sample means for the two margins, using the category scores $\{u_i\}$ in the model. Denote the sample means for the rows (X) and columns (Y) by $\bar{x} = \sum_i u_i p_{i+}$ and $\bar{y} = \sum_i u_i p_{+i}$. One forms a chi-squared test statistic with $df = 1$ by the ratio of the square of $(\bar{x} - \bar{y})$ to its estimated variance, which equals

$$\left(\frac{1}{n}\right)\left[\sum_i \sum_j (u_i - u_j)^2 p_{ij} - (\bar{x} - \bar{y})^2\right].$$

This test essentially analyzes paired difference scores for the n matched pairs. It is a good test for detecting differences between true marginal means, even if the ordinal quasi-symmetry model exhibits lack of fit.

For large samples, these ordinal likelihood-ratio, Wald, and efficient score tests of marginal homogeneity usually have similar results. Unless the ordinal quasi-symmetry model fits quite poorly, tests based on it are more powerful than tests such as $G^2(S \mid QS)$. In utilizing the ordinality, the tests are directed toward narrower alternatives and have $df = 1$.

We illustrate the ordinal tests of marginal homogeneity using Table 9.6 on opinions about premarital and extramarital sex. The symmetry model has $G^2 = 402.2$ ($df = 6$), and the ordinal quasi-symmetry model has $G^2 = 2.1$ ($df = 5$). The likelihood-ratio statistic for testing marginal homogeneity is $402.2 - 2.1 = 400.1$, with $df = 1$, providing extremely strong evidence of heterogeneity ($P < .0001$).

Strong evidence also results from the Wald test statistic $(\hat{\beta}/ASE)^2 = (-2.857/0.420)^2 = 46.4$ or the score statistic $[(\bar{x} - \bar{y})/ASE]^2 = [(1.28 - 2.69)/0.0563]^2 = 629.5$. The estimate $\hat{\beta} = -2.857$ or the difference in sample marginal means indicates that responses were considerably more conservative on extramarital sex than premarital sex.

9.4.3   A Proportional Odds Comparison of Margins*

The marginal comparisons presented so far in this section utilize loglinear models. The analyses also result from generalizations of the Rasch logit model (9.2.2). The statistic $G^2(S \mid QS)$ based on the quasi-symmetry model relates to a baseline-category logit model having subject effects and margin effects. The statistic based on the ordinal quasi-symmetry model relates to an adjacent-categories logit model having subject effects and having margin effects that are the same for each pair of adjacent response categories. Alternatively, for ordinal responses, one can compare the margins with a model using logits of cumulative probabilities. For instance, a proportional odds generalization of the Rasch model expresses cumulative logits (Section 8.2.1) for each margin in terms of subject effects and margin effects that are the same for each binary collapsing of the response.

Let β denote a parameter such that for each subject, the odds that the row response falls in category j or below (instead of above category j) are $\exp(\beta)$ times the odds for the column response. An estimate of this parameter is

$$\hat{\beta} = \log \left(\frac{\sum \sum_{i<j} (j - i) n_{ij}}{\sum \sum_{i>j} (i - j) n_{ij}} \right). \tag{9.4.1}$$

The numerator sum refers to cells above the main diagonal, weighted by their distance $(j - i)$ from that diagonal. The denominator sum refers to cells below the main diagonal.

For Table 9.6, the numerator of (9.4.1) equals

$$1(2 + 2 + 1) + 2(0 + 0) + 3(0) = 5,$$

and the denominator equals

$$1(33 + 14 + 25) + 2(84 + 29) + 3(126) = 676.$$

Thus, $\hat{\beta} = \log(5/676) = -4.91$, and the estimated odds ratio is $\exp(\hat{\beta}) = 5/676 = 0.0074$. For instance, for each subject, the estimated odds of response "always wrong" (instead of the other three categories) on premarital sex are .0074 times the estimated odds for extramarital sex. The estimate $\hat{\beta} = -4.91$ falls far from 0 and indicates a substantial difference between the marginal distributions.

One can base an ordinal test of marginal homogeneity ($\beta = 0$) on this margin effect. Estimator (9.4.1) of β has estimated variance

$$\frac{\sum\sum_{i<j}(j-i)^2 n_{ij}}{\left[\sum\sum_{i<j}(j-i)n_{ij}\right]^2} + \frac{\sum\sum_{i>j}(i-j)^2 n_{ij}}{\left[\sum\sum_{i>j}(i-j)n_{ij}\right]^2}. \tag{9.4.2}$$

The ratio $\hat{\beta}/ASE$, where ASE is the square root of (9.4.2), is an approximate standard normal test statistic, and its square is chi-squared with $df = 1$. For Table 9.6, $\hat{\beta} = -4.91$ has $ASE = 0.45$. The chi-squared statistic $(-4.91/0.45)^2 = 118.2$ also provides strong evidence against the null hypothesis of marginal homogeneity.

The result of this ordinal test for a proportional odds model is usually similar to the tests discussed in Section 9.4.2 of $\beta = 0$ in the ordinal quasi-symmetry model. The two models have different forms and are not equivalent, their interpretations referring to different types of odds ratios. Both are sensitive, though, to detecting marginal differences for which responses tend to be higher in one margin than the other.

9.5   ANALYZING RATER AGREEMENT*

Table 9.7 shows ratings by two pathologists, labeled X and Y, who separately classi-fied 118 slides regarding the presence and extent of carcinoma of the uterine cervix. The rating scale has the ordered categories (1) negative, (2) atypical squamous hyper-plasia, (3) carcinoma *in situ*, (4) squamous or invasive carcinoma. This table illustrates

Table 9.7   Diagnoses of Carcinoma, with Adjusted Residuals for Independence Model

Pathologist X	Pathologist Y				Total
	1	2	3	4	
1	22	2	2	0	26
	(8.5)	(−0.5)	(−5.9)	(−1.8)	
2	5	7	14	0	26
	(−0.5)	(3.2)	(−0.5)	(−1.8)	
3	0	2	36	0	38
	(−4.1)	(−1.2)	(5.5)	(−2.3)	
4	0	1	17	10	28
	(−3.3)	(−1.3)	(0.3)	(5.9)	
Total	27	12	69	10	118

Source: N. S. Holmquist, C. A. McMahon, and O. D. Williams, *Arch. Pathol., 84:* 334–345 (1967). Reprinted with permission of the American Medical Association. See also J. R. Landis and G. G. Koch, *Biometrics, 33:* 363–374 (1977).

another type of matched-pairs data, referring to separate ratings of a sample by two observers using the same categorical scale.

Let $\pi_{ij} = P(X = i, Y = j)$ denote the probability that observer X classifies a slide in category i and observer Y classifies it in category j. Their ratings of a particular subject *agree* if they classify the subject in the same category. In the square table, the main diagonal $\{i = j\}$ represents observer agreement. The term π_{ii} is the probability that they both place a subject in category i, and $\sum_i \pi_{ii}$ is the total probability of agreement. Perfect agreement occurs when $\sum_i \pi_{ii} = 1$.

Many categorical scales are quite subjective, and perfect agreement is rare. This section presents ways of measuring strength of agreement and detecting patterns of disagreement. We distinguish between measuring *agreement* and measuring *association*. Strong agreement requires strong association, but strong association can exist without strong agreement. If observer X consistently rates subjects one level higher than observer Y, the strength of agreement is poor even though the association is strong.

9.5.1   Quasi Independence

A common way of evaluating agreement compares the cell counts $\{n_{ij}\}$ to the values $\{n_{i+}n_{+j}/n\}$ predicted by the loglinear model of independence (6.1.1) for the two-way table. That model provides a baseline, showing the degree of agreement expected if no association existed between the ratings. One would normally expect it to fit poorly if there is even mild agreement, but its cell adjusted residuals (Section 2.4.5) provide information about patterns of agreement and disagreement.

Cells with positive adjusted residuals have frequencies that are higher than expected under independence. Ideally, large positive adjusted residuals occur on the main diagonal, and large negative adjusted residuals occur off that diagonal. The sizes are influenced, however, by the sample size n, larger values tending to occur as n increases.

The independence model fits Table 9.7 poorly ($G^2 = 118.0$, $df = 9$), as one would expect. Table 9.7 reports the adjusted residuals in parentheses. The large positive adjusted residuals on the main diagonal indicate that agreement for each category is greater than expected by chance, especially for the first category. The off-main-diagonal residuals are primarily negative. Disagreements occurred less than expected under independence, though the evidence of this is weaker for categories closer together. Inspection of cell counts reveals that the most common disagreements refer to observer Y choosing category 3 and observer X instead choosing category 2 or 4.

More complex loglinear models add components that relate to agreement beyond that expected under independence. One useful generalization of independence is the *quasi-independence model*,

$$\log \mu_{ij} = \lambda + \lambda_i^X + \lambda_j^Y + \delta_i I(i = j), \tag{9.5.1}$$

where the indicator $I(i = j)$ equals 1 when $i = j$ and equals 0 when $i \neq j$. This model adds to the independence model a parameter δ_1 for the cell in row 1 and column 1, a parameter δ_2 for the cell in row 2 and column 2, and so forth. When $\delta_i > 0$, more agreements regarding outcome i occur than would be expected under independence.

Because of the addition of the $\{\delta_i\}$ parameters when $\{i = j\}$, the quasi-independence model treats the main diagonal differently from the rest of the table. The *ML* fit in those cells is perfect, with $\hat{\mu}_{ii} = n_{ii}$ for all i. For the remaining cells, the independence model still applies. In other words, conditional on observer disagreement, the rating by X is independent of the rating by Y. This model implies that odds ratios equal 1.0 for all rectangularly formed 2×2 tables in which all cells fall off the main diagonal. For 4×4 tables, for instance, the model implies that odds ratios such as

$$\frac{\mu_{13}\mu_{24}}{\mu_{14}\mu_{23}} = \frac{\mu_{31}\mu_{42}}{\mu_{41}\mu_{32}} = 1.0.$$

The independence model is the special case of quasi independence in which $\delta_1 = \cdots = \delta_I = 0$. The quasi-independence model has I more parameters, so its residual $df = (I - 1)^2 - I$. It is a special case of the quasi-symmetry model (9.3.4) in which $\lambda_{ij} = 0$ when $i \neq j$. The two models are equivalent when $I = 3$. One can fit it using iterative methods in loglinear model software.

For Table 9.7, the quasi-independence model has $G^2 = 13.2$ and $X^2 = 11.5$, with $df = 5$. It fits much better than the independence model, though some lack of fit remains. Table 9.8 displays the fit. The fitted counts have the same main-diagonal values and the same row and column totals as the observed data, but satisfy independence for cells not on the main diagonal.

For Table 9.5 on choice of coffee brand at two occasions, the quasi-independence model has $G^2 = 13.8$ with $df = 11$. This is a dramatic improvement over independence, which has $G^2 = 346.4$ with $df = 16$. Given a change in brands, the new choice of coffee brand is plausibly independent of the original choice. The quasi-independence model does not give a significantly poorer fit than the quasi-symmetry model, which has $G^2 = 10.0$ with $df = 6$, the change in G^2 of 3.8 being based on $df = 11 - 6 = 5$. Its interpretation is simpler.

9.5.2 Summarizing Agreement

For a pair of subjects, consider the event that each observer classifies one subject in category a and one subject in category b. The odds that the raters agree rather than disagree on which subject is in category a and which is in category b equal

$$\tau_{ab} = \frac{\pi_{aa}\pi_{bb}}{\pi_{ab}\pi_{ba}} = \frac{\mu_{aa}\mu_{bb}}{\mu_{ab}\mu_{ba}}. \tag{9.5.2}$$

As τ_{ab} increases, the observers are more likely to agree on which subject receives each designation.

For the quasi-independence model, the odds (9.5.2) summarizing agreement for categories a and b equal

$$\tau_{ab} = \exp(\delta_a + \delta_b).$$

These increase as the diagonal parameters increase, so larger $\{\delta_i\}$ represent stronger agreement. For instance, categories 2 and 3 in Table 9.7 have $\hat{\delta}_2 = 0.6$ and $\hat{\delta}_3 = 1.9$. The estimated odds that one observer's rating is category 2 rather than 3 are $\exp(0.6 + 1.9) = 12.3$ times as high when the other observer's rating is 2 than when it is 3. The degree of agreement seems fairly strong, which also happens for the other pairs of categories.

9.5.3 Quasi Symmetry and Agreement Modeling

For Table 9.7, the quasi-independence model shows some lack of fit. This model is normally inadequate for ordinal scales, which almost always exhibit a positive association between ratings. Conditional on observer disagreement, a tendency usually remains for high (low) ratings by one observer to occur with relatively high (low) ratings by the other observer.

For rater agreement data, the quasi-symmetry model (9.3.4) often fits much better than the quasi-independence model. For Table 9.7, it has $G^2 = 1.0$ and $X^2 = 0.6$, based on $df = 2$. Table 9.8 displays the fit. It is not unusual for observer agreement tables to have many empty cells. When $n_{ij} + n_{ji} = 0$ for any pair (such as categories 1 and 4 in Table 9.7), the ML fitted values in those cells must also be zero. One should ideally eliminate those cells from the fitting process to get the proper residual df value.

For the quasi-symmetry model, one can estimate the agreement odds (9.5.2) by substituting the fitted values $\{\hat{\mu}_{ij}\}$ into (9.5.2), or equivalently using $\exp(\hat{\lambda}_{aa} + \hat{\lambda}_{bb} - \hat{\lambda}_{ab} - \hat{\lambda}_{ba})$. For categories 2 and 3 of Table 9.7, for instance, the estimate equals 10.7.

Loglinear models directly address the association component of agreement. The quasi-symmetry model also yields information about similarity of marginal distri-

Table 9.8 Fitted Values for Quasi-Independence and Quasi-Symmetry Models

Pathologist X	Pathologist Y			
	1	2	3	4
1	22	2	2	0
	$(22)^a$	(0.7)	(3.3)	(0.0)
	$(22)^b$	(2.4)	(1.6)	(0.0)
2	5	7	14	0
	(2.4)	(7)	(16.6)	(0.0)
	(4.6)	(7)	(14.4)	(0.0)
3	0	2	36	0
	(0.8)	(1.2)	(36)	(0.0)
	(0.4)	(1.6)	(36)	(0.0)
4	0	1	17	10
	(1.9)	(3.0)	(13.1)	(10)
	(0.0)	(1.0)	(17.0)	(10)

[a]Quasi-independence model.
[b]Quasi-symmetry model.

butions. The simpler symmetry model that forces the margins to be identical fits Table 9.7 poorly, with $G^2 = 39.2$ and $X^2 = 30.3$, based on $df = 5$. The statistic $G^2(S \mid QS) = 39.2 - 1.0 = 38.2$, with $df = 3$, provides strong evidence of marginal heterogeneity. The lack of perfect agreement reflects differences in marginal distributions, which Table 9.7 reveals to be substantial in each category but the first.

The ordinal quasi-symmetry model uses the category orderings. This model fits Table 9.7 poorly, partly because ratings do not tend to be consistently higher by one observer than the other.

9.5.4 Kappa Measure of Agreement

An alternative approach describes strength of agreement using a single summary index, rather than a model. For nominal scales, the most popular index is *Cohen's kappa*. It compares the agreement to that expected if the ratings were independent. The probability of agreement equals $\sum_i \pi_{ii}$. If the observers' ratings were independent, then $\pi_{ii} = \pi_{i+}\pi_{+i}$, and the probability of agreement equals $\sum_i \pi_{i+}\pi_{+i}$.

Cohen's kappa is defined by

$$\kappa = \frac{\sum \pi_{ii} - \sum \pi_{i+}\pi_{+i}}{1 - \sum \pi_{i+}\pi_{+i}}.$$

The numerator compares the probability of agreement to that expected under independence. The denominator replaces $\sum \pi_{ii}$ by its maximum possible value of 1, corresponding to perfect agreement. Kappa equals 0 when the agreement merely equals that expected under independence, and it equals 1.0 when perfect agreement occurs. The stronger the agreement, for a given pair of marginal distributions, the higher the value of kappa.

For Table 9.7, the sample estimate of $\sum \pi_{ii}$ equals $(22 + 7 + 36 + 10)/118 = .636$, and the sample estimate of $\sum \pi_{i+}\pi_{+i}$ equals $[(26)(27) + (26)(12) + (38)(69) + (28)(10)]/(118)^2 = .281$. The sample value of kappa equals $(.636 - .281)/(1 - .281) = .493$. The difference between the observed agreement and that expected under independence is about 50% of the maximum possible difference.

Controversy surrounds the utility of kappa, primarily because its value depends strongly on the marginal distributions. The same diagnostic rating process can yield quite different values of kappa, depending on the proportions of cases of the various types. We prefer to construct models describing the structure of agreement and disagreement, rather than to depend solely on this summary index.

9.6 BRADLEY–TERRY MODEL FOR PAIRED PREFERENCES*

Table 9.9 summarizes results of matches among five women tennis players during the 1989–1990 season. For instance, Steffi Graf won 3 of the 5 matches that she and Monica Seles played. This section presents a model that applies to data of this sort, in which observations consist of pairwise comparisons that result in a preference for

Table 9.9   Results of 1989–1990 Tennis Matches for Women Players, with Fit of Bradley–Terry model

Winner	Loser				
	Seles	Graf	Sabatini	Navratilova	Sanchez
Seles	—	2	1	3	2
		(2.01)	(0.69)	(3.66)	(1.64)
Graf	3	—	6	3	7
	(2.99)		(6.92)	(2.10)	(6.99)
Sabatini	0	3	—	1	3
	(0.31)	(2.08)		(1.24)	(3.38)
Navratilova	3	0	2	—	3
	(2.34)	(0.90)	(1.76)		(2.99)
Sanchez	0	1	2	1	—
	(0.36)	(1.01)	(1.62)	(1.01)	

one category over another. The fitted model provides a ranking of the players, and gives estimates of the probabilities of each result for matches between each pair.

The model is often applied in product comparisons. For instance, a wine-tasting session comparing several brands of sauvignon blanc might consist of a series of pairwise competitions. For each pair of wines, raters taste each wine and indicate a preference for one of them. Using results of several pairwise evaluations, one could use the model to establish a ranking of the wines.

9.6.1   The Bradley–Terry Model

The Bradley–Terry model is a logit model for paired preference data. For Table 9.9, let Π_{ij} denote the probability that player i is the victor when i and j play. The probability that player j wins is $\Pi_{ji} = 1 - \Pi_{ij}$; ties cannot occur. For instance, when Seles (player 1) and Graf (player 2) play, Π_{12} is the probability that Seles wins and $\Pi_{21} = 1 - \Pi_{12}$ is the probability that Graf wins.

The Bradley–Terry model has player parameters $\{\beta_i\}$ such that

$$\text{logit}(\Pi_{ij}) = \log(\Pi_{ij}/\Pi_{ji}) = \beta_i - \beta_j. \qquad (9.6.1)$$

The probability that player i wins equals $\frac{1}{2}$ when $\beta_i = \beta_j$ and exceeds $\frac{1}{2}$ when $\beta_i > \beta_j$. One parameter is redundant, and software imposes a constraint such as setting the last one equal to 0.

One can fit this logit model by treating each separate pair of cell counts (n_{ij}, n_{ji}) as an independent binomial variate. For instance, from Seles's perspective, the (Seles, Graf) results correspond to 2 successes and 3 failures in 5 trials. The logit model has one parameter for each player and no intercept term. To fit the model to Table 9.9, one sets up five artificial explanatory variables, corresponding to the coefficients of those parameters in the model. For the logit of Π_{ij} for a given match, the variable for player i is 1, the variable for player j is -1, and the variables for the other players equal 0. (Table A.16 in the Appendix illustrates this representation in SAS code. One of the five explanatory variables is redundant, corresponding to the redundant parameter.)

Interestingly, the fit of logit model (9.6.1) to the square table of results is identical to the fit of the quasi-symmetry model (9.3.4). For that model, $\log(\mu_{ij}/\mu_{ji})$ has form $\beta_i - \beta_j$, when one identifies β_i with $(\lambda_i^X - \lambda_i^Y)$. One can estimate parameters in the logit model by fitting the quasi-symmetry model and calculating $\{\hat{\beta}_i = \hat{\lambda}_i^X - \hat{\lambda}_i^Y\}$. After fitting the logit model or the quasi-symmetry model, one can estimate $\{\hat{\Pi}_{ij}\}$ using $\hat{\Pi}_{ij} = \exp(\hat{\beta}_i - \hat{\beta}_j)/[1 + \exp(\hat{\beta}_i - \hat{\beta}_j)]$. Or, one can use $\hat{\Pi}_{ij} = \hat{\mu}_{ij}/(\hat{\mu}_{ij} + \hat{\mu}_{ji})$ from the fit of either model.

9.6.2 Ranking Women Tennis Players

For Table 9.9, the goodness-of-fit statistics for testing the Bradley-Terry model are $G^2 = 4.6$ and $X^2 = 3.2$, with $df = 6$. The table also contains the fitted values. With software that sets $\beta_5 = 0$ (for Sanchez), the estimates of the other parameters are 1.53 for Seles, 1.93 for Graf, 0.73 for Sabatini, and 1.09 for Navratilova. Graf ranked highest of the players, and Sanchez ranked lowest.

As well as providing a player ranking, the model fit yields estimates of the probability of each outcome. To illustrate, when Graf played Seles in 1993, model (9.6.1) estimates the probability of a Graf win to be

$$\hat{\Pi}_{21} = \frac{\exp(\hat{\beta}_2 - \hat{\beta}_1)}{1 + \exp(\hat{\beta}_2 - \hat{\beta}_1)} = \frac{\exp(0.40)}{1 + \exp(0.40)} = .60.$$

Or, from the fitted values,

$$\hat{\Pi}_{21} = \frac{\hat{\mu}_{21}}{\hat{\mu}_{21} + \hat{\mu}_{12}} = \frac{2.99}{2.99 + 2.01} = .60$$

For such small data sets, the model smoothing provides estimates that are more pleasing and realistic than the sample proportions. For instance, Seles beat Sanchez in both their matches, but the Bradley–Terry model estimates the probability of a Seles victory to be .82 rather than 1.00.

To check whether the difference between two players is "statistically significant," one can compare $(\hat{\beta}_i - \hat{\beta}_j)$ to its *ASE*. From the covariance matrix of parameter estimates, the *ASE* equals the square root of $[\text{Var}(\hat{\beta}_i) + \text{Var}(\hat{\beta}_j) - 2\,\text{Cov}(\hat{\beta}_i, \hat{\beta}_j)]$. For instance, for comparing Graf and Seles, $\hat{\beta}_2 - \hat{\beta}_1 = 0.400$ has *ASE* $= 0.669$, indicating an insignificant difference. Estimates are imprecise for this small data set. The only two comparisons showing fairly good evidence of a true difference in the ranking are those placing both Graf and Seles higher than Sanchez.

A confidence interval for $\beta_i - \beta_j$ translates directly to one for Π_{ij}. For Graf and Seles, a 95% confidence interval for $\beta_2 - \beta_1$ is $0.400 \pm 1.96(0.669)$, or $(-0.91, 1.71)$. This translates to $(.29, .85)$ for the probability Π_{21} of a Graf win (e.g., $\exp(1.71)/[1 + \exp(1.71)] = .85$).

The assumption of independent, identical trials that leads to the binomial distribution and the usual fit of the logit model may be overly simplistic for this application. For instance, the probability Π_{ij} that i beats j may vary according to whether the court is clay, grass, or hard, and it may vary somewhat over time.

PROBLEMS

9.1. Apply McNemar's test to Table 9.3. Interpret.

9.2. Table 9.10 refers to a sample of juveniles convicted of a felony in Florida in 1987. Matched pairs were formed using criteria such as age and the number of prior offenses. For each pair, one subject was handled in the juvenile court and the other was transferred to the adult court. The response of interest was whether the juvenile was rearrested by the end of 1988.

Table 9.10

| Adult | Juvenile Court | |
Court	Rearrest	No Rearrest
Rearrest	158	515
No Rearrest	290	1134

Source: Based on a study at the University of Florida by D. Bishop, C. Frazier, L. Lanza-Kaduce, and L. Winner. Thanks to Dr. Larry Winner for showing me these data.

a. Test the hypothesis that the true proportions rearrested were identical for the adult and juvenile court assignments. Use a two-sided alternative.

b. Find a 90% confidence interval for the difference between the true proportions of rearrest for the adult and juvenile court assignments. Interpret.

9.3. Refer to the previous problem. Specify a logit model for the probability of rearrest, using court assignment as a predictor. Explain how to estimate and interpret the effect of court assignment.

9.4. Explain the following analogy: McNemar's test is to binary data as the paired difference t test is to normally distributed data.

9.5. Refer to Table 6.12, and treat the data as matched pairs on opinion, stratified by gender.

a. For each gender, test equality of the true proportions supporting government action for the two items, and construct a 90% confidence interval for the difference between the true proportions of support. Interpret.

b. Construct a 90% confidence interval for the difference between males and females in their differences of proportions of support for each item. (*Hint*: The gender samples are independent, so the variance of the difference is the sum of the variances.)

c. Estimate the odds ratio $\exp(\beta)$ for logit model (9.2.1) for each gender. Interpret.

d. Explain why a test of independence for the 2×2 table using entries (6, 160) from Table 6.12 in row 1 and entries (11, 181) in row 2 compares the odds ratios in **(c)** for the genders.

9.6. Refer to Table 9.1. Suppose sample proportions of approval of .59 and .55 were based on *independent* samples of size 1600 each. Construct a 95% confidence interval for the true difference of proportions. Compare to the result in Section 9.1.2, and comment on how the use of dependent samples can improve precision.

9.7. A crossover experiment with 100 subjects compares two drugs used to treat migraine headaches. The response scale is success $(+)$ or failure $(-)$. Half the study subjects, randomly selected, used drug A the first time they get a migraine headache and drug B the next time. For them, 6 had responses $(A+, B+)$, 25 had responses $(A+, B-)$, 10 had responses $(A-, B+)$, and 9 had responses $(A-, B-)$. The other 50 subjects took the drugs in the reverse order. For them, 10 were $(A+, B+)$, 20 were $(A+, B-)$, 12 were $(A-, B+)$, and 8 were $(A-, B-)$.

a. Ignoring treatment order, compare the success probabilities for the two treatments. Interpret.

b. McNemar's test uses only the pairs of responses that differ. For this study, Table 9.11 shows such data from both treatment orders. Explain why a test of independence for this table tests the hypothesis that success rates are identical for the two treatments. Analyze these data, and interpret.

Table 9.11

Treatment Order	Treatment That Is Better	
	First	Second
A then B	25	10
B then A	12	20

9.8. Fitting a logistic regression model to difference scores, estimate β in model (9.2.1) applied to Table 9.1. Interpret.

9.9. A case-control study has 8 pairs of subjects. The cases have colon cancer, and the controls are matched with the cases on gender and age. A possible explanatory variable is the extent of red meat in a subject's diet, measured as "low" or "high." For three pairs, both the case and the control were high; for one pair, both the case and the control were low; for three pairs, the case was high and the control was low; for one pair, the case was low and the control was high.

a. Display the data in a 2×2 cross-classification of diet for the case against diet for the control. Display the $2 \times 2 \times 8$ table with partial tables relating diet to response (case, control) for the pairs. Successive parts refer to these as Table A and Table B.

b. Calculate the McNemar z^2 statistic for Table A and the *CMH* statistic for Table B. Compare.

   c. Show that the Mantel–Haenszel estimate of a common odds ratio for the eight partial tables in Table B is identical to n_{12}/n_{21} for Table A.

   d. For Table B, show that the *CMH* statistic and the Mantel–Haenszel odds ratio estimate do not change if one deletes pairs from the data set in which both the case and the control had the same diet.

   e. This sample size is too small for these large-sample tests. Use the binomial distribution with Table A to find the exact P-value for testing marginal homogeneity against the alternative hypothesis of a higher incidence of colon cancer for the "high" red meat diet.

9.10. The estimated variance for the conditional ML estimate $\hat{\beta} = \log(n_{12}/n_{21})$ of β in model (9.2.1) is $(1/n_{12} + 1/n_{21})$. Show that a 95% confidence interval for the odds ratio $\exp(\beta)$ for Table 9.1 equals $(1.34, 2.27)$.

9.11. Refer to Table 6.3. Viewing the table as matched triplets, one can compare the proportion of "yes" responses among the three drugs.

   a. Construct the marginal distribution for each drug, and compute the three sample proportions of "yes" responses.

   b. Representing the data with a 3×2 partial table of drug-by-response for each subject, use a generalized *CMH* procedure (Cochran's Q) to test marginal homogeneity.

   c. Repeat the test after eliminating the $911 + 279$ subjects who make the same response for every drug. What effect do such observations have on the test?

9.12. Refer to the previous problem. Using conditional ML, estimate the marginal parameters in logit model (9.2.2), and interpret using subject-specific odds ratios.

9.13. Refer to Table 9.6. For the symmetry model, compute and interpret the adjusted residual for the pair of categories (1,4).

9.14. Table 9.12, from the 1991 General Social Survey, reports subjects' religious affiliation in 1991 and at age 16, for categories (1) Protestant, (2) Catholic, (3) Jewish, (4) None or other.

Table 9.12

Affiliation at Age 16	Religious Affiliation Now			
	1	2	3	4
1	863	30	1	52
2	50	320	0	33
3	1	1	28	1
4	27	8	0	33

Source: 1991 General Social Survey.

   a. Test whether the proportions classified as Catholic differed in 1991 and at age 16. Construct a 90% confidence interval for the change in the proportion classified Catholic. Interpret.

   b. Show that the symmetry model has $G^2 = 32.2$ with $df = 6$. Use residuals to analyze transition patterns between pairs of religions.

   c. Show that the quasi-symmetry model has $G^2 = 2.0$ with $df = 3$. Interpret.

   d. Test marginal homogeneity. Show that the small P-value mainly reflects the large sample size and is due to a small decrease in the proportion classified Catholic and an increase in the proportion classified None or Other, with no evidence of change for other categories.

   e. Fit the quasi-independence model, and interpret.

9.15. Table 9.13, from the 1991 General Social Survey, reports respondents' current region of residence and region of residence at age 16. Analyze these data.

Table 9.13

Residence	Residence in 1991			
at Age 16	Northeast	Midwest	South	West
Northeast	245	16	40	20
Midwest	12	333	31	51
South	14	31	321	16
West	3	51	12	309

Source: 1991 General Social Survey.

9.16. Refer to Table 9.6. Fit the ordinal quasi-symmetry model, using $u_1 = 1$ and $u_4 = 4$ and picking u_2 and u_3 to represent sensible choices for distances between categories that are unequally spaced. Compare results and interpretations to those given in Sections 9.3.4 and 9.4.2.

9.17. Refer to Table 9.6. Using a generalized *CMH* procedure, test marginal homogeneity.

9.18. Table 9.14 is from the 1989 General Social Survey. Subjects were asked their opinion on early teens (age 14–16) having sex relations and on a man and a woman having sex relations before marriage. The response categories are $1 = $ always wrong, $2 = $ almost always wrong, $3 = $ wrong only sometime, $4 = $ not wrong at all. Analyze these data.

Table 9.14

Teen	Premarital Sex			
Sex	1	2	3	4
1	141	34	72	109
2	4	5	23	38
3	1	0	9	23
4	0	0	1	15

Source: 1989 General Social Survey.

9.19. Refer to the previous problem. Using the ordinal quasi-symmetry model or a proportional odds model, estimate the marginal effect. Interpret.

9.20. Refer to Table 6.17. The two-way table relating responses for the environment (as rows) and cities (as columns) has cell counts, by row, $(108, 179, 157/21, 55, 49/5, 6, 24)$.

a. Fit the symmetry and quasi-symmetry models. Interpret.

b. Test marginal homogeneity, and interpret.

c. Conduct an analysis that utilizes the ordinality, and interpret.

d. Fit the quasi-independence model, and interpret.

e. Compare the margins using a proportional odds model.

f. Use kappa to describe agreement between responses.

9.21. Refer to Table 6.17. The two-way table relating health (as rows) and law enforcement (as columns) has cell counts, by row, $(292, 117, 25/72, 60, 6/14, 12, 9)$. Analyze these data.

9.22. Refer to all four items in Table 6.17.

a. Generalize the symmetry model, so that the cell probability is identical for all cells with the same set of response outcomes; that is, μ_{hijk} is identical for all permutations of the subscript. Fit this *complete symmetry model*, and interpret. (*Hint:* Fit a model with a factor that has a different level for each distinct combination of the four indices.)

b. Fit a generalized quasi-symmetry model by using different main effect factors for the four items. Test homogeneity of the four margins by comparing the fits of the complete symmetry and quasi-symmetry models. Interpret.

c. Fit a generalized ordinal quasi-symmetry model by adding a quantitative main effect term for each margin to the complete symmetry model. Use this model to test marginal homogeneity in a way that uses the ordering.

9.23. An alternative ordinal model for square tables, called the *conditional symmetry model*, has form $\log(\mu_{ij}/\mu_{ji}) = \beta$, for all $i < j$. Show that symmetry is the special case $\beta = 0$. Comparing the fit of this model to the symmetry model provides an alternative ordinal test of marginal homogeneity. Do this for Table 9.6, and interpret. (The ordinal quasi-symmetry model has the advantage of extending naturally to multiway tables, as discussed in the previous problem.)

9.24. Table 9.15 refers to a case-control study investigating a possible relationship between cataracts and the use of head coverings during the summer. Each case

Table 9.15

	Control			
	Always or			
Cataract Case	Almost Always	Frequently	Occasionally	Never
Always or almost always	29	3	3	4
Frequently	5	0	1	1
Occasionally	9	0	2	0
Never	7	3	1	0

Source: J. M. Dolezal et al., *Am. J. Epidemiol., 129:* 559–568 (1989).

reporting to a clinic for care for a cataract was matched with a control of the same sex and similar age not having a cataract. The row and column categories refer to the frequency with which the subject used head coverings. Analyze these data using models presented in this chapter. Interpret results.

9.25. Refer to Problem 8.18 with Table 8.13. For both the control and treatment groups, use methods of this chapter to compare the beginning and ending cholesterol levels. Compare the changes in cholesterol levels for the two groups. Interpret.

9.26. Refer to Table 9.13. Fit the independence model and the quasi-independence model. Describe lack of fit. What can you say about the numbers of people who moved from the Northeast to the South and from the Midwest to the West, relative to what quasi independence predicts?

9.27. Table 9.16 displays diagnoses of multiple sclerosis for two neurologists. The categories are (1) Certain multiple sclerosis, (2) Probable multiple sclerosis, (3) Possible multiple sclerosis, (4) Doubtful, unlikely, or definitely not multiple sclerosis.

Table 9.16

| | Neurologist B | | | |
Neurologist A	1	2	3	4
1	38	5	0	1
2	33	11	3	0
3	10	14	5	6
4	3	7	3	10

Source: Based on data in J. R. Landis and G. Koch, *Biometrics, 33:* 159–174 (1977).
Reprinted with permission of the Biometric Society.

a. Test marginal homogeneity (i) by comparing symmetry and quasi-symmetry models, (ii) with a test that uses the ordering of response categories. Interpret.

b. Use the independence model and residuals to study the pattern of agreement. Interpret.

c. Use more complex models to study the pattern and strength of agreement between the neurologists. Interpret results.

d. Use kappa to describe agreement. Interpret.

9.28. Refer to Table 9.5. Fit the quasi-independence model. Calculate the fitted odds ratio for the four cells in the first two rows and the last two columns. Interpret. Analyze the data from the perspective of describing agreement between choice of coffee at the two times.

9.29. Refer to Table 9.7. Based on the adjusted residuals, explain why the linear-by-linear association model (7.2.1) might fit these data well. Fit that model, and use the fit to describe the association between the diagnoses.

9.30. In 1990, a sample of psychology graduate students at the University of Florida made blind, pairwise preference tests of three cola drinks. For 49 comparisons of Coke and Pepsi, Coke was preferred 29 times. For 47 comparisons of Classic Coke and Pepsi, Classic Coke was preferred 19 times. For 50 comparisons of Coke and Classic Coke, Coke was preferred 31 times. Comparisons resulting in ties are not reported. Fit the Bradley–Terry model, analyze the quality of fit, and establish a ranking of the drinks. Estimate the probability that Coke is preferred to Pepsi, using the model fit, and compare to the sample proportion.

9.31. Table 9.17 refers to journal citations among four statistical theory and methods journals (*Biometrika, Communications in Statistics, Journal of the American Statistical Association, Journal of the Royal Statistical Society Series B*) during 1987–1989. The more often that articles in a particular journal are cited, the more prestige that journal accrues. For citations involving a pair of journals X and Y, view it as a "victory" for X if it is cited by Y and a "defeat" for X if it cites Y. Fit the Bradley–Terry model. Interpret the fit, and give a prestige ranking of the journals. For citations involving *Comm. Stat.* and *JRSS-B*, estimate the probability that the *Comm. Stat.* article cites the *JRSS-B* article.

Table 9.17

| | Cited Journal | | | |
Citing Journal	*Biometrika*	*Comm. Stat.*	*JASA*	*JRSS-B*
Biometrika	714	33	320	284
Comm. Stat.	730	425	813	276
JASA	498	68	1072	325
JRSS-B	221	17	142	188

Source: Based on Table 4 in S. M. Stigler, Citation patterns in the journals of statistics and probability, *Statist. Sci., 9:* 94–108 (1994). Reprinted with permission of the Institute of Mathematical Statistics.

9.32. Table 9.18 refers to tennis matches for several men players during 1989 and 1990.

Table 9.18

| | Loser | | | | |
Winner	Edberg	Lendl	Agassi	Sampras	Becker
Edberg	—	5	3	2	4
Lendl	4	—	3	1	2
Agassi	2	0	—	1	3
Sampras	0	1	2	—	0
Becker	6	4	2	1	—

a. Fit the Bradley–Terry model. Analyze the lack of fit.

b. Estimate the probability of Edberg beating Sampras. Compare the model estimate to the sample proportion.

c. Construct a 90% confidence interval for the probability that Edberg beats Sampras. Interpret.

d. Which pairs of players are significantly different according to .10-level tests?

9.33. Fit the Bradley–Terry model to Table 9.9.

a. Estimate the probability of Graf beating Sanchez, and construct a 90% confidence interval for this probability. Interpret.

b. By comparing the fit of this model and the simpler model, $\text{logit}(\Pi_{ij}) = 0$ all i and j, show that the likelihood-ratio statistic for testing $H_0 : \beta_1 = \cdots = \beta_5$ equals 11.5 with $df = 4$. Interpret.

9.34. When the Bradley–Terry model holds, explain why it is not possible that A could be preferred to B (i.e., $\Pi_{AB} > 1/2$) and B could be preferred to C, yet C could be preferred to A.

9.35. For the quasi-symmetry model, show that

$$\log(\mu_{ij}/\mu_{ji}) = (\lambda_i^X - \lambda_i^Y) - (\lambda_j^X - \lambda_j^Y).$$

Thus, the main effect parameters determine departures from symmetry. Explain the connection with the Bradley–Terry model (9.6.1).

CHAPTER 10

A Twentieth-Century Tour of Categorical Data Analysis*

We conclude by providing a historical overview of the evolution of methods for categorical data analysis (CDA). The beginnings of CDA were often shrouded in controversy. Key figures in the development of statistical science made groundbreaking contributions, but these statisticians were often in heated disagreement with one another.

10.1   THE PEARSON–YULE ASSOCIATION CONTROVERSY

Much of the early development of methods for CDA took place in England, and it is fitting that we begin our tour in London at the beginning of the twentieth century. The year 1900 is an apt starting point, since in that year Karl Pearson introduced his chi-squared statistic (X^2) and G. Udny Yule presented the odds ratio and related measures of association.

By 1900, Karl Pearson (1857–1936) was already well known in the statistical community. Head of a statistical laboratory at University College in London, his work the previous decade included developing a family of skewed probability distributions (called *Pearson curves*), obtaining the product-moment estimate of the correlation coefficient and finding its standard error, and extending work by Francis Galton on linear regression. In fact, Pearson was a true renaissance man, writing on a wide variety of topics that included art, religion, philosophy, socialism, women's rights, physics, genetics, eugenics, and evolution. Pearson's motivation for developing the chi-squared test included testing whether outcomes on a roulette wheel in Monte Carlo varied randomly, checking the fit to various data sets of Pearson curves, and testing statistical independence in two-way contingency tables.

Much of the literature on CDA in the early 1900s consisted of vocal debates about appropriate summary indices for describing association. Pearson's approach assumed that continuous bivariate distributions underlie cross-classification tables. He argued that one should describe association by approximating a measure, such as the correlation, for the underlying continuum. In 1904, Pearson introduced the term *contingency* as a "measure of the total deviation of the classification from independent probability," and he introduced measures to describe its extent. The *tetrachoric correlation* is a ML estimate of the correlation for a normal distribution assumed to underlie counts in 2×2 tables. The *mean square contingency* and the *contingency coefficient* are normalizations of X^2 to the $(0, 1)$ scale.

George Udny Yule (1871–1951), an English contemporary of Pearson's, took an alternative approach. Having completed pioneering work developing multiple regression models and related multiple and partial correlation coefficients, Yule turned his attention between 1900 and 1912 to the study of association in contingency tables. Yule believed that many categorical variables are inherently discrete. He defined indices directly in terms of the cell counts, without assuming an underlying continuum. His measures included the odds ratio θ and a transformation of it to the $[-1, +1]$ scale, $Q = (\theta - 1)/(\theta + 1)$, now called *Yule's Q*. Discussing one of Pearson's measures that assumes underlying normality, Yule stated "at best the normal coefficient can only be said to give us in cases like these a hypothetical correlation between supposititious variables. The introduction of needless and unverifiable hypotheses does not appear to me a desirable proceeding in scientific work." Yule also showed the potential discrepancy between marginal and conditional associations in contingency tables, later noted by E. H. Simpson in 1951 and now called *Simpson's paradox*.

In the first quarter of the twentieth century, Karl Pearson was the rarely challenged leader of statistical science in England. Pearson's strong personality did not take kindly to criticism, and he reacted negatively to Yule's ideas. In particular, he argued that Yule's own coefficients were unsuitable. For instance, he claimed that their values were unstable, since different collapsings of $I \times J$ tables to 2×2 tables could produce quite different values of the measures. In 1913, Pearson and D. Heron filled more than 150 pages of Pearson's journal (*Biometrika*) with a scathing reply to Yule's criticism. In a passage critical also of Yule's well-received book *An Introduction to the Theory of Statistics* (London: Griffin, (1911)), they stated

> If Mr. Yule's views are accepted, irreparable damage will be done to the growth of modern statistical theory.... [Yule's Q] has never been and never will be used in any work done under his [Pearson's] supervision.... We regret having to draw attention to the manner in which Mr. Yule has gone astray at every stage in his treatment of association, but criticism of his methods has been thrust on us not only by Mr Yule's recent attack, but also by the unthinking praise which has been bestowed on a text-book which at many points can only lead statistical students hopelessly astray.

Pearson and Heron attacked Yule's "half-baked notions" and "specious reasoning" and concluded that Yule would have to withdraw his ideas "if he wishes to maintain any reputation as a statistician."

In retrospect, Pearson and Yule both had valid points. Some classifications, such as most nominal variables, have no apparent underlying continuous distribution. On the other hand, many applications relate naturally to an underlying continuum, and it can be useful to direct model-building and inference toward that continuum. The ordinal models presented in Sections 7.2 and 8.2 provide a sort of reconciliation between Yule and Pearson, since Yule's odds ratio characterizes models that fit well when underlying distributions are approximately normal.

Half a century after the Pearson-Yule controversy, Leo Goodman and William Kruskal of the University of Chicago surveyed the development of measures of association for contingency tables and also made many contributions of their own. Their book, *Measures of Association for Cross Classifications* (1979), reprinted their four influential articles on this topic. One of their articles contains the following quote from an article by M. H. Doolittle in 1887, which illustrates the lack of precision in early attempts to quantify association even in 2×2 tables.

> Having given the number of instances respectively in which things are both thus and so, in which they are thus but not so, in which they are so but not thus, and in which they are neither thus nor so, it is required to eliminate the general quantitative relativity inhering in the mere thingness of the things, and to determine the special quantitative relativity subsisting between the thusness and the soness of the things.

10.2   R. A. FISHER'S CONTRIBUTIONS

Pearson's disagreements with Yule were minor compared to his later ones with Ronald A. Fisher (1890–1962). In 1922, Fisher introduced the concept of *degrees of freedom*, using a geometric representation. The df index characterizes the family of chi-squared distributions, and Fisher claimed that for tests of independence in $I \times J$ tables, X^2 had $df = (I - 1)(J - 1)$. By contrast, in 1900 Pearson had argued that for any application of his statistic, df equalled the number of cells minus 1, or $IJ - 1$ for two-way tables. Fisher pointed out, however, that estimating hypothesized cell probabilities using estimated row and column probabilities resulted in an additional $(I - 1) + (J - 1)$ constraints on the fitted values, thus affecting the distribution of X^2.

Not surprisingly, Pearson reacted critically to Fisher's suggestion that his formula for df was incorrect. He stated

> I hold that such a view [Fisher's] is entirely erroneous, and that the writer has done no service to the science of statistics by giving it broad-cast circulation in the pages of the *Journal of the Royal Statistical Society*.... I trust my critic will pardon me for comparing him with Don Quixote tilting at the windmill; he must either destroy himself, or the whole theory of probable errors, for they are invariably based on using sample values for those of the sampled population unknown to us.

Pearson claimed that using row and column sample proportions to estimate unknown probabilities had negligible effect on large-sample distributions. Fisher was unable to get his rebuttal published by the Royal Statistical Society, and he ultimately resigned his membership.

Statisticians soon realized that Fisher was correct, but he maintained much bitterness over this and other dealings with Pearson. In a later volume of his collected works, he remarked that his 1922 article "had to find its way to publication past critics who, in the first place, could not believe that Pearson's work stood in need of correction, and who, if this had to be admitted, were sure that they themselves had corrected it." Writing about Pearson, he stated "If peevish intolerance of free opinion in others is a sign of senility, it is one which he had developed at an early age." In an article in 1926, he was able to dig the knife a bit deeper into the Pearson family using 12,000 2×2 tables randomly generated by Karl Pearson's son, E. S. Pearson. Fisher showed that the sample mean of X^2 for these tables was 1.00001, much closer to the 1.0 predicted by his formula for $E(X^2)$ of $df = (I - 1)(J - 1) = 1$ than Pearson's $IJ - 1 = 3$.

Fisher's preeminent reputation among statisticians today accrues primarily from his theoretical work (introducing concepts such as sufficiency, information, and optimal properties of ML estimators) and his methodological contributions to areas such as the design of experiments and the analysis of variance. Though not so well known for work in CDA, he did make other interesting contributions to its history. Moreover, he made good use of the methods in his applied work. For instance, Fisher was also a famed geneticist. In one article, he used Pearson's goodness-of-fit test to test Mendel's theories of natural inheritance. Calculating a summary P-value from the results of several of Mendel's experiments, he obtained an unusually large value ($P = 0.99996$) for the right-tail probability of the reference chi-squared distribution. In other words X^2 was so small that the fit seemed *too* good, leading Fisher in 1936 to comment "the general level of agreement between Mendel's expectations and his reported results shows that it is closer than would be expected in the best of several thousand repetitions.... I have no doubt that Mendel was deceived by a gardening assistant, who knew only too well what his principal expected from each trial made." In a letter written at the time, he stated "Now, when data have been faked, I know very well how generally people underestimate the frequency of wide chance deviations, so that the tendency is always to make them agree too well with expectations."

Fisher realized the limitations of large-sample statistical methods for laboratory work, and he was at the forefront of advocating specialized small-sample procedures. He was among the first to promote the work by W. S. Gosset (pseudonym "Student") on the t distribution, and the fifth edition of his classic text, *Statistical Methods for Research Workers* (Edinburgh: Oliver and Boyd (1934)) introduced "Fisher's exact test" for 2×2 contingency tables. In his book *The Design of Experiments* (Edinburgh: Oliver and Boyd (1935)), Fisher described the tea-tasting experiment (Section 2.6.2) based on his experience at an afternoon tea break while employed at Rothamsted Experiment Station.

The mid 1930s finally saw some work on model-building for CDA. For instance, Chester Bliss popularized the probit model for applications in toxicology dealing with a binary response. In the appendix of one of Bliss's articles in 1935, Fisher provided an algorithm for obtaining ML estimates of parameters in the probit model. That algorithm was a Newton–Raphson type method, today commonly called *Fisher scoring*.

The definition for homogeneous association (no three-factor interaction) in contingency tables originates in an article by the British statistician Maurice Bartlett in 1935, concerning $2 \times 2 \times 2$ tables. Bartlett showed how to calculate ML estimates of cell probabilities satisfying the property of equality of odds ratios between two variables at each level of the third. He attributed this idea to Fisher.

In 1940 Fisher developed canonical correlation methods for contingency tables, showing how to assign scores to rows and columns of a contingency table in order to maximize the correlation. His work relates to the later development, particularly in France, of *correspondence analysis* methods.

10.3 LOGISTIC REGRESSION AND LOGLINEAR MODELS

In a book of statistical tables published in 1938, R. A. Fisher and Frank Yates suggested $\log[\pi/(1 - \pi)]$ as one of several possible transformations of a binomial parameter for analyzing binary data. In 1944, the physician and statistician Joseph Berkson introduced the term "logit" for this transformation. Berkson showed that the logistic regression model fitted similarly to the probit model, and his subsequent work did much to popularize the model. In 1951, Jerome Cornfield, another statistician with a medical background, showed the use of the odds ratio for approximating relative risks in case-control studies with this model.

Sir David R. Cox (currently at Oxford University) also had considerable influence in popularizing logistic regression, both through a 1958 article and a 1970 book, *The Analysis of Binary Data*. About the same time, an article by the Danish statistician and mathematician Georg Rasch sparked an enormous literature on item response models, the most important of which is the logit model with subject and item parameters, now called the *Rasch model* (Section 9.2.4). This work was highly influential in the psychometric community of northern Europe (especially in Denmark, the Netherlands, and Germany) and spurred many generalizations in the educational testing community in the United States.

The quarter century following the end of World War II saw strong theoretical advances in CDA. For instance, general expressions were derived by H. Cramér and by C. R. Rao for large-sample distributions of parameter estimators in models for categorical data. In 1949, the Berkeley-based statistician Jerzy Neyman, who had already performed fundamental work on hypothesis testing and confidence interval methods with E. S. Pearson, introduced the family of *best asymptotically normal* (BAN) estimators. These estimators have the same optimal large-sample properties as ML estimators. The BAN family includes estimators obtained by minimizing chi-squared-type measures comparing observed proportions to proportions predicted by the model. This type of estimator itself includes some *weighted least squares* (WLS) estimators, which generalize ordinary least squares to permit non-constant variance. These are simpler to compute than ML estimators, which was an important consideration before the advent of modern computing.

In the early 1950s, William Cochran published work dealing with a variety of important topics in CDA. Scottish-born, Cochran spent most of his career at American universities: Iowa State, North Carolina State, Johns Hopkins, and Harvard. He introduced a generalization (Cochran's Q) of McNemar's test for comparing proportions in several matched samples. He showed how to partition chi-squared statistics into components that described various aspects of association, such as a linear trend in binomial proportions across quantitatively-defined rows of an $I \times 2$ table. He developed sample size guidelines for chi-squared approximations to work well for the X^2 statistic. Cochran also proposed a test of conditional independence for $2 \times 2 \times K$ tables, similar to the one later proposed by Mantel and Haenszel in 1959.

In the 1950s and early 1960s, Bartlett's work on interaction structure in contingency tables was finally extended to multiway tables in articles by J. N. Darroch, I. J. Good, L. Goodman, H. O. Lancaster, N. Mantel, R. L. Plackett, and S. Roy, among others. These articles as well as some influential articles by Martin W. Birch in 1963–1965 were the genesis of research work on loglinear models. Birch's work was part of a never-submitted Ph.D. thesis at the University of Glasgow. He showed how to obtain ML estimates of cell probabilities in three-way tables, under various conditions. Birch showed the equivalence of those ML estimates for Poisson and multinomial sampling. He also extended earlier theoretical results of Cramér and Rao on large-sample distributions for categorical-data models. The articles by Birch and others stimulated a great deal of research on loglinear models between about 1965 and 1975. A survey article by the French statistician Henri Caussinus (see Caussinus (1966)) provides a good glimpse of the state of the art of CDA just before this decade of advances. In that article, Caussinus introduced the quasi-symmetry model for square tables.

Much of the work in the next decade on loglinear and logit modeling took place at three American universities: Chicago, Harvard, and North Carolina. At the University of Chicago, Leo Goodman wrote a series of groundbreaking articles, dealing with such topics as partitionings of chi-squared, models for square tables (e.g., quasi-independence), stepwise logit and loglinear model-building procedures, latent class models (CDA analogs of factor analysis methods), and specialized models for ordinal data. For surveys of Goodman's early work, see Goodman ((1968), a R. A. Fisher memorial lecture) and (1970). Goodman also wrote a stream of articles for social science journals that had a substantial impact on popularizing loglinear and logit methods for applications.

Over the past forty years, Goodman (now at the University of California at Berkeley) has been the most prolific contributor to the advancement of CDA methodology. The field owes tremendous gratitude to his steady and impressive body of work. In addition, some of Goodman's students at Chicago have also made fundamental contributions. In 1970, for instance, Shelby Haberman completed a Ph.D. dissertation making substantial theoretical contributions to loglinear modeling. Among topics he considered were residual analyses (introducing adjusted residuals), loglinear models for ordinal variables, and theoretical results for models (such as the Rasch model) for which the number of parameters grows with the sample size.

Simultaneously, related research on ML methods for loglinear-logit models occurred at Harvard University by students of Frederick Mosteller (such as Stephen Fienberg) and William Cochran. Much of this research was inspired by problems arising in analyzing large, multivariate data sets in the National Halothane Study. That study investigated whether halothane was more likely than other anaesthetics to cause death due to liver damage. A presidential address by Mosteller to the American Statistical Association (Mosteller, (1968)) describes early uses of loglinear models for smoothing multidimensional discrete data sets. Fienberg and his own students

Karl Pearson   G. Udny Yule

Ronald A. Fisher   Leo Goodman

Figure 10.1   Four leading figures in the development of categorical data analysis.

advanced this work further. A landmark book in 1975 by him with Yvonne Bishop and Paul Holland, *Discrete Multivariate Analysis*, was largely responsible for introducing loglinear models to the general statistical community.

Research at the University of North Carolina by Gary Koch and several students and coworkers has been highly influential in the biomedical sciences. Their research developed WLS methods for categorical data models. An article in 1969 by Koch with J. Grizzle and F. Starmer popularized this approach, which was extended in later articles to an impressive variety of problems. In particular, Koch and colleagues applied WLS to problems for which ML methods are awkward to use, such as the analysis of repeated categorical measurement data (Koch et al. (1977)). In 1966, Vasant Bhapkar showed that the WLS estimator is often identical to one of Neyman's BAN estimators that minimizes a chi-squared metric. For large samples with fully categorical data, WLS estimators have similar properties as ML.

10.4   LATER DEVELOPMENTS

It is surely unwise to guess which contributions from the final quarter of this century are most important. We mention here only some work that has already been very useful in applications:

1. the modeling of ordinal data, articles by Leo Goodman in 1979 on the loglinear approach and by Peter McCullagh in 1980 on the proportional odds approach being particularly influential,

2. the development of efficient algorithms for implementing exact small-sample methods, available in the StatXact and LogXact software, by Cyrus Mehta, Nitin Patel, and colleagues at Harvard,

3. the development of graphical models for multi-way contingency tables (related to the association graphs discussed in Section 7.1), spurred by an article in 1980 by John Darroch, Steffen Lauritzen, and Terry Speed,

4. conditional likelihood methods for modeling odds ratios in case-control studies by Norman Breslow, Ross Prentice, and colleagues at the University of Washington (see Breslow and Day (1980)).

5. methodology for longitudinal and multivariate categorical responses using general estimating equations by Kung-Yee Liang and Scott Zeger and colleagues at Johns Hopkins and by biostatisticians at Harvard University and the University of Washington (see Diggle et al. (1994)).

Perhaps the most far-reaching contribution has been the introduction by British statisticians John Nelder and R. W. M. Wedderburn in 1972 of the concept of *generalized linear models*. This unifies the primary categorical modeling procedures—logistic and probit regression models for binomial data and loglinear models for Poisson data—with long-established regression and ANOVA methods for normal-response data. Interestingly, the algorithm they used to fit GLMs is Fisher scoring, which R. A. Fisher introduced in 1935 for ML fitting of probit models.

And so, it is fitting that we end this brief survey by giving yet further credit to R. A. Fisher for his influence on the practice of modern statistical science. The biography, *R. A. Fisher: The Life of a Scientist*, by Fisher's daughter, Joan Fisher Box (New York: Wiley, 1978), gives a fascinating account of R. A. Fisher's impressive contributions to statistics and genetics. Besides this biography, sources for our brief historical tour include *The History of Statistics* by S. Stigler (Cambridge, MA: Harvard, 1986), *Studies in the History of Probability and Statistics* edited by E. S. Pearson and M. G. Kendall (London: Griffin, 1970), an article on Fisher by S. Fienberg (1980b), and personal conversations over the years with several statisticians, including Henri Caussinus, William Cochran, Sir David Cox, John Darroch, Leo Goodman, Gary Koch, Frederick Mosteller, John Nelder, and C. R. Rao. Full references for all work and articles cited appear in Agresti (1990).

SAS and SPSS for Categorical Data Analysis

Most major statistical software packages have procedures for categorical data analyses. In particular, SAS, SPSS, S-Plus, GLIM, and BMDP can perform nearly all the large-sample analyses presented in this text. StatXact and its companion package LogXact (Cytel Software, 675 Massachusetts Avenue, Cambridge, MA 02139) perform small-sample analyses; SPSS Exact Tests also provides many of the StatXact analyses. This appendix illustrates software use for the analyses presented in this text.

There is insufficient space to discuss all the major packages, and the examples show only SAS code (release 6.10). For more detailed discussion of the use of SAS for categorical data analyses, see Stokes, Davis, and Koch (1995). We also briefly discuss SPSS, which is simple to use in a menu-driven windows environment. See Agresti (1990) for GLIM and BMDP examples. We focus on basic fitting of models rather than the great variety of options provided by various procedures. The material is organized by chapter of presentation in this text. The tables and the full data sets are available from StatLib. The file is available on the World Wide Web at http://lib.stat.cmu.edu/datasets/agresti. For information about StatLib, send the e-mail message

send index

to statlib@lib.stat.cmu.edu.

CHAPTER 2: TWO-WAY CONTINGENCY TABLES

Table A.1 illustrates SAS for analyzing two-way tables, using data from Table 2.3. The @@ symbol indicates that each line of data contains more than one observation. PROC FREQ conducts chi-squared tests of independence using the CHISQ option, and provides the estimated expected frequencies for the test with the EXPECTED option. The MEASURES option provides a wide assortment of measures of association and their ASE values. For 2×2 tables this option provides confidence intervals for the odds ratio (labeled "case-control" on output) and the relative risk. One can

Table A.1   SAS for Analyzing Table 2.3

```
data aspirin;
input group mi count @@;
cards;
    1  1  189    1  2  10845
    2  1  104    2  2  10933
;
proc freq; weight count;
        tables group*mi / chisq expected measures ;
proc genmod;
        model count = group mi / dist=poi link=log obstats residuals;
run;
```

also perform chi-squared tests using PROC GENMOD (discussed in the Chapter 4 section of this Appendix), as shown in Table A.1; its OBSTATS and RESIDUALS options provide cell residuals. (The output labeled "StReschi" is the adjusted residual (2.4.4).)

Table A.2 shows SAS code for analyzing Table 2.7. The option CMH1 in PROC FREQ provides a "nonzero correlation" statistic that is ordinal statistic (2.5.1). Table A.2 uses scores (0, .5, 1.5, 4.0, 7.0) for alcohol consumption. The option SCORES = RIDIT with CMH1 performs the analysis with midrank scores.

For tables having small cell counts, the EXACT option in PROC FREQ performs an exact test of independence that treats the variables as nominal. For 2×2 tables, this is Fisher's exact test. Table A.3 shows SAS code for performing this test with the tea-tasting data of Table 2.8.

In SPSS, one can obtain standard chi-squared tests and the correlation statistic (2.5.1), which SPSS calls the *Mantel–Haenszel test for linear association*, using the CROSSTABS procedure, available under the "Summarize" option in the Statistics menu. CROSSTABS also provides the adjusted residuals, Fisher's exact test, and several measures of association, such as the odds ratio and relative risk.

StatXact provides nominal and ordinal exact tests for two-way tables, as does SPSS Exact Tests.

Table A.2   SAS for Analyzing Table 2.7

```
data infants;
input malform alcohol count @@;
cards;
    1   0   17066    1   0.5   14464    1   1.5   788    1   4.0   126    1   7.0   37
    2   0      48    2   0.5      38    2   1.5     5    2   4.0     1    2   7.0    1
;
proc freq; weight count;
     tables malform*alcohol / chisq cmh1;
proc freq; weight count;
     tables malform*alcohol / cmh1 scores=ridit;
run;
```

Table A.3   SAS for Analyzing Table 2.8

```
data tea;
input poured guess count @@;
cards;
    1   1   3      1   2   1
    2   1   1      2   2   3
;
proc freq; weight count;
        tables poured*guess / exact;
run;
```

CHAPTER 3: THREE-WAY CONTINGENCY TABLES

The *CMH* option in PROC FREQ in SAS provides the Cochran–Mantel–Haenszel statistic (3.2.1), the Mantel–Haenszel estimate (3.2.2) of a common odds ratio and its confidence interval, and the Breslow–Day statistic (3.2.3). Table A.4 shows SAS code for analyzing Table 3.3. FREQ treats the first variable listed in the "TABLES" directive (Center) as the control variable; the CHISQ option yields chi-square tests of independence for each partial table.

StatXact provides exact tests of conditional independence and homogeneous association for $2 \times 2 \times K$ tables.

CHAPTER 4: GENERALIZED LINEAR MODELS

PROC GENMOD in SAS fits generalized linear models. GENMOD (available beginning with version 6.08 of SAS) specifies the distribution of the random component in the DIST option ("poi" for Poisson, "bin" for binomial, "nor" for normal) and specifies the link in the LINK option (including "logit", "probit", and "identity").

Table A.5 shows SAS (GENMOD) analyses for Table 4.1. For binomial models, the response in the model statements must have the form of the number of "successes" divided by the number of cases. In Table A.5, the variable labeled TOTAL contains

Table A.4   SAS for Analyzing Table 3.3

```
data cmh;
input center smoke cancer count @@;
cards;
    1   1   1   126      1   1   2   100      1   2   1   35      1   2   2   61
    . . .
    8   1   1   104      8   1   2   89      8   2   1   21      8   2   2   36
;
proc freq; weight count;
        tables center*smoke*cancer / cmh chisq;
run;
```

Table A.5   SAS for Analyzing Table 4.1

```
data glm;
input snoring disease total;
cards;
    0   24   1379
    2   35    638
    4   21    213
    5   30    254
;
proc genmod; model disease/total = snoring / dist=bin link=identity ;
proc genmod; model disease/total = snoring / dist=bin link=logit ;
proc genmod; model disease/total = snoring / dist=bin link=probit;
run;
```

the number of cases at each level of snoring, and DISEASE contains the number of "yes" responses on heart disease. The predictor uses scores $(0, 2, 4, 5)$ for snoring. The models for the probability of heart disease use identity, logit, and probit links.

Table A.6 shows SAS (GENMOD) analyses for Poisson regression and logistic regression modeling of the data in Table 4.2, in which each observation refers to a single crab. Using width as the predictor in Poisson regression, the first model statement fits the model with log link (model (4.3.1)), and the second model statement fits the model with identity link. The third model statement fits the logistic regression model (4.2.2) to a constructed binary variable Y that equals 1 when a crab has satellites and equals 0 when she does not. (Chapter 5 discusses this model.) The response in the

Table A.6   SAS for Analyzing Table 4.2

```
data crabs;
input color spine width weight satell;
if satell > 0 then y = 1; if satell = 0 then y = 0; n=1;
cards;
    2    3    28.3    3.05    8
    3    3    26.0    2.60    4
    . . .
    2    2    24.5    2.00    0
;
proc genmod; model satell = width / dist=poi link=log ;
proc genmod; model satell = width / dist=poi link=identity ;
proc genmod; model y/n = width / dist=bin link=logit obstats ;
proc genmod; class color;
     model y/n = color width / dist=bin link=logit; contrast 'a-d' color 1 0 0 -1;
proc genmod; model y/n = color width / dist=bin link=logit;
proc genmod; class color spine ;
     model y/n = color spine width weight / dist=bin link=logit type3;
proc logistic; model y = width / lackfit;
proc logistic; model y = color weight width / selection=backward;
run;
```

Table A.7   SAS for Analyzing Table 4.3

```
data crabs;
input width cases satell; log_case = log(cases);
cards;
     22.69    14       14
     23.84    14       20
     24.77    28       67
     25.84    39      105
     26.79    22       63
     27.74    24       93
     28.67    18       71
     30.41    14       72
 ;
proc genmod;
     model satell = width / dist=poi link=log offset=log_case obstats residuals ;
run;
```

model statement is the number of successes divided by the number of cases (1, in this case). The OBSTATS option in GENMOD provides various "observation statistics," including predicted values and their confidence limits.

Table A.6 also uses PROC LOGISTIC to fit logistic regression models to these data. This procedure orders the levels of the response variable alphanumerically, forming the logit, for instance, as $\log[P(Y = 0)/P(Y = 1)]$. One can use the DESCENDING option to reverse the order, invoking the procedure using the statement PROC LOGISTIC DESCENDING;.

Table A.7 uses SAS (GENMOD) to fit the Poisson regression model (4.3.3) with log link for the grouped data in Table 4.3. It models the total number of satellites at each weight level, using the log of the number of cases as the offset. The OBSTATS option provides Pearson residuals, and the RESIDUALS option provides the adjusted residuals (labeled "StReschi" on output), which adjust the Pearson residuals to have approximate standard normal distributions.

SPSS has a general loglinear modeling procedure (GENLOG) having a wide variety of loglinear and logit modeling capabilities. (See Chapter 6 in *SPSS 6.1 for Windows Update*, SPSS Inc., Chicago, 1994.) GENLOG can fit Poisson regression models (pp. 95–101 of the SPSS publication). In the dialog box, one specifies a Poisson distribution for the cell counts, specifies the predictors as terms in a customized model, and specifies the index used for the offset in the "cell structure."

CHAPTER 5: LOGISTIC REGRESSION

One can fit logistic regression models either using software for generalized linear models or specialized software for logistic regression. Table A.8 applies SAS (PROC GENMOD and PROC LOGISTIC) to Table 5.1. In the code, "satell" refers to the number of crabs that had satellites at the given width level. In GENMOD, the WALDCI option provides the ordinary large-sample confidence intervals for parameters. The

Table A.8   SAS for Analyzing Table 5.1

```
data crabs;
input width cases satell;
cards;
    22.69    14    5
    23.84    14    4
    24.77    28    17
    25.84    39    21
    26.79    22    15
    27.74    24    20
    28.67    18    15
    30.41    14    14
;
proc genmod; model satell/cases = width / dist=bin link=logit obstats
    residuals waldci lrci;
proc logistic; model satell/cases = width / influence;
    output out=predict p=pi_hat lower=LCL upper=UCL;
proc print data=predict;
run;
```

LRCI option presents an alternative set of intervals, likelihood-based, called *profile likelihood intervals*. These are beyond the scope of this text; a substantial difference between the two sets of intervals is a warning that large-sample inference may be inadequate. In that case, one can instead use software for small-sample inference, such as LogXact, which provides conditional ML fitting and exact inference for parameters in logistic regression models.

Like GENMOD, LOGISTIC can also apply some other links, such as the probit. The INFLUENCE option in LOGISTIC provides regression diagnostics. Following the model fit, Table A.8 requests predicted probabilities and lower and upper 95% confidence limits for the true probabilities.

Table A.9 uses SAS (GENMOD) to fit a logit model with qualitative predictors to Table 5.5. When one inputs characters rather than numbers for levels of variables, the variables have an accompanying $ label in the INPUT statement. One sets up dummy variables for the factors in GENMOD by declaring them in a CLASS statement. The

Table A.9   SAS for Analyzing Table 5.5

```
data aids;
input race $ azt $ yes no @@;
cases = yes + no;
cards;
    white    y    14    93        white    n    32    81
    black    y    11    52        black    n    12    43
;
proc genmod order=data; class race azt;
    model yes/cases = race azt / dist=bin link=logit type3 obstats residuals ;
run;
```

parameter estimate for the last level of each factor equals 0. SAS lists the category levels in alphabetical order unless one states ORDER=DATA in the PROC directive, in which case the levels have the order in which they occur in the input data. In models with multiple predictors, the TYPE3 option provides likelihood-ratio tests for testing the significance of each individual predictor in the model.

The fourth and fifth GENMOD statements in Table A.6 use both color and width as predictors for the crab data; color is qualitative in the fourth model (because of the CLASS statement) and quantitative in the fifth. The sixth GENMOD statement in Table A.6 fits the main effects model using all the predictors from Table 4.2. One can use a CONTRAST option in GENMOD to test contrasts of parameters, such as testing whether parameters for two levels of a factor are identical; when color is qualitative, for instance, the contrast statement shown contrasts the first and fourth levels of color.

PROC LOGISTIC has options for stepwise selection of variables, as shown in the final model statement in Table A.6. The LACKFIT option in this procedure yields the Hosmer–Lemeshow statistic.

In SPSS, one can fit logistic regression models using the LOGISTIC REGRES-SION procedure, which is available as a regression option on the Statistics menu in the windows environment. One identifies the response (dependent) variable and the explanatory predictors (covariates), and identifies qualitative predictors using the "categorical" option. This program also has options for stepwise model selection procedures, such as backward elimination, and can provide a wide variety of regression diagnostics. Among several options for setting up dummy variables for categorical predictors, the "simple" contrast constructs them as in this text, using the final category as a baseline. One can also fit such models using the logit option in the general loglinear (GENLOG) procedure, identifying qualitative predictors as factors and quantitative predictors as cell covariates.

CHAPTER 6: LOGLINEAR MODELS FOR CONTINGENCY TABLES

One can fit loglinear models using either software for generalized linear models or specialized software for loglinear models. Table A.10 uses SAS (PROC GENMOD

Table A.10   SAS for Analyzing Table 6.3

```
data drugs;
input a c m count @@;
cards;
  1  1  1  911    1  1  2  538    1  2  1  44    1  2  2  456
  2  1  1    3    2  1  2   43    2  2  1   2    2  2  2  279
;
proc genmod; class a c m;
     model count = a c m a*m c*m / dist=poi link=log obstats residuals ;
proc catmod; weight count;
     model a*c*m = _response_ ; loglin a|m c|m ;
run;
```

and PROC CATMOD) to fit model (AM, CM) to Table 6.3. The CLASS statement in GENMOD generates dummy variables for the classification factors. The A-M association is represented by $A*M$ in GENMOD and by $A|M$ in CATMOD. CATMOD codes estimates for a factor so they sum to zero, whereas GENMOD sets the estimate for the last level equal to 0. The OBSTATS and RESIDUALS options in GENMOD provides diagnostics, including Pearson and adjusted residuals.

In SPSS, one can fit standard loglinear models using the GENLOG procedure. With this procedure, one can also conduct chi-squared tests of independence for two-way tables and display standardized and adjusted residuals for model fits. One enters the factors and relevant interactions in a customized (unsaturated) model, and weights each cell by the cell count using the "Weight cases" option in the "Data" menu.

CHAPTER 7: BUILDING AND APPLYING LOGIT AND LOGLINEAR MODELS

One can fit ordinal loglinear models using software for generalized linear models. Table A.11 uses SAS (GENMOD) to fit the independence model and the linear-by-linear association model (7.2.1) to Table 7.3. The defined variable "assoc" represents the cross-product of row and column scores, which has β parameter as coefficient in the latter model.

GENLOG in SPSS can also fit the linear-by-linear association model. One creates a variable having the cross-product $u_i v_j$ of the row and column scores, and identifies it as a cell covariate in a customized model that contains the row and column classifications as factors.

With the *CMH* option in PROC FREQ, SAS provides the generalized *CMH* tests of conditional independence discussed in Section 7.3. The statistic for the "general association" alternative treats X and Y as nominal, the statistic for the "row mean scores differ" alternative treats X as nominal and Y as ordinal (this statistic is (7.3.2)

Table A.11   SAS for Analyzing Table 7.3

```
data sex;
input premar birth count @@; assoc = premar*birth;
cards;
    1   1    38    1   2    60    1   3   68    1   4   81
    2   1    14    2   2    29    2   3   26    2   4   24
    3   1    42    3   2    74    3   3   41    3   4   18
    4   1   157    4   2   161    4   3   57    4   4   36
;
proc genmod; class premar birth;
    model count = premar birth / dist=poi link=log;
proc genmod; class premar birth;
    model count = premar birth assoc / dist=poi link=log;
run;
```

Table A.12   SAS for Analyzing Table 7.5

```
data cmh;
input gender $ income satisf count @@;
cards;
    F    3   1   1     F    3   3   3     F    3   4   11     F    3   5   2
    F   10   1   2     F   10   3   3     F   10   4   17     F   10   5   3
    F   20   1   0     F   20   3   1     F   20   4   8      F   20   5   5
    F   35   1   0     F   35   3   2     F   35   4   4      F   35   5   2
    . . .
    M   35   1   0     M   35   3   1     M   35   4   9      M   35   5   6
;
proc freq; weight count;
     tables gender*income*satisf / cmh;
run;
```

when $K = 1$), and the statistic for the "nonzero correlation" alternative treats X and Y as ordinal (statistic (2.5.1) when $K = 1$). Table A.12 shows SAS code for Table 7.5, using scores $(1, 3, 4, 5)$ for $Y =$ job satisfaction and $(3, 10, 20, 35)$ for $X =$ income in the ordinal test statistics. For average-rank scores, one uses the "SCORES = RIDIT" option.

PROC LOGISTIC in SAS has a built-in check of whether ML estimates for logistic regression models exist.

CHAPTER 8: MULTI-CATEGORY LOGIT MODELS

SAS can fit generalized logit models directly using PROC CATMOD. For nominal responses, CATMOD uses the final response category as the default baseline for the logits. Table A.13 uses CATMOD to fit model (8.1.1) to Table 8.1. The DI-RECT statement in CATMOD identifies predictors to be treated as quantitative. The PRED=PROB and PRED=FREQ options provide predicted probabilities and fitted values and their standard errors.

Table A.13   SAS for Analyzing Table 8.1

```
data gator;
input length choice $ @@;
cards;
   1.24   I     1.30   I     1.30   I     1.32   F     1.32   F     1.40   F     1.42   I     1.42   F
   . . .
   3.68   O     3.71   F     3.89   F
;
proc catmod; response logits; direct length;
     model choice = length / pred=prob pred=freq;
run;
```

When the number of response categories exceeds two, PROC LOGISTIC in SAS provides ML fitting of the proportional odds version of cumulative logit models. Table A.14 uses it to fit the proportional odds model (8.2.1) to Table 8.6. PROC CATMOD has options (CLOGIT and ALOGIT) for fitting cumulative logit and adjacent-categories logit models to ordinal responses; however, those options provide weighted least squares (WLS) rather than ML fits. For large samples with categorical predictors, WLS and ML fits are practically identical. Table A.14 uses CATMOD for the WLS fit of model (8.2.1) and the adjacent-categories logit model (8.3.2).

SPSS can fit baseline-category logit models using the logit option in the general loglinear program (GENLOG). (See pp. 71–78 of *SPSS 6.1 for Windows Update*, SPSS Inc., Chicago, 1994.) It is also simple to fit continuation-ratio logit models with this procedure.

When all predictors are categorical, one can fit the logit models of Sec. 8.1 and 8.3 using any software that fits corresponding loglinear models. For instance, to fit an adjacent-categories logit model with an ordinal categorical predictor, one could use generalized linear model software (such as GENMOD) to fit the corresponding linear-by-linear association model for Poisson loglinear models. One can fit the continuation-ratio logit model using ordinary logistic regression software for each separate binary logit.

CHAPTER 9: MODELS FOR MATCHED PAIRS

Table A.15 refers to Table 9.6. For square tables, the AGREE option in PROC FREQ provides the McNemar chi-squared statistic for binary responses, the X^2 test of fit of the symmetry model (also called *Bowker's test*), and Cohen's kappa and its ASE value. One can use *CMH* procedures in PROC FREQ to conduct tests of marginal homogeneity (Sections 9.2.1 and 9.2.4). (For details, see Section 6.4 of Stokes et al. (1995.))

Table A.14   SAS for Analyzing Table 8.6

```
data politics;
input party ideology count @@;
cards;
  1  1  80    1  2  81    1  3  171    1  4  41    1  5  55
  0  1  30    0  2  46    0  3  148    0  4  84    0  5  99
;
proc logistic; weight count;
     model ideology = party;
proc catmod; weight count; response clogits;
     model ideology = _response_ party;
proc catmod; weight count; response alogits;
     model ideology = _response_ party;
run;
```

Table A.15   SAS for Analyzing Table 9.6

```
data sex;
input premar extramar symm qi count @@;
cards;
  1  1  1  1   144    1  2  2  5   2    1  3  3  5   0    1  4   4  5   0
  2  1  2  5    33    2  2  5  2   4    2  3  6  5   2    2  4   7  5   0
  3  1  3  5    84    3  2  6  5  14    3  3  8  3   6    3  4   9  5   1
  4  1  4  5   126    4  2  7  5  29    4  3  9  5  25    4  4  10  4   5
;
proc freq; weight count;
      tables premar*extramar / agree;
proc genmod; class symm;
      model count = symm / dist=poi link=log;
proc genmod; class extramar premar symm;
      model count = symm extramar premar / dist=poi link=log;
proc genmod; class symm;
      model count = symm extramar premar / dist=poi link=log;
proc genmod; class extramar premar qi;
      model count = extramar premar qi / dist=poi link=log;
data sex2; input score below above @@; trials = below + above;
cards;
  1  33  2    1  14  2    1  25  1    2  84  0    2  29  0    3  126  0
;
proc genmod;
      model above/trials = score / dist=bin link=logit noint;
proc genmod;
      model above/trials = / dist=bin link=logit noint;
run;
```

Table A.15 also uses SAS (GENMOD) to fit the symmetry, quasi-symmetry, ordinal quasi-symmetry, and quasi-independence models to Table 9.6. The defined "symm" factor indexes the various pairs of cells that have the same association terms in the symmetry and quasi-symmetry models. For instance, "symm" takes the same value for cells $(1, 2)$ and $(2, 1)$, another value for cells $(1, 3)$ and $(3, 1)$, and so forth. Including this term as a factor in the model represents a parameter λ_{ij} satisfying $\lambda_{ij} = \lambda_{ji}$. The first model fits this factor alone, providing the symmetry model.

In Table A.15, the second model statement (for quasi symmetry) looks like the third model statement (for ordinal quasi symmetry). The difference is that the second model statement identifies "premar" and "extramar" as class variables and the third model statement does not. The main effects are qualitative (nominal) in model two and quantitative (ordinal) in model three. The fourth model statement fits the quasi-independence model. The "qi" factor represents the δ_i parameters in that model. It takes a separate level for each cell on the main diagonal, and a common value for all other cells; the last level (for cells off the main diagonal) is redundant, and SAS sets its coefficient equal to 0. Deleting the "qi" term from the final model yields the ordinary independence loglinear model.

Table A.16   SAS for Analyzing Table 9.9

```
data tennis;
input wins matches seles graf sabat navrat sanchez;
cards;
    2    5    1    -1    0    0    0
    1    1    1     0   -1    0    0
    3    6    1     0    0   -1    0
    2    2    1     0    0    0   -1
    6    9    0     1   -1    0    0
    3    3    0     1    0   -1    0
    7    8    0     1    0    0   -1
    1    3    0     0    1   -1    0
    3    5    0     0    1    0   -1
    3    4    0     0    0    1   -1
;
proc genmod;
     model wins/matches = seles graf sabat navrat sanchez / dist=bin link=logit
     noint covb;
run;
```

The bottom of Table A.15 shows how to fit the symmetry and ordinal quasi-symmetry models as logit models. The pairs of cell counts (n_{ij}, n_{ji}), labeled as "above" and "below" with reference to their positions relative to the main diagonal, are treated as six sets of binomial counts. The variable defined as "score" is the distance $(u_j - u_i) = j - i$ between the column and row indices. Neither model contains an intercept term, indicated by the NOINT option, and the ordinal model uses "score" as the predictor.

One can fit the Bradley–Terry model by fitting the quasi-symmetry model, or directly using logit models. Table A.16 uses SAS (GENMOD) for logit fitting of the tennis data of Table 9.9, by forming an artificial explanatory variable for each player. For a given observation, the variable for player i is 1 if she wins, -1 if she loses, and 0 if she is not one of the players for that match. Each observation lists the number of wins ("wins") for the player with variate-level equal to 1 out of the number of matches ("matches") against the player with variate-level equal to -1. One fits the logit model having these artificial variates (one of which is redundant) as explanatory variables, deleting the intercept term using the NOINT option. The COVB option provides the estimated covariance matrix of the model parameter estimators.

In SPSS, one can fit models for matched pairs using the GENLOG procedure by creating appropriate factors and variates for loglinear models, as just discussed. A simple way to fit quasi independence is to set up a dummy variable that equals 1 for the cells to which the independence structure applies and 0 for the other cells and specify that dummy variable under "cell structure." The kappa measure of agreement and its standard error are available as part of CROSSTABS in SPSS.

Chi-Squared Distribution Values for Various Right-Tail Probabilities

			Right-Tail Probability				
df	0.250	0.100	0.050	0.025	0.010	0.005	0.001
1	1.32	2.71	3.84	5.02	6.63	7.88	10.83
2	2.77	4.61	5.99	7.38	9.21	10.60	13.82
3	4.11	6.25	7.81	9.35	11.34	12.84	16.27
4	5.39	7.78	9.49	11.14	13.28	14.86	18.47
5	6.63	9.24	11.07	12.83	15.09	16.75	20.52
6	7.84	10.64	12.59	14.45	16.81	18.55	22.46
7	9.04	12.02	14.07	16.01	18.48	20.28	24.32
8	10.22	13.36	15.51	17.53	20.09	21.96	26.12
9	11.39	14.68	16.92	19.02	21.67	23.59	27.88
10	12.55	15.99	18.31	20.48	23.21	25.19	29.59
11	13.70	17.28	19.68	21.92	24.72	26.76	31.26
12	14.85	18.55	21.03	23.34	26.22	28.30	32.91
13	15.98	19.81	22.36	24.74	27.69	29.82	34.53
14	17.12	21.06	23.68	26.12	29.14	31.32	36.12
15	18.25	22.31	25.00	27.49	30.58	32.80	37.70
16	19.37	23.54	26.30	28.85	32.00	34.27	39.25
17	20.49	24.77	27.59	30.19	33.41	35.72	40.79
18	21.60	25.99	28.87	31.53	34.81	37.16	42.31
19	22.72	27.20	30.14	32.85	36.19	38.58	43.82
20	23.83	28.41	31.41	34.17	37.57	40.00	45.32
25	29.34	34.38	37.65	40.65	44.31	46.93	52.62
30	34.80	40.26	43.77	46.98	50.89	53.67	59.70
40	45.62	51.80	55.76	59.34	63.69	66.77	73.40
50	56.33	63.17	67.50	71.42	76.15	79.49	86.66
60	66.98	74.40	79.08	83.30	88.38	91.95	99.61
70	77.58	85.53	90.53	95.02	100.4	104.2	112.3
80	88.13	96.58	101.8	106.6	112.3	116.3	124.8
90	98.65	107.6	113.1	118.1	124.1	128.3	137.2
100	109.1	118.5	124.3	129.6	135.8	140.2	149.5

Source: Calculated using *StaTable*, software from Cytel Software, Cambridge, MA.

Bibliography

Agresti, A. (1984). *Analysis of Ordinal Categorical Data*. New York: Wiley.

Agresti, A. (1990). *Categorical Data Analysis*. New York: Wiley.

Andersen, E. B. (1980). *Discrete Statistical Models with Social Science Applications*. Amsterdam: North Holland.

Bishop, Y. V. V., S. E. Fienberg, and P. W. Holland. (1975). *Discrete Multivariate Analysis*. Cambridge, MA: MIT Press.

Breslow, N., and N. E. Day. (1980). *Statistical Methods in Cancer Research*, Vol. I: *The Analysis of Case-Control Studies*. Lyon: IARC.

Caussinus, H. (1966). Contribution à l'analyse statistique des tableaux de correlation, *Ann. Fac. Sci. Univ. Toulouse*, **29**, 77–182.

Clogg, C. C. and E. S. Shihadeh. (1994). *Statistical Models for Ordinal Variables*. Thousand Oaks, CA: Sage.

Collett, D. (1991). *Modelling Binary Data*. London: Chapman & Hall.

Cox, D. R. (1970). *Analysis of Binary Data*. London: Chapman and Hall (2nd ed. 1989 with E. J. Snell).

Diggle, P. J., K.-Y. Liang, and S. L. Zeger. (1994). *Analysis of Longitudinal Data*. Oxford: Oxford Univ. Press.

Fahrmeir, L. and G. Tutz. (1994) *Multivariate Statistical Modelling Based on Generalized Linear Models*. Berlin: Springer.

Fienberg, S. E. (1980a). *The Analysis of Cross-Classified Categorical Data*, 2nd ed. Cambridge, MA: MIT Press.

Fienberg, S. E. (1980b). Fisher's contributions to the analysis of categorical data. Pp. 75–84 in *R. A. Fisher: An Appreciation*, (S. E. Fienberg and D. V. Hinkley, eds.). Berlin: Springer-Verlag.

Fisher, R. A. (1970). *Statistical Methods for Research Workers*, 14th ed. (Originally published 1925). Edinburgh: Oliver and Boyd, Ltd.

Fleiss, J. L. (1981). *Statistical Methods for Rates and Proportions*, 2nd ed. New York: Wiley.

Goodman, L. A. (1968). The analysis of cross-classified data: independence, quasi-independence, and interactions in contingency tables with or without missing entries. *J. Amer. Statist. Assoc.*, **63**, 1091–1131.

Goodman, L. A. (1970). The multivariate analysis of qualitative data: interaction among multiple classifications. *J. Amer. Statist. Assoc.*, **65**, 226–256.

Goodman, L. A. and W. H. Kruskal. (1979). *Measures of Association for Cross Classifications.* New York: Springer-Verlag. (Contains articles appearing in *J. Amer. Statist. Assoc.* in 1954, 1959, 1963, 1972.)

Grizzle, J. E., C. F. Starmer and G. G. Koch. (1969). Analysis of categorical data by linear models. *Biometrics,* **25**, 489–504.

Hastie, T. and R. Tibshirani. (1990). *Generalized Additive Models.* London: Chapman and Hall.

Hosmer, D. W. and S. Lemeshow. (1989). *Applied Logistic Regression.* New York: Wiley.

Imrey, P. B., G. G. Koch, and M. E. Stokes. (1981). Categorical data analysis: Some reflections on the log linear model and logistic regression. Part I: Historical and methodological overview. *Int. Statist. Rev.,* **49**, 265–283.

Koch, G. G., J. R. Landis, J. L. Freeman, D. H. Freeman, and R. G. Lehnen. (1977). A general methodology for the analysis of experiments with repeated measurement of categorical data. *Biometrics,* **33**, 133–158.

Lindsey, J. K. (1993). *Models for Repeated Measurements.* Oxford: Oxford Univ. Press.

Mantel, N. and W. Haenszel. (1959). Statistical aspects of the analysis of data from retrospective studies of disease. *J. Natl. Cancer Inst.,* **22**, 719–748.

McCullagh, P. (1980). Regression models for ordinal data (with discussion). *J. Roy. Statist. Soc.,* **B 42**, 109–142.

McCullagh, P. and J. A. Nelder. (1989). *Generalized Linear Models.* 2nd ed. London: Chapman and Hall.

Morgan, B. J. T. (1992). *Analysis of Quantal Response Data.* London: Chapman & Hall.

Mosteller, F. (1968). Association and estimation in contingency tables. *J. Amer. Statist. Assoc.,* **63**, 1–28.

Nelder, J. and R. W. M. Wedderburn. (1972). Generalized linear models. *J. Roy. Statist. Soc.,* **A 135**, 370–384.

Pearson, K. (1900). On a criterion that a given system of deviations from the probable in the case of a correlated system of variables is such that it can reasonably supposed to have arisen from random sampling. *Philos. Mag.,* Ser. 5, **50**, 157–175.

Rasch, G. (1961). On general laws and the meaning of measurement in psychology. *Proc. 4th Berkeley Symp. Math. Statist. Prob.* (J. Neyman, ed.), **4**, 321–333. Berkeley: Univ. of California Press.

Read, T. R. C. and N. A. C. Cressie. (1988). *Goodness-of-Fit Statistics for Discrete Multivariate Data.* New York: Springer-Verlag.

Santner, T. J. and D. E. Duffy. (1989). *The Statistical Analysis of Discrete Data.* Berlin: Springer-Verlag.

Stigler, S. (1986). *The History of Statistics: The Measurement of Uncertainty Before 1900.* Cambridge, MA: Harvard Univ. Press.

Stokes, M. E., C. S. Davis, and G. G. Koch. (1995). *Categorical Data Analysis Using the SAS System.* Cary, NC: SAS Institute Inc.

Whittaker, J. (1990). *Graphical Models in Applied Multivariate Statistics.* New York: Wiley.

Wickens, T. D. (1989). *Multiway Contingency Table Analysis for the Social Sciences.* Hillsdale, NJ: Lawrence Erlbaum.

Yule, G. U. (1900). On the association of attributes in statistics. *Philosohical Transactions of the Royal Society of London,* **A 194**, 257–319.

Index of Examples

283

Subject Index

289

WILEY SERIES IN PROBABILITY AND STATISTICS

*Now available in a lower priced paperback edition in the Wiley Classics Library.

*Now available in a lower priced paperback edition in the Wiley Classics Library.

*Now available in a lower priced paperback edition in the Wiley Classics Library.

*Now available in a lower priced paperback edition in the Wiley Classics Library.